Early Development of *Xenopus laevis*

A LABORATORY MANUAL

Early Development of *Xenopus laevis*

A LABORATORY MANUAL

Hazel L. Sive
Whitehead Institute for Biomedical Research, Cambridge

Robert M. Grainger
University of Virginia, Charlottesville

Richard M. Harland
University of California, Berkeley

COLD SPRING HARBOR LABORATORY PRESS
Cold Spring Harbor, New York

Early Development of *Xenopus laevis*
A LABORATORY MANUAL

©2000 by Cold Spring Harbor Laboratory Press, Cold Spring Harbor, New York
Paperback edition 2010

Developmental Editor: Siân Curtis
Project Coordinator: Mary Cozza
Production Editor: Dorothy Brown

Desktop Editor: Danny deBruin
Book Designer: Denise Weiss
Cover Design: Tony Urgo

Front cover (printed hardcover): Photograph of a transgenic *X. laevis* froglet expressing green fluorescent protein in the lens under the control of a lens-specific crystallin promoter. (*Top*) Normal early blastula. (*Middle*) Posterior neural tissue (dorsal view, early neurula): HoxD1. (*Bottom*) Retinoic-acid-treated series.

Back cover (printed hardcover): (*Left*) Forebrain/cement gland (head-on view, mid-neurula): (*Pale blue*) otx2; (*purple*) XCG-1. (*Middle*) Cement gland/hatching gland (head-on view, late neurula): (*Red*) XCG-1; (*brown*) XA-1. (*Right*) En-2 expression in the tadpole brain: (*purple*) En-2; (*brown*) pigment.

Library of Congress Cataloging-in-Publication Data

Sive, Hazel L.
    Early development of Xenopus laevis : a laboratory manual / by
Hazel L. Sive, Robert M. Grainger, Richard M. Harland.
       p. cm.
    Includes bibliographical references and index.
    ISBN 978-087969-504-0 (printed hardcover)-- ISBN 978-087969-578-1 (cloth)--
      ISBN 978-087969-942-0 (pbk)
      1. Xenopus laevis–Development–Laboratory manuals. I. Grainger,
Robert M. II. Harland, Richard M. III. Title.
QL668.E265S55 1998
571.8´1786–dc21   98-25174
                                                CIP

For a complete catalog of all Cold Spring Harbor Laboratory Press publications, visit our website at www.cshlpress.org.

Contents

Preface, vii
Acknowledgments, ix

*For color art in Chapters 2 and 9, see Color Plates between pages 102 and 103.

Preface

Experiments using amphibian embryos have led the way in understanding signaling events in the early development of vertebrates. The power of amphibian embryology stems from the ability to obtain embryos of all stages and from the large size of the embryos. These attributes have made detailed fate maps possible, and allow embryos to be microinjected and micromanipulated relatively easily. In addition, many amphibian species develop rapidly, so that the interesting stages of axis formation and tissue differentiation are accessible in a relatively short time. Over the years, *Xenopus* has come to dominate experimental embryology in amphibians: *Xenopus* is easy to keep, and ovulates at any time of year in response to simple hormone injection.

This is an exciting time in the *Xenopus* field. During the past several years, many new techniques have been devised or adapted for *Xenopus*. These techniques include whole mount in situ hybridization and immunocytochemistry that allow visualization of gene expression in the intact embryo. Transgenic technology has for the first time led to the correctly regulated expression of promoters in *Xenopus* embryos, eliminating previous problems of mosaic and inefficient expression from DNA expression vectors. Expression cloning has led to the isolation of powerful inducing molecules. Inducible gene expression systems can control the timing of gene expression in gain-of-function assays, whereas dominant-negative proteins have been instrumental in eliminating gene function. Together, these techniques have led to sophisticated new understanding of early *Xenopus* development and continue to ensure *Xenopus* a prominent position in the group of "model organisms."

Given the growth of techniques available for work with *Xenopus*, it is time to collect a comprehensive series of protocols in one place. This book arose from a course first taught at Cold Spring Harbor Laboratory in April 1991. Various in-house protocols were cobbled together, and in subsequent years, the lab manual became more extensive, and more accurate. However, to turn this from a collection of informal protocols into something more comprehensive and comprehensible took more than a rash promise by the authors. Much of the credit for the completion of the manual goes to the marathon efforts and cajoling of Mary Cozza and Siân Curtis at Cold Spring Harbor Laboratory Press.

H.L. Sive, R.M. Grainger, and R.M. Harland

Acknowledgments

The contribution of many people to entries in this book cannot be over-stated. Many sections in the manual have been written by our colleagues, and the procedures have been road-tested and modified by many. Enrique Amaya and Kris Kroll contributed Chapter 11; Sally Moody contributed extensively to Chapter 9; Nancy Papalopulu contributed much of Chapter 2. In addition, we thank Vladimir Apekin, David Bentley, Leila Bradley, Marietta Dunaway, Tabitha Doniach, Rick Elinson, Janet Heasman, Laura Gammill, Josh Gamse, John Gerhart, Jeremy Green, Ali Hemmati-Brivanlou, Ray Keller, Chris Kintner, Anne Knecht, Peggy Kolm, Mike Klymkowsky, Paul Krieg, Martin Offield, Charles Sagerström, Bill Smith, David Turner, Daniel Wainstock, and Paul Wilson. We also thank students of the course, members of our laboratories, and lecturers in the course for suggestions and criticism.

Thanks for permission to use figures goes to Rick Elinson, Jonathan Slack, Sally Moody, Tabitha Doniach, and Ray Keller.

We thank Terri Grodzicker and Jim Watson for their support in starting the *Xenopus* course, the laboratories of Hollis Cline and Nick Tonks for special help with emergency course supplies, and Clifford Sutkavich for finding essential equipment. We would also like to thank members of the CSH community for help in this project. H.L.S. thanks Andrew Lassar for his support and for looking after two infants while she coorganized this course.

While writing and revising the manual, we have been supported by our respective institutions and the National Institutes of Health. The course was supported by the National Science Foundation, the National Institute of Child Health and Human Development, and the Howard Hughes Medical Institute.

CHAPTER 1

Introduction

Welcome to the world of *Xenopus*. This manual is designed to introduce developmental biologists to the use of *Xenopus* as a model system. However, it is not a comprehensive volume and should be used in conjunction with Kay and Peng (1991) and Nieuwkoop and Faber (1994). Two recent and useful studies of *Xenopus* rearing are described by Hilken et al. (1994, 1995). Another useful volume by Hausen and Riebesell (1991) contains an excellent histology of early embryos. The companion video series "Manipulating the Early Embryo of *Xenopus laevis*" presents very helpful illustrations of many of the developmental processes and procedures described here. The reader is encouraged to use these video demonstrations to complement the material presented in this manual.

At several instances in this manual, more than one protocol is presented for a single procedure. The protocol of choice generally depends on the facilities available in the laboratory and on personal preference. There is rarely one "correct" method.

Xenopus laevis is a gentle, freshwater animal that can be induced by simple hormone injection to lay eggs repeatedly. These features, coupled with the large size of the embryos, which allows micromanipulation and microinjection, and their rapid rate of development, make *Xenopus* an excellent animal for analyzing early vertebrate development. The chief disadvantages of *X. laevis* are long generation times (1–2 years) and tetraploidy. The diploid *Xenopus (Silurana) tropicalis* does not have these disadvantages. This species has a generation time of 5 months or less and so may provide a useful alternative to *X. laevis* for future research (Amaya et al. 1995).

WHERE TO OBTAIN *XENOPUS*

The two major suppliers in the United States are NASCO and Xenopus I (5654 Merkel Road, Dexter, Michigan 48130). Although adult frogs are quite expensive, late juveniles are less costly and will usually yield eggs, although in smaller numbers than adults. Frogs are usually shipped in peat moss (or equivalent), but they cannot be shipped in extreme summer heat or in frigid winters without special produres. After receiving the frogs, they should be allowed a minimum recovery period of 2 weeks prior to experimentation.

nuptial pads

cloaca

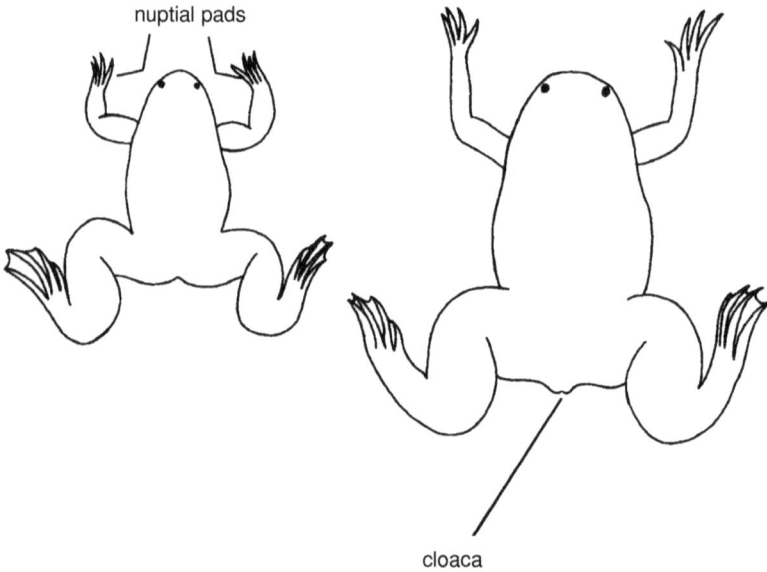

Figure 1.1 Relative sizes of male (*left*) and female (*right*) *Xenopus laevis*. Males are small-er than females of equivalent age. Males can also be distinguished from females by the absence of a cloaca, i.e., the fleshy protuberence between the legs, and by the presence of "nuptial pads," black rough regions on the inner side of the forelimb. These nuptial pads assist the male in grasping the female during mating (may not be obvious in males that are not ready to mate). In females, the cloaca becomes red and swollen when the animal is ready to lay eggs. Note the slightly rounded shape of both animals, although females may be some-what plumper than males due to the size of their ovaries. Animals should not be excessively skinny, which could indicate illness, or excessively bloated, which can be a sign of poor health. (Courtesy of Mark Curtis.)

Females are larger than males, with a prominent cloaca. Males have rough, dark pads on the inside wrist, used to clasp the female during amplexus (see Figure 1.1). Animals may be endemically infected with nematodes and can be treated prophylactically on arrival (see below, Diseases, Preventions, and Cures). Thin animals tend to be sickly and should not be used.

HOUSING AND FEEDING

Good animal husbandry is vital for maintaining a healthy frog population. This requires some effort but is generally rewarded by high-quality egg and embryo production. A healthy frog is placid, with moderately slimy skin and a nice pear shape. Jumpy frogs, frogs with dry or excessively slimy skin, bloated frogs, and frogs that look gray and thin or reddish are not healthy and should not be used for egg collection, as this would lead to further deterioration of the animals' condition, and the resulting eggs would be generally unsuitable for experimental purposes.

Containers

Xenopus never leaves the water and so it is unnecessary to provide dry areas in their containers. Frogs may be housed in different containers depending on their age and number and the quality of the water available in the laboratory.

Standing Water Tanks

For fewer than 50 or so animals, it is a simple matter to house frogs in tanks of still water. Four females or six males can be accommodated comfortably in approximately 16 liters of water. The water should be 12–20 cm deep. Plastic tanks are the most convenient containers, since their opaque sides approximate pond conditions. Frogs are talented jumpers (up to 45 cm) and thus their tanks should be covered with a heavy lid, either Plexiglas with half-inch holes or a stainless steel mesh. An opaque pipe (12 cm long and 8 cm in diameter) provides a comfortable hiding place for the frogs. The water in standing tanks should be replaced at least three times each week.

Drip-through Systems

For a larger number of frogs, it is more cost-effective to house animals in a system that is at least partially self-cleaning. One possibility is a system in which fresh water drips in and out continuously, thus preventing accumulation of wastes. This is probably the optimal way to keep frogs, since levels of toxic wastes are kept low, and solid waste (in suspension) can be drained continuously. The disadvantages of this system are that it uses a lot of water and the quality of input water must be monitored constantly. The chlorine content, pH, and salinity of tap water can vary widely with the season and the whim of the local water authority. Alternatively, distilled water, supplemented with NaCl or Instant Ocean (available from local aquarium suppliers) to approximately 20 mM, can be used as the input water supply. However, this system can use excessive quantities of distilled water. The water should drip in at a rate of about 10–50 ml per minute depending on tank size. A faster drip rate uses more water than is necessary.

Recirculating Systems

Many investigators use a continuously recirculating system that incorporates a biological filter. The water in these systems is generally collected into a common drain, with U-tubes between the tank and the common drain to prevent mixing dirty water between tanks. The water then flows into a reservoir in which the gross particulate matter is removed by an inexpensive air-conditioner filter that is replaced on a daily to weekly basis. From the reservoir, the water is pumped through a biological (bacterial) filter to remove ammonia and nitrites. It then flows through a sand filter to remove finer particles and across a series of high-capacity UV lights to kill bacteria and other potential pathogens, before being pumped back to the tanks. The sand filter is backwashed every day, with some fresh water entering the system, such that water in the entire system is replaced every 5 days. Again, the rate of flow should be limited to less than 100 ml of water per minute.

Water quality is monitored by analyzing pH and the concentration of ammonia, nitrates, and nitrites at least once a week. Chlorine and chloramine should be undetectable, and ammonia should be less than 5 ppm; nitrates and nitrites should also be undetectable. Bacterial counts are carried out from the water that goes back into the tanks (the count should be essentially zero). This type of system requires more maintenance than the drip-through system, but uses less water—a particular advantage in laboratories where water quality is a problem and supplies of distilled water are limited. In a system that has a capacity of approximately 400 frogs and contains about 2000 liters of water, maintenance should take less than 1 hour each day. The health of frogs in such a controlled environment can be excellent.

Water

Water quality is very important, particularly with respect to pH, chlorine, and ammonia content. The water must be completely dechlorinated before use, which can be achieved by exposure to the air for several days in standing tubs. However, many water authorities also add chloramine to the water. This compound is extremely stable, but can be removed by running the water through a carbon filter (obtainable as a cartridge type from most aquarium dealers, e.g., Barnstead). The chloramine content of the water determines how often the cartridge should be changed, but it should be changed before chloramine is detected in the output water. Chlorine and chloramine levels can be monitored with easy-to-use kits (Hach Co.). Deaminating liquids (e.g., Prime) can be purchased from a local aquarium supplier and used according to the manufacturers' instructions for removal of chloramine and ammonia. In addition, water should be filtered through a dirt/rust or particle filter, also obtainable through aquarium dealers. Add NaCl or Instant Ocean to a final concentration of 20 mM. Alternatively, use rock salt at 1 g/liter of tap water, which is less expensive than NaCl.

Frogs can also be maintained in distilled or reverse-osmosis water supplemented with Instant Ocean to 20 mM; NaCl alone is not sufficient. This is advisable if the quality of the tap water in the laboratory is poor or variable. Frogs must not be kept in unsupplemented distilled water. This will cause their skin to flake and may lead to the development of stress-related diseases. The pH of distilled and reverse-osmosis water should be adjusted to 6.5 with SeaKem Neutralization Buffer (available from That Fish Place, 237 Centerville Road, Lancaster, Pennsylvania 17603), with soda lime, or with NaOH. A conductivity of 1.0 ms/cm ± 0.1 units is optimal.

pH

A pH of 6.5 is optimal for *Xenopus*. At low pH (below 7.0), ammonia waste is present as ammonium ions, which are nontoxic, but as the pH increases, free ammonia, which is toxic, forms rapidly. Even in the absence of ammonia, a change of 1 pH unit (e.g., from 6.5 to 7.5) can lead to the loss of the frogs' protective mucus, a higher susceptibility to pathogen attack, and other stress-related conditions.

Light and Temperature

Frogs must be kept on a regular light/dark cycle. They respond well to 12–14 hours of light and 12–10 hours of dark. Daylight spectrum fluorescent lighting can be used, but some investigators believe that egg quality improves when frogs are exposed to sunlight-equivalent light levels. Despite normally living in dark ditches, *Xenopus* appears to enjoy light conditions. However, it is important to have dark areas, such as opaque plastic pipes, so that the frogs can escape the light if they want to.

Temperatures of 16–20°C are optimal for *Xenopus* at all stages of development. Above 25°C, egg quality declines precipitously. After fertilization, it should be noted that the higher the temperature, the faster the rate of development. The effect of temperature on development is discussed in more detail in Appendix 2.

Food

Adult frogs must be fed a minimum of three times each week, several hours before changing the water. Frogs feed well on floating food (e.g., Trout Chow

Pellets; Purina), which is inexpensive and convenient. An alternative includes Frog Brittle (NASCO). If frogs have previously been fed liver, it may take them 1 week or more to eat the pellets. They can be encouraged to accept the new type of food by housing a frog that already eats pellets with those that do not in order to initiate a feeding frenzy. Each adult frog will eat 5–10 pellets per feeding. Feeding regimes for tadpoles are discussed below (see Raising Tadpoles and Frogs).

Seasonal Variation

Many investigators report seasonal variation in the quality of embryos, even for animals that have been kept for years or even bred without seeing seasonal light/dark changes. Variation may be due to changes in water quality or more significantly water temperature. High summer temperatures (above 26°C) will adversely affect egg production. Injection of frogs with 50 units of pregnant mare serum (PMS) a few weeks before inducing ovulation may help to reduce the effects of high temperatures. Whole serum is not commercially available, but an alternative to whole PMS is chorionic gonadotropin (CG) obtainable from Sigma. Injecting the frogs with 50 units of human chorionic gonadotropin (hCG) also may help to maintain egg production.

RAISING TADPOLES AND FROGS

Given sufficient space and time, raising *Xenopus* is not difficult. Tadpoles require approximately 1 liter of water each, whereas fully grown frogs need up to 4 liters each.

Hundreds of tadpoles can be generated by fertilizing several dishes of eggs in vitro as described in Chapter 5. Alternatively, unused embryos from several days of experiments can be pooled and reared through metamorphosis to adulthood. It is important not to overcrowd dejellied embryos (see Protocol 6.1). Overcrowding leads to abnormal development, usually as a result of gastrulation defects thought to be caused by anoxia, and dejellied embryos should be separated from each other by at least one embryo diameter at all times. Swimming tadpole-stage embryos should be transferred to a tank with the appropriate volume (1 liter per individual) of good-quality water containing 20 mM NaCl.

The most efficient method of generating large numbers of tadpoles is to induce natural matings between several females and males. Each male is injected with approximately 300 units of hCG and each female is injected with approximately 800 units of hCG (concentration 1000–2000 units/ml; Sigma) and placed in a large tank (80 liters) containing good-quality water supplemented with 20 mM NaCl at a depth of approximately 15 cm. Equal numbers of males and females should be used during mating. After 6–8 hours, the males clasp the females around the hips (amplexus) and the eggs are fertilized as they emerge from the cloaca. It is important that the frogs are not disturbed during mating, which may result in the release of amplexus and a decrease in fertilization efficiency. For this reason, it is generally preferable to inject the frogs in the afternoon so that they will have most of the night to mate undisturbed. After 24 hours or so, the males release the females and mating ceases. At this time, the energy-depleted frogs are removed from the tank to prevent them from eating their own eggs. After 3–4 days, the embryos will hatch and begin to colonize the sides of the tank at the water's edge. Unfertilized eggs must be removed promptly since they will decay and allow bacteria and fungi to contaminate the water. If the fertilization frequency is particularly low (<20%), healthy embryos should be screened as soon as possible and transferred to fresh water supplemented with penicillin and streptomycin or gentamycin (0.05 mg/ml). This procedure is time consuming and very laborious, but worthwhile if embryos are in limited supply.

After 1 week, the tadpoles should be free-swimming and ready to begin feeding, as evidenced by the rhythmic opening of their mouths. The tadpoles thrive on very fine food. They can be fed routinely on nettle powder mixture, i.e., a combination of nettle powder, active dry yeast, and powdered bone meal, mixed at 7:2:1, respectively. These ingredients can be obtained at health food stores. Tadpole Brittle (NASCO) can also be used according to the manufacturer's instructions. Food should be delivered to the tank as a water suspension through a pipette. This method decreases the amount of surface scum that forms when the food is added directly to the tank. It is important not to overfeed the tadpoles, because they will become anoxic, presumably due to food-clogged gills. Death due to overfeeding is common, particularly in young tadpoles. In general, tadpoles should clear the water within 2 hours of feeding. Ideally, they should be fed daily, with a weekly supplement of fresh whole milk added dropwise until the water in the tank is slightly cloudy. The milk and bone meal add calcium and phosphate to the water, which should eliminate skeletal deformities in the metamorphosing frogs. Although not essential, oxygenating the water with submerged aerators appears to enhance the growth rate of tadpoles. With proper feeding, the water in tadpole tanks need only be changed once every 1–2 weeks.

Tadpoles do not develop synchronously, even among siblings. Some metamorphose after 8 weeks, whereas others require 4–6 months to mature.

During metamorphosis, the water level in the tank should be no more than 30 cm deep because the newly metamorphosed froglets may have difficulty reaching the surface to breathe. Froglets cannot eat tadpole food and so must be fed small Trout Chow Pellets (Purina, pellet size 4) or Frog Brittle (NASCO). The small pieces of food sink to the bottom where the froglets prefer to eat. There is no need to separate froglets and sibling tadpoles. They can coexist for several months until all the tadpoles metamorphose. In fact, it can be rather convenient to have a mixture of tadpoles and froglets in the tank because the tadpoles tend to keep the water clean. Once one third of the tadpoles have metamorphosed, it is no longer necessary to feed the remaining tadpoles nettle powder mixture since the froglets break up the Trout Chow sufficiently for the tadpoles to filter the scraps. This simplifies the feeding regimen to:

- nettle powder mixture only, up to the emergence of the first froglets

- nettle powder mixture and small Trout Chow Pellets, until one third of the froglets have emerged

- Trout Chow Pellets or Frog Brittle exclusively beyond this time

Given sufficient space, froglets reach sexual maturity after 1 year and full size after 3 years.

DISEASES, PREVENTIONS, AND CURES

The two most common diseases in *Xenopus* are bacterial septicemia (the culprit bacterium varies widely) and nematode infestation. Fungal infections may also occur. In general, veterinarians are not familiar with *Xenopus,* and thus, treatment must often be carried out by the investigator. Many diseases are stress-induced and commonly occur after induction of ovulation or egg collection. It is therefore imperative to minimize stress and to observe animals carefully during and after these procedures. Females should be handled with care and isolated for 24 hours after egg collection in water supplemented with 20 mM NaCl and 5 µg/ml gentamycin (i.e., 10x lower than the dose given to embryos). As an additional precaution, oxytetracycline, which targets bacteria different from those killed by gentamycin, can be included to a final concentration of 50 mg/ml, replaced every day. (Never leave a frog in a bucket of dirty water containing excess food or especially rotting eggs. After a few days, it will get septicemia and die a horrible death.)

It should be noted that the treatment of a disease with drugs may not remove the cause of the disease. Bacterial, nematode, and fungal infections often arise from pathogens endemic to the frogs that become a problem only when frogs are stressed. Such stresses can include overcrowding, a change in pH or water quality, or careless frog handling. It is important to establish the root cause of a problem before it inflicts significant damage on the colony.

Frogs should have a moist but not excessively slimy skin. The skin should not be visibly flaking (although they do shed some skin normally) and the pigmentation should not be patchy (normal pigmentation is mottled, but the mottling should cover the entire body). Animals should be fat but not bloated (the difference is easily recognized when the frogs are handled) or excessively thin. The skin should not be red, which may indicate subcutaneous hemorrhaging.

Nematode Infection (Capillariasis)

The symptoms of nematode infection include sloughing of skin, patchy pigmentation, grayish and thin skin, and weight loss. Redness is not usually apparent. Examination of skin scrapings will reveal nematode presence. This infection is often precipitated by stress.

Many frogs carry latent nematode infections. Animals can be treated prophylactically with Ivermectin (PRO-VET; available from local aquarium suppliers) on arrival. Apply treatment as soon as symptoms are recognized. Infected animals must be isolated, since both eggs and adult nematodes are shed into the water. Administer Ivermectin via injection into the dorsal lymph sac, as described in Protocol 5.2. Deliver two injections of 2 µg/g of body weight 2 weeks apart, each in a volume of approximately 100 µl. Although the drug can also be administered by oral gavage, it is difficult to carry out without severely stressing the frogs. This treatment is very effective, and frogs will recover if it is administered promptly. If the frogs are very thin with fragile skin, recovery can take several months (see Cromeens et al. 1987; Stephens et al. 1987).

Red Leg (Bacterial Septicemia)

The symptoms of red leg are cutaneous hemorrhages, especially on flexor surfaces of thighs and foot webs, dull discoloration of skin, subcutaneous edema, and neurological disorders (trembling, initially of limbs). This disease can be caused by a number of gram-negative bacteria, primarily *Enterobacteria*, particularly *Aeromona, Pseudomonas*, and *Citrobacter*, and is precipitated by stress.

Infected individuals must be isolated since the disease is highly contagious. Treatment can be effective if administered promptly. The most convenient antibacterial treatment is to add oxytetracycline to the water, to a final concentration of 100 µg/ml. Change the water every day and administer antibiotic for 7 days. Other antibiotics can be administered by oral gavage, but this is difficult to do without severely stressing the frogs. Oxytetracycline is given orally at a dose of 200 µg/g of body weight per day for 5 days, and chloramphenicol at a dose of 330 µg/g of body weight per day, also for 5 days. In conjunction with these drugs, increasing the salt content of the water to 1 g/liter may be helpful.

Fungal Infections

The symptoms of fungal infections are thread-like, generally whitish bodies around injection sites. Fungus should be confirmed by microscopy.

Maroxy solution (available from a local aquarium supplier) is generally effective in the treatment of fungal infections. It should be added to the water as recommended by the manufacturer. The antifungal permanganate treatment prescribed by Wu and Gerhart (1991) can be fatal. In many cases, the fungus involved is nonpathogenic and an opportunistic infection may have arisen as a result of damage to the protective mucus of the skin. In these instances, application of antifungal agents such as Maroxy may help, but the cause of the skin damage must also be eliminated. See comments on stress at the beginning of this section.

REFERENCES

Amaya E., Offield M.F., and Grainger R.M. 1998. Frog genetics: *Xenopus tropicalis* jumps into the future. *Trends Genet.* **14:** 253–255.

Cromeens D.M., Robbins V.W., and Stephens L.C. 1987. Diagnostic exercise: Cutaneous lesions in frogs. *Lab. Anim. Sci.* **37:** 58–59.

Hausen P. and Riebesell M., eds. 1991. *The early development of* Xenopus laevis: *An atlas of the histology*. Springer-Verlag, Berlin.

Hilken G., Dimigen J., and Iglauer F. 1995. Growth of *Xenopus laevis* under different laboratory rearing conditions. *Lab. Anim.* **29:** 152–162.

Hilken G., Willmann F., and Dimigen J., and Iglauer F. 1994. Preference of *Xenopus laevis* for different housing conditions. *Scand. J. Lab. Anim. Sci.* **21:** 71–80.

Kay B.K. and Peng H.B., eds. 1991. *Methods in cell biology*, vol. 36. Xenopus laevis: *Practical Uses in cell and molecular biology*. Academic Press, New York.

Nieuwkoop P.K. and Faber J., eds. 1994. *Normal table of* Xenopus laevis *(Daudin): A systematical and chronological survey of development from the fertilized egg till the end of metamorphosis*. Garland, New York.

Stephens L.C., Cromeens D.M., Robbins V.W., Stromberg P.C., and Jardine J.H. 1987. Epidermal capillariasis in South African clawed frogs (*Xenopus laevis*). *Lab. Anim. Sci.* **37**: 341–344.

Wu M. and Gerhart J. 1991. Raising *Xenopus* in the laboratory. *Methods Cell Biol.* **36**: 3–18.

WWW RESOURCES

http://vize222.zo.utexas.edu Peter Vize (University of Texas, Austin) has set up a *Xenopus* Web Site, which contains a wealth of useful information.

http://www.library.wisc.edu/guides/Biology/demo/frog2/mainmenu.html Another good Web site is provided by the University of Wisconsin, Amphibian Development Tutorial. It contains a glossary and movies of developmental events.

http://minerva.acc.Virginia.edu/~develbio/trop/ This Web site contains much useful information regarding *X. tropicalis*.

CHAPTER 2

Morphology of *Xenopus* Embryos and Adults

The morphology of *Xenopus laevis* has been well studied in some regards. Comprehensive treatises on the developmental stages of *Xenopus* describe the external morphology of embryos (Nieuwkoop and Faber 1994) and the histology of the early stages of development (Hausen and Riebesell 1991).

Other areas of *Xenopus* morphology, for example, later neural development and organogenesis, are poorly documented. For a thorough histological description of later developmental stages of other anurans, see Rugh (1951). Athough no such text exists specifically for *Xenopus*, an increasing amount of information on *Xenopus* is currently being gleaned from identification of morphological domains in embryos, using in situ hybridization and antibody staining. Several examples of advances made using these techniques are presented in this chapter. A useful resource, illustrating a large number of genes that show tissue-specific expression patterns in *Xenopus,* is "The *Xenopus* Molecular Marker Resource" web site (*http://vize222.zo.utexas.edu/*). Comparative studies of gene expression in tissues from different vertebrates are likely to provide extremely useful information in the future (see, e.g., Fernandez et al. 1998).

This chapter briefly summarizes the features of early development and some important aspects of neural development. Also presented is a brief discussion on adult morphology as it pertains to the experimental manipulations described in this manual.

EARLY DEVELOPMENT

The development of the oocyte and the stages of egg maturation are extensively discussed by Hausen and Riebesell (1991). The stages of embryonic development are described by Nieuwkoop and Faber (1994), who devised the staging system used universally for *X. laevis*.

Cleavage and Blastula Formation

Shortly after fertilization, a transient contraction of the highly pigmented animal cap occurs, and a region of denser pigmentation on the ventral side of the embryo becomes discernible. The first three cleavages occur during a 2-hour interval; the asymmetry of dorsoventral pigmentation persists through these and later stages (Figure 2.1).

Subsequent rapid cleavages result in the formation of the blastula (stages 7–8) in 4 to 5 hours postfertilization (hpf; note that all times refer to development at 22°C; for the effect of temperature on rates of development, see Appendix 2), and gastrulation commences at 9 hpf (stage 10) when the blastopore is first discernible (Figure 2.2). At these stages, the animal cap is that region of the embryo fated to become epidermis. The marginal zone will become neural and mesodermal tissue, and the vegetal yolk mass will form endodermal tissues.

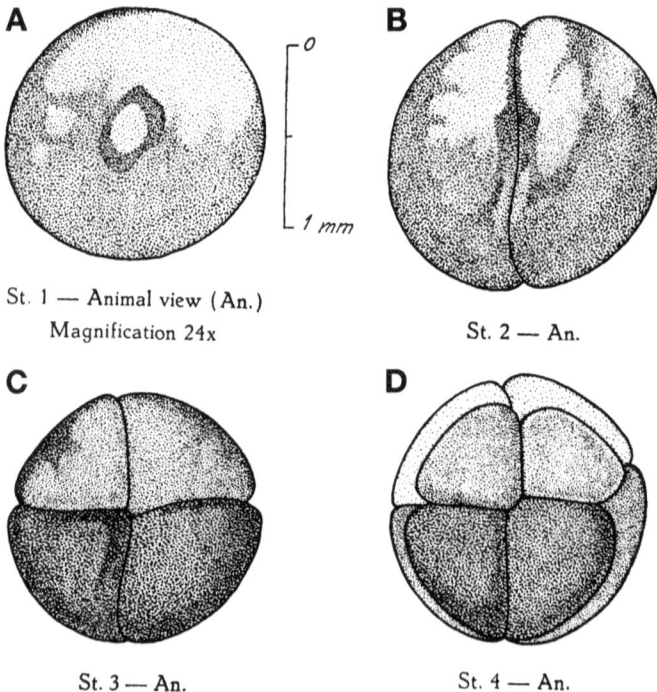

A

St. 1 — Animal view (An.)
Magnification 24x

B

St. 2 — An.

C

St. 3 — An.

D

St. 4 — An.

Figure 2.1. Early cleavage in *X. laevis*. (*A*) Embryo shortly after fertilization (stage 1); (*B*) at the 2-cell stage (stage 2); (*C*) at the 4-cell stage (stage 3); (*D*) at the 8 cell stage (stage 4). All views are from the animal pole. Note the asymmetry of the dorsoventral pigmentation that persists through this and later stages. (Reprinted, with permission, from Nieuwkoop and Faber 1994.)

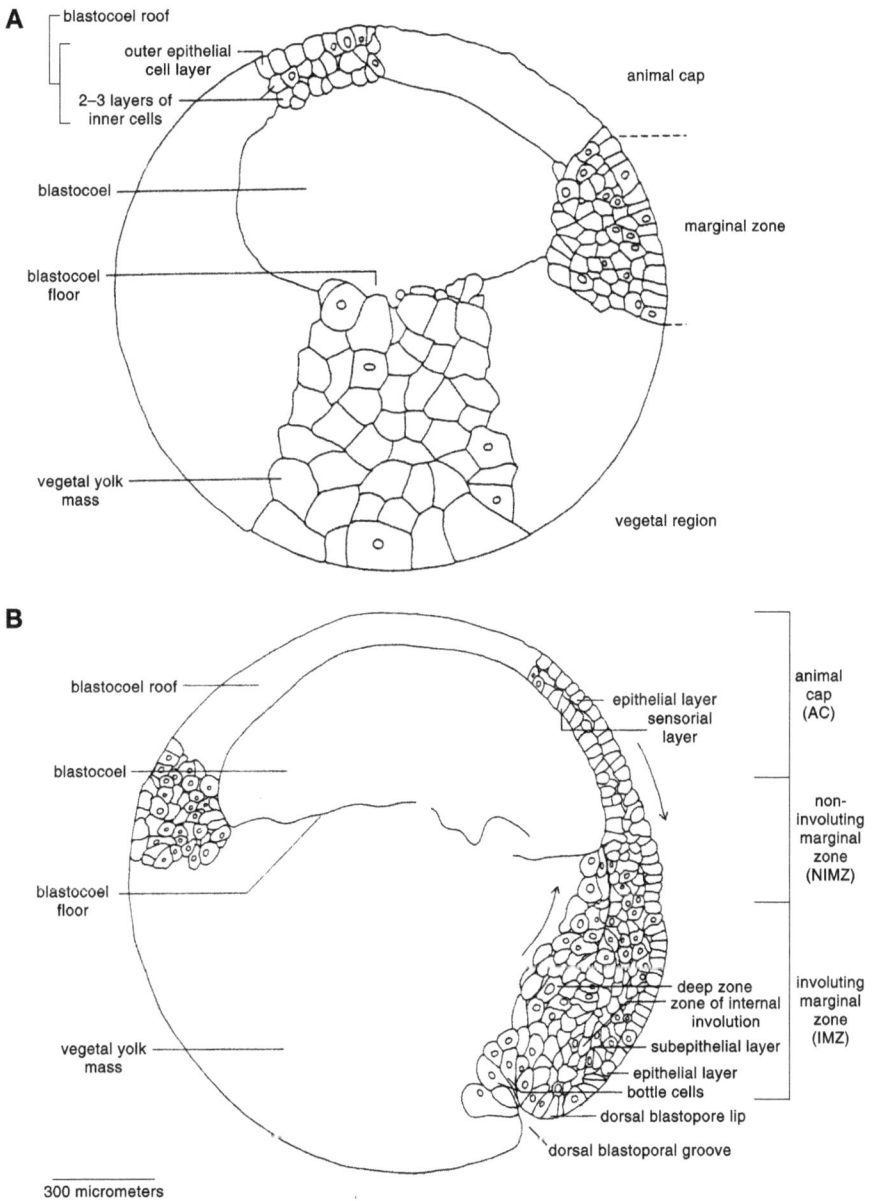

Figure 2.2. Cross-sections through blastula (*A*, stage 9) and early gastrula stage (*B*, stage 10+) embryos, indicating the nomenclature used for describing parts of the embryo at these stages. (Reprinted, with permission, from Hausen and Riebesell 1991.)

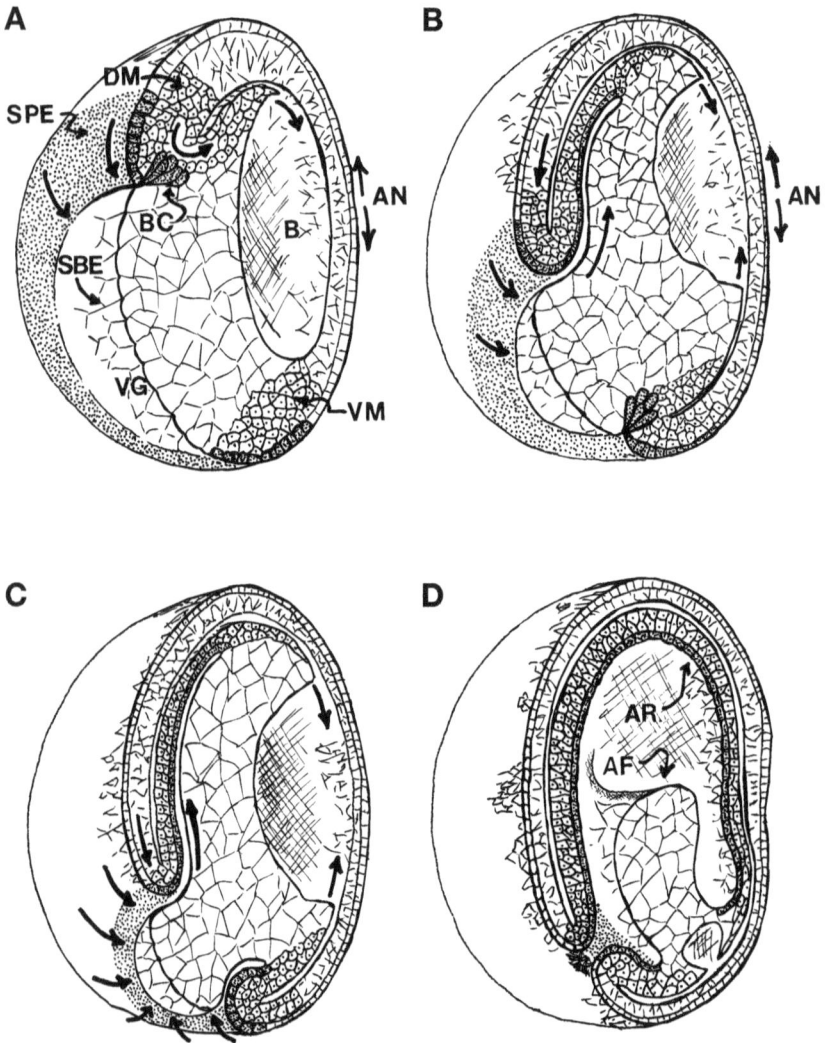

Figure 2.3. Gastrulation in *X. laevis* viewed in mid-sagittal sections. The vegetal pole (VG) faces the observer. (*A*) Early gastrula; (*B,C*) middle gastrula; (*D*) late gastrula. (AN) Animal pole; (AF) archenteron floor; (AR) archenteron roof; (B) blastocoel; (BC) bottle cells; (DM) dorsal mesoderm; (SBE) subblastoporal endoderm; (SPE) suprablastoporal endoderm; (VM) ventral mesoderm. (Reprinted, with permission, from Keller 1986.)

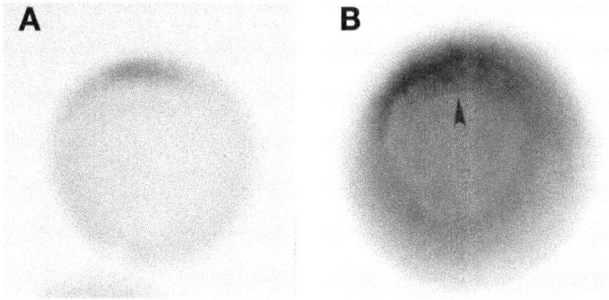

Figure 2.4. (*See Color Plate 1*) Genes expressed in dorsal mesoderm at the beginning of gastrulation. (*A*) Goosecoid (Reprinted, with permission, from Cho et al. 1991 [© Cell Press]); (*B*) Noggin (Reprinted, with permission, from Smith and Harland 1992 [© Cell Press]). Both images are views from the vegetal pole of the embryo.

Gastrulation

The process of gastrulation occurs over a period of several hours (ending at stage 12) and entails an extremely complex set of movements (Figure 2.3; for review, see Keller 1991). Briefly, the first sign of gastrulation is the appearance of a condensed area of pigmentation on the dorsal side of the embryo where endodermally derived bottle cells are elongating (at early stage 10, sometimes referred to as stage 10–). Subsequently, involution of dorsal marginal zone tissue begins. The dorsal marginal zone includes both dorsal mesoderm (the deep involuting layer) and endodermal cells (the superficial involuting area). Involution then progresses to lateral tissues and finally to the ventral side of the embryo. The end result of this extensive rearrangement is to move the mesoderm and endoderm inside the embryo, displacing the blastocoel and forming a new body cavity, the archenteron. Tissues along the anteroposterior embryonic axis become elongated due to movements of cells from ventral and lateral regions of the embryo toward the dorsal midline.

Morphological boundaries between and within germ layers are not obvious at the beginning of gastrulation, but they can be defined by domains of gene expression. For example, Goosecoid (Cho et al. 1991) and Noggin (Smith and Harland 1992) are activated at the beginning of gastrulation in dorsal mesoderm that will develop into the notochord (Figure 2.4).

Neurulation and Organogenesis

Shortly after the end of gastrulation, the neural plate becomes progressively more prominent on the dorsal side of the embryo. It comprises a thickened,

Stage 15

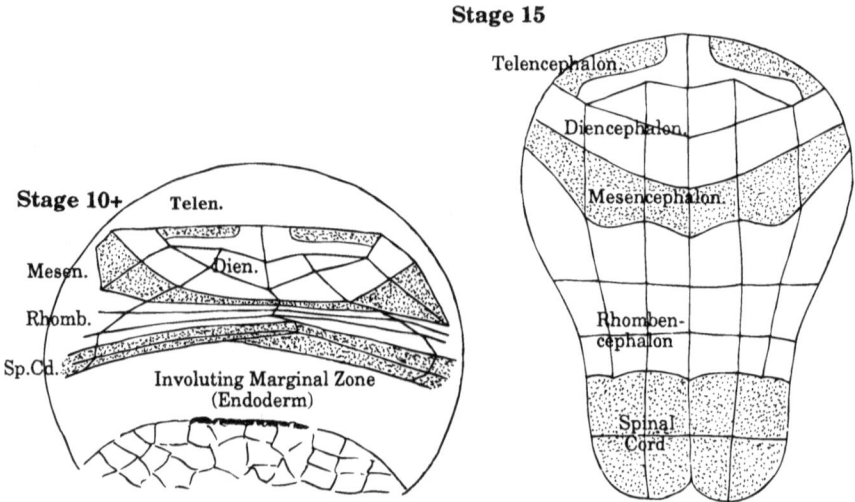

Figure 2.5. Rearrangements within ectoderm during gastrulation define the shape of the neural plate. (*Left*) At stage 10+ (advanced stage 10), the regions of ectoderm fated to form the different regions of the neural plate, telencephalon (Telen.), diencephalon (Dien.), mesencephalon (Mesen.), rhombencephalon (Rhomb.), and spinal cord (Sp.Cd.), are spread broadly around the embryo (dorsal view). (*Right*) By stage 15, when the shape of the definitive neural plate is clear, these regions have been stretched along the anteroposterior axis due to cell rearrangements. (Reprinted, with permission, from Keller et al. 1992a [© Wiley-Liss.])

flat region of ectoderm that will form all of the components of the central nervous system (Eagleson and Harris 1990). The edges of the neural plate, the neural folds, will give rise to all of the neural crest (Baker and Bronner-Fraser 1997). The neural plate is formed by extensive cell rearrangements within the ectoderm, resulting in narrowing (convergence) and elongation (extension) of dorsal ectoderm during gastrulation. This process can be seen by comparing the fate maps of regions of the brain at stage 10 and stage 15 (Figure 2.5; for further discussion of fate mapping at these stages, see Chapter 9). Although both the neural ectoderm and the underlying mesoderm undergo convergence and extension, the mechanisms in these two germ layers differ (see Keller 1986; Keller et al. 1992a,b; Elul et al. 1997). Between stages 14 (16 hpf) and 20 (22 hpf), the process of neurulation occurs, during which the neural plate folds to become the neural tube (see Figures 2.6 and 2.7) (for review, see Smith and Schoenwolf 1997).

Following closure of the neural tube, the period of organogenesis commences. Organogenesis is illustrated in the views of neurula (stage 20) and early-tailbud-stage (stage 26) *Xenopus* embryos shown in Figures 2.8, 2.9,

A

B

St. 14 — Post.-dors.   St. 16 — Post.-dors.

C

D

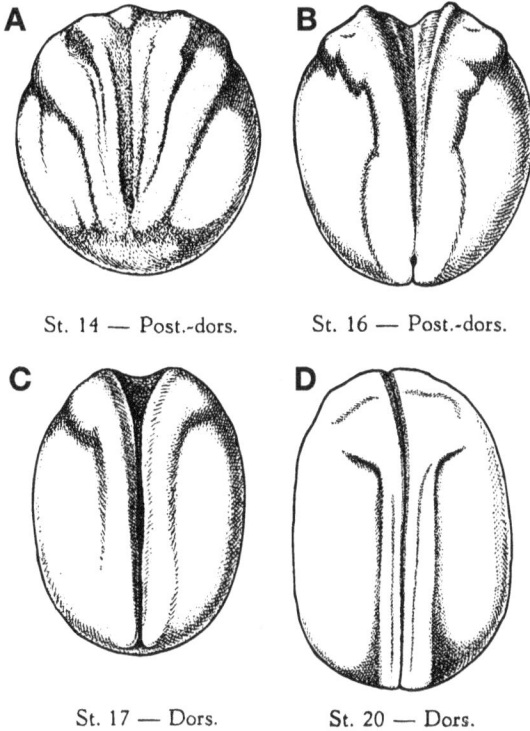

Figure 2.6. Neurulation in *Xenopus*. All views are from the dorsal side, anterior at the top. (*A*) Stage-14 embryo; (*B*) stage-16 embryo; (*C*) stage-17 embryo; (*D*) stage-20 embryo. (Reprinted, with permission, from Nieuwkoop and Faber 1994.)

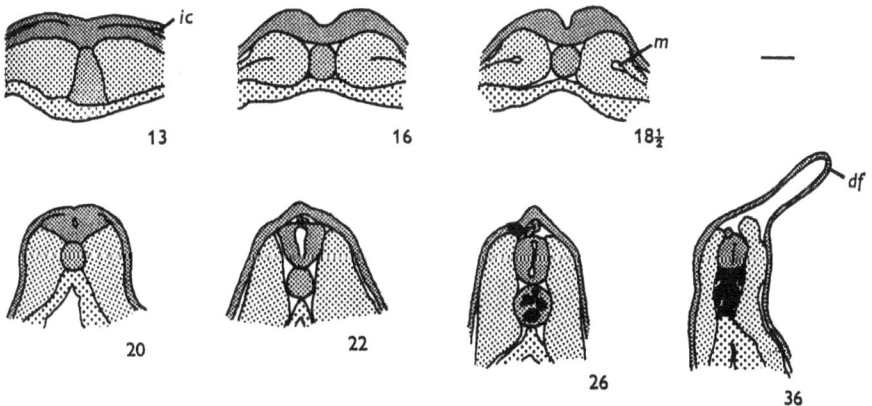

St. 17 — Dors.   St. 20 — Dors.

Figure 2.7. Cross-sectional view of neurulation in *Xenopus* showing the folding of the neural plate, fusion of the lateral edges to form the dorsal part of the neural tube, and the covering of the neural tube by epidermis. Stippling: (*finest*) ectoderm; (*fine*) notochord; (*medium*) somatic mesoderm; (*coarse*) endoderm. (df) Dorsal fin; (ic) interectodermal cleft; (m) myocoelic space. Numbers refer to stages of Nieuwkoop and Faber (1994). (Reprinted, with permission, from Schroeder 1970 [© Company of Biologists].)

| Neural tube stage
Embryo stage 20
21 h 45 min p.f. | dorsal
↑
posterior ←――┼――→ anterior
↓
ventral | sagittal section |

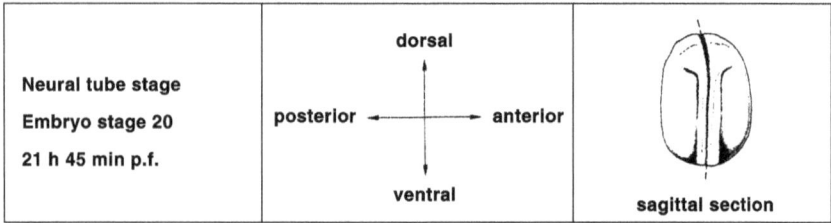

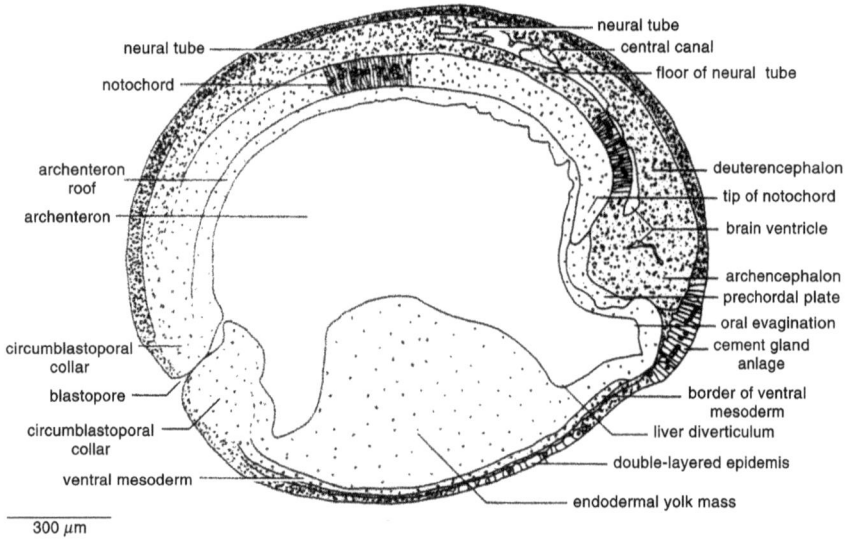

Figure 2.8. Sagittal section of a stage-20 embryo. Subdivisions of the neural tube, mesoderm, and endoderm are denoted. (Reprinted, with permission, from Hausen and Riebesell 1991.)

and 2.10. No systematic review of the later stages of organogenesis in *Xenopus* has been published, but some of the features of these stages are illustrated in diagrams of other anuran embryos (see Figure 2.11). Although it is often difficult to identify a particular tissue anlage, or regions within tissues, at these stages based on histology alone, once again, studies of gene expression have been very informative. Some examples to illustrate this point are presented in Figure 2.12: *xWT1* (Carroll and Vize 1996) serves as a marker for pronephric mesoderm, *XTin1* is a marker for the presumptive heart (Evans et al. 1995; Cleaver et al. 1996), and *xHB9* is a marker for the presumptive motor neuron area of the spinal cord (Saha et al. 1997).

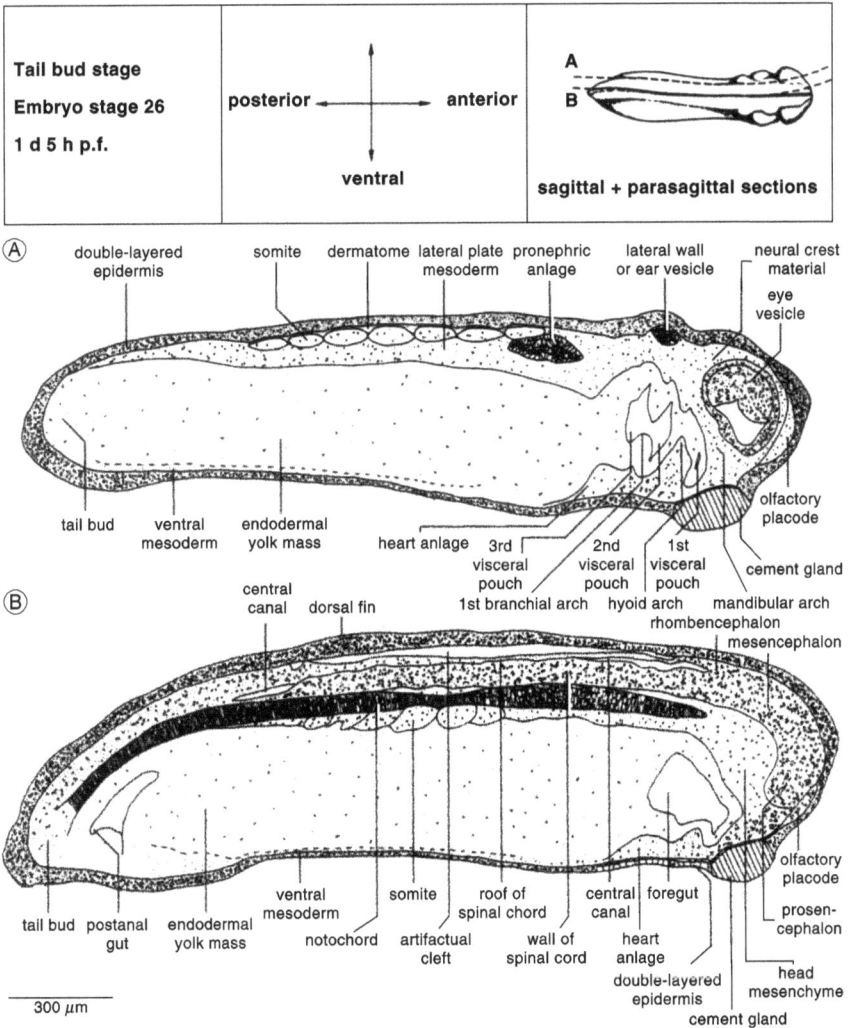

Figure 2.9. Sagittal and parasagittal sections of a stage-26 embryo. Further regionalization of the germ layers along the anteroposterior axis becomes apparent at this stage. (Reprinted, with permission, from Hausen and Riebesell 1991.)

NEURAL DEVELOPMENT

The vertebrate central nervous system can be divided into spinal cord, hindbrain (or rhombencephalon), midbrain (or mesencephalon), and forebrain (or prosencephalon). The forebrain is further subdivided into diencephalon and telencephalon. The *Xenopus* central nervous system and its subdivisions

| Tail bud stage

Embryo stage 26

1 d 5 h p.f. | dorsal
↑
↓
ventral | A   B
transversal sections |

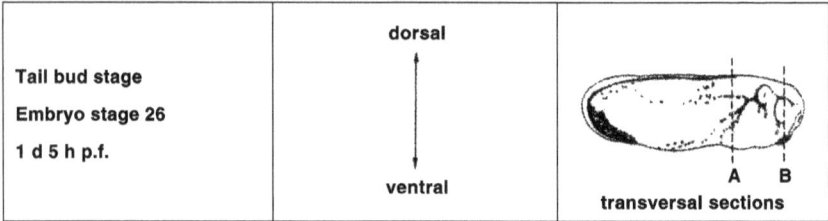

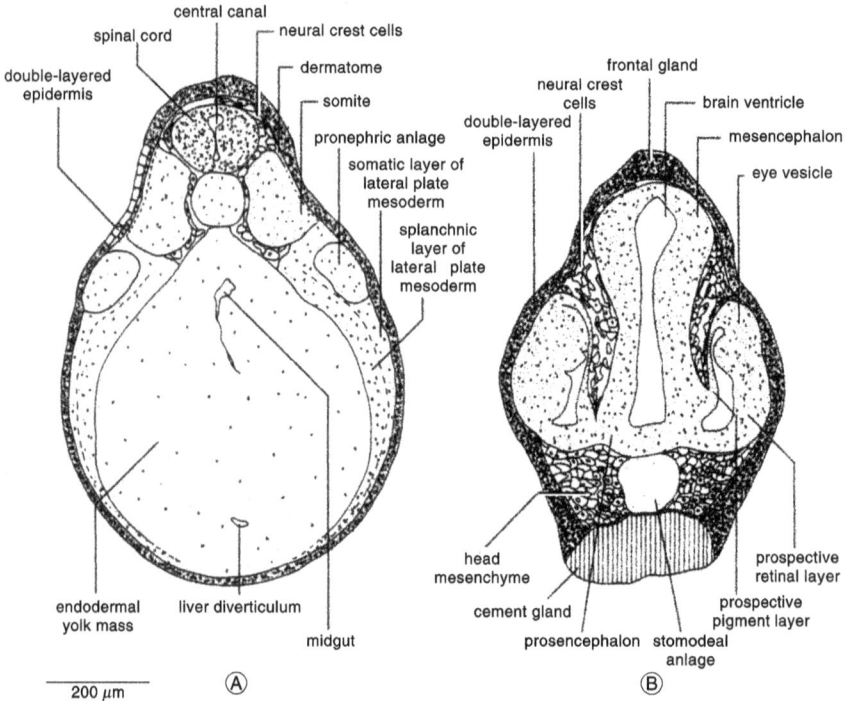

Figure 2.10. Transverse sections of a stage-26 embryo. Extensive regionalization of the germ layers along the dorsoventral and left-right axes is apparent at this stage. (Reprinted, with permission, from Hausen and Riebesell 1991.)

are beginning to be determined in embryos undergoing gastrulation, when the neural ectoderm is induced on the dorsal side of the embryo by signals originating from the dorsal mesoderm (for review, see Hemmati-Brivanlou and Melton 1997). Fate mapping in *Xenopus* (Keller et al. 1992a; see also the discussion of fate mapping in Chapter 9) has shown that in the early gastrula, the prospective hindbrain and spinal cord extend well into the lateral regions of the embryo (Figure 2.13A). During gastrulation and neurulation, cells in the posterior neural plate undergo dramatic convergence toward the

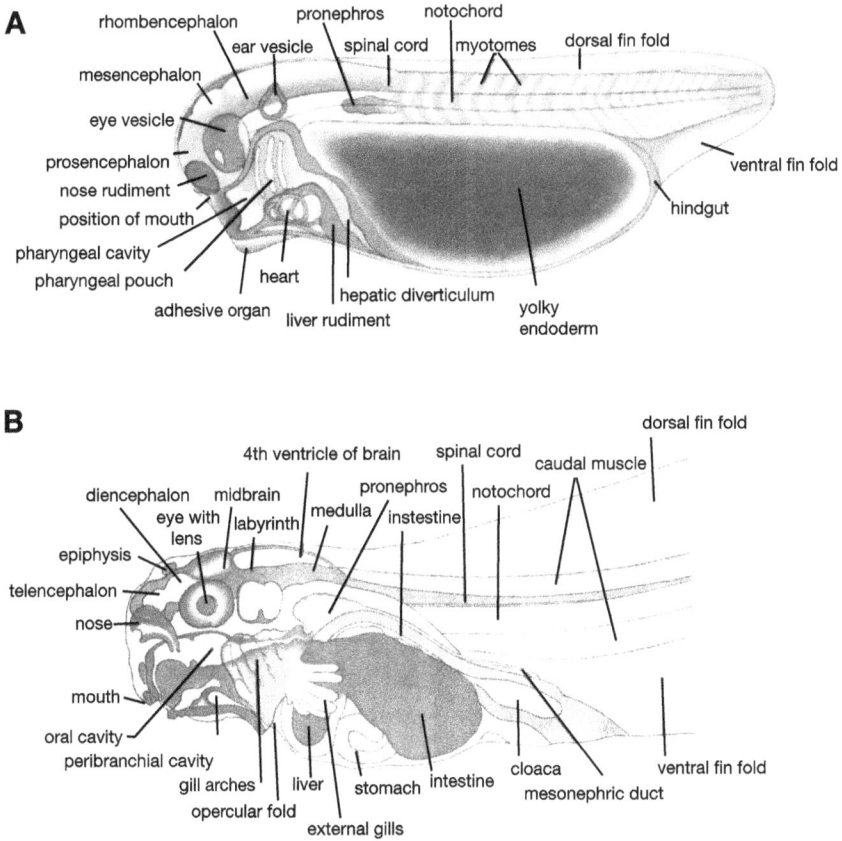

Figure 2.11. Tissue differentiation during tailbud (*A*) and tadpole (*B*) stages of frog embryos. At these stages, overt tissue differentiation is clearly apparent throughout the embryo. (Reprinted, with permission, from Balinsky 1981 [© Saunders College Publishing].)

midline, and anteroposterior extension (Keller et al. 1992a,b; Elul et al. 1997). Thus, material from the lateral sides of the gastrula is brought toward the midline, and the neural plate transiently adopts a keyhole shape (Figure 2.13B). This is the first stage at which the neural ectoderm displays a clear mediolateral and anteroposterior (or rostrocaudal) axis. The neural plate can be subdivided along the anteroposterior axis into two regions: the prechordal neural plate, which gives rise to the forebrain, and the epichordal neural plate, which gives rise to the midbrain, hindbrain, and spinal cord. The epichordal neural plate takes its name from the fact that it lies over the notochord, which forms along the midline of the underlying dorsal mesodermal mantle during gastrulation.

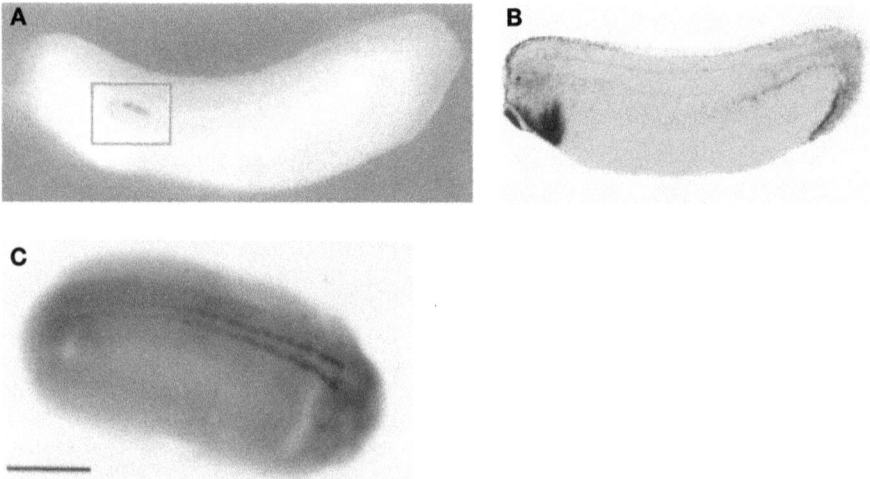

Figure 2.12. (*See Color Plate 2*) Regionalized gene expression at neural tube and tailbud stages. (*A*) *xWT1* defines the pronephric region. (Reprinted, with permission, from Carroll and Vize 1996 [© Wiley-Liss].) (*B*) *XTin1* delineates the presumptive heart region. (Reprinted, with permission, from Cleaver et al. 1996 [© Company of Biologists].) (*C*) *xHB9* is expressed in the presumptive motor neuron area of the spinal cord (Reprinted, with permission, from Saha et al. 1997.)

During neurulation, the midline of the neural plate deepens and the lateral edges rise toward each other, eventually fusing to create the dorsal midline, as seen in Figures 2.6 and 2.7. The lateral-to-medial axis of the neural plate becomes the dorsoventral axis of the neural tube. A large number of genes that are active during neurulation have been identified, and many have expression domains that define highly restricted regions within the neural plate. Some examples of these genes are shown in Figure 2.14. Figure 2.14A illustrates expression of a neuron-specific β-tubulin gene (N-tubulin) in three stripes in the posterior neural plate defining domains along the lateral-to-medial (future dorsoventral) axis where primary neurons will form (Chitnis et al. 1995). Early *Xash-3* and *F-cadherin* expression define the sulcus limitans, the boundary separating the dorsal and ventral parts of the neural tube (Zimmerman et al. 1993; Espeseth et al. 1995). Figure 2.14B shows early *Xash-3* expression. *Rx* gene expression, seen in Figure 2.14C, is a marker for the presumptive retina (Mathers et al. 1997).

Neurulation is more complicated in the anterior neural plate; in addition to lateral folding of the neural edges, there is also downward rotation of the anteroposterior axis, by approximately 90°, at the cephalic flexure (Figure 2.13C). The complex movements of the anterior neural plate and ridge during neurulation are described by Eagleson et al. (1995). At later embryonic

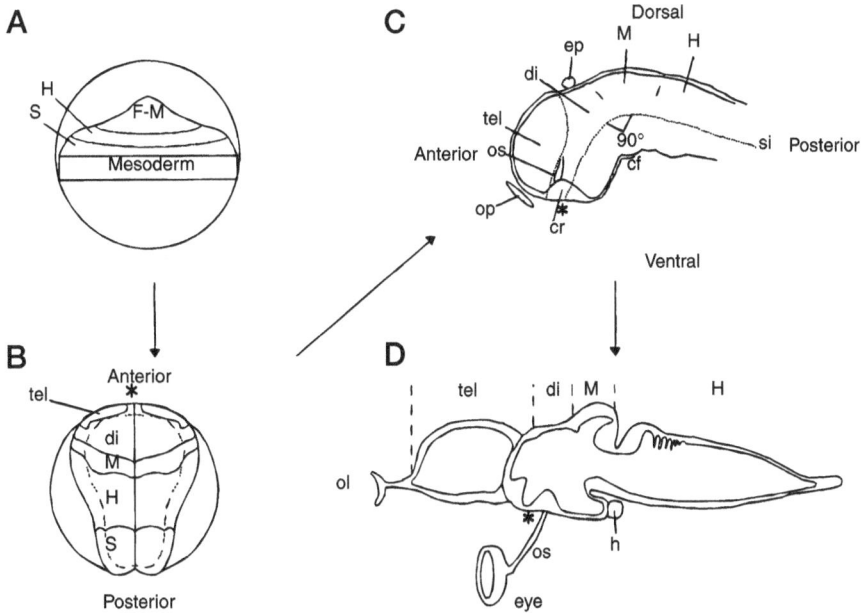

Figure 2.13. Diagrammatic series of brain development. (*A*) Early gastrula stage, dorsal view; (*B*) neural plate stage, dorsal view; (*C*) stage-29–30 side view of developing brain structures (N. Papapulu, unpubl.); (*D*) side view of stage-46 brain structures. The positions of the boundaries between different brain regions are approximate. The asterisk shows the position on the neural tube of the original anterior end of the neural plate. (cr) Chiasmatic ridge; (cf) cephalic flexure; (di) diencephalon; (ep) epiphysis; (H) hindbrain; (h) hypophysis; (M) midbrain; (ol) olfactory organ; (op) olfactory placode; (os) optic stalk; (S) spinal cord; (sl) sulcus limitans; (tel) telencephalon. (Figure compiled, with permission, from Eagleson and Harris 1989 [© John Wiley & Sons], and Keller et al. 1992a [© Wiley-Liss].)

stages (after stage 30), the forebrain gradually acquires a more straightened appearance, due to growth and elongation of tissue in front of the cephalic flexure (Figure 2.13D). Figure 2.15 shows the major parts of the central nervous system of a frog tadpole.

The neural plate is surrounded by prospective neural crest and placodes. Cranial neural crest material segregates on the lateral edges of the anterior neural plate at stage 15 and starts migrating after stage 20. This progression of events is illustrated in Figure 2.16. The cranial neural crest contributes to a variety of tissues, including the skeleton of the head and the branchial (gill) arches. *Xtwi* (Hopwood et al. 1989) and *Xslu* (Mayor et al. 1995) are good markers for neural crest at neural plate stages (*Xslu* is shown in Figure 2.17). Placodes develop as thickenings of the deep layer of the ectoderm, visible from about stage 22. The olfactory, adenohypophyseal, otic, epibrachial, and

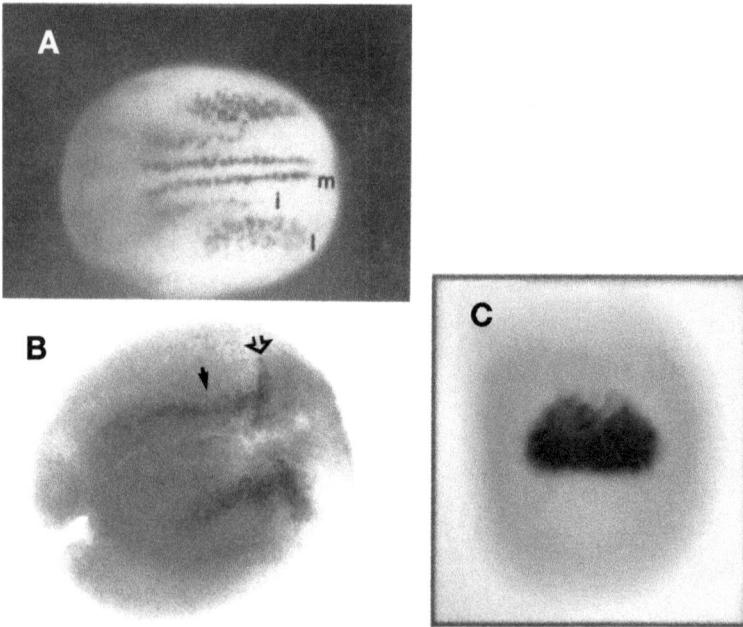

Figure 2.14. (*See Color Plate 3*) Genes expressed in particular domains of neural tissue during neurulation. (*A*) Expression of *N-tubulin* in stripes corresponding to the sites where primary neurons will form (dorsal view). (Reprinted, with permission, from Chitinis et al. 1995 [© Macmillian Magazines].) (*B*) Lateral domains of *Xash-3* expression demarcate the future sulcus limitans, the border between dorsal and ventral parts of the spinal cord (dorsal view). (Reprinted, with permission, from Zimmerman et al. 1993 [© Company of Biologists].) (*C*) *Rx* expression in the domain of the future retina (anterior view). (Reprinted, with permission, from Mathers et al. 1997 [© Macmillian Magazines.])

dorsolateral placodes give rise to neuronal cells (Figure 2.18), whereas the lens placode gives rise to the lens of the eye; see Figure 2.19.

Both cranial neural crest and placodes contribute to the formation of the cranial ganglia. Cranial nerves, in addition to other neural structures, can be visualized by hybridization to tanabin or N-tubulin mRNA (expression of tanabin is shown in Figure 2.20) (Hemmati-Brivanlou et al. 1992; Chitnis et al. 1995) or immunostaining with anti-acetylated tubulin antibody (Hartenstein 1993). The adenohypophysis originates from a very anterior placode at the neural plate stage (for an early molecular marker, see Mathers et al. 1995) but translocates posteriorly during neurulation and establishes a connection with the posterior diencephalon, which also originates from an anterior position at the neural plate stage (Eagleson et al. 1995 and references therein). Just outside the anterior neural plate lies the primordium of the cement gland, a mucus-secreting ectodermal organ that helps the tadpole

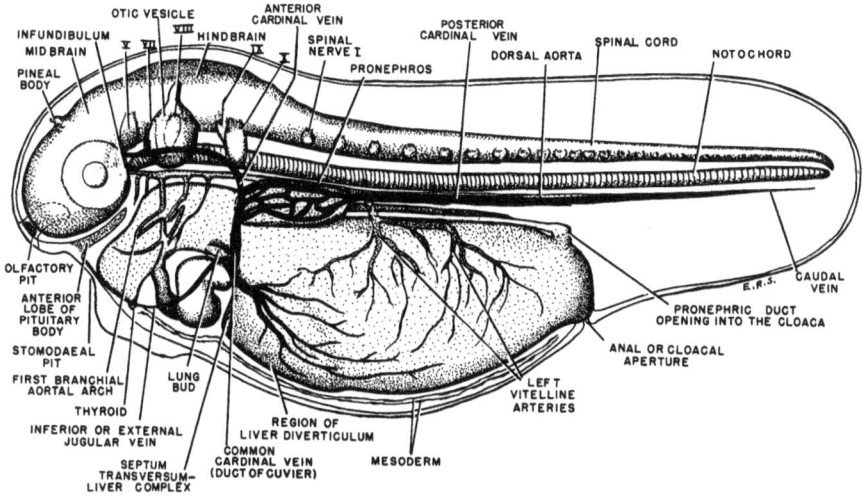

Figure 2.15. Frog tadpole nervous system. This illustration highlights the key features of the tadpole nervous system, including the main divisions of the brain and the location of cranial nerves. (Reprinted from Nelson 1953.)

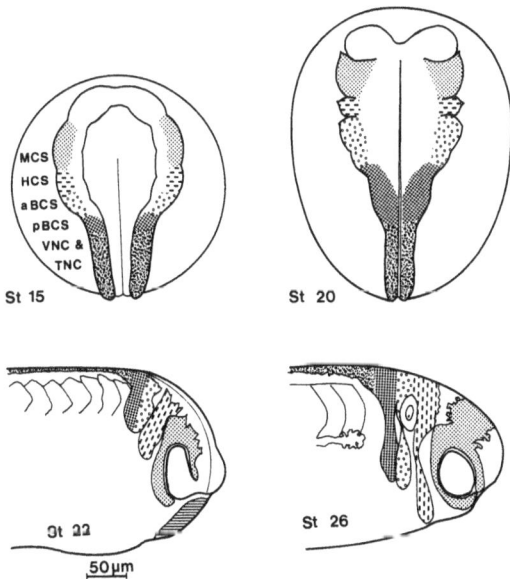

Figure 2.16. Diagrammatic series of neural crest development. (MCS) Mandibular crest segment; (HCS) hyoid crest segment; (aBCS) anterior branchial crest segment; (pBCS) posterior branchial crest segment; (VNC and TNC) vagal and trunk neural crest. (Reprinted, with permission, from Sadaghiani and Thibaud 1987.)

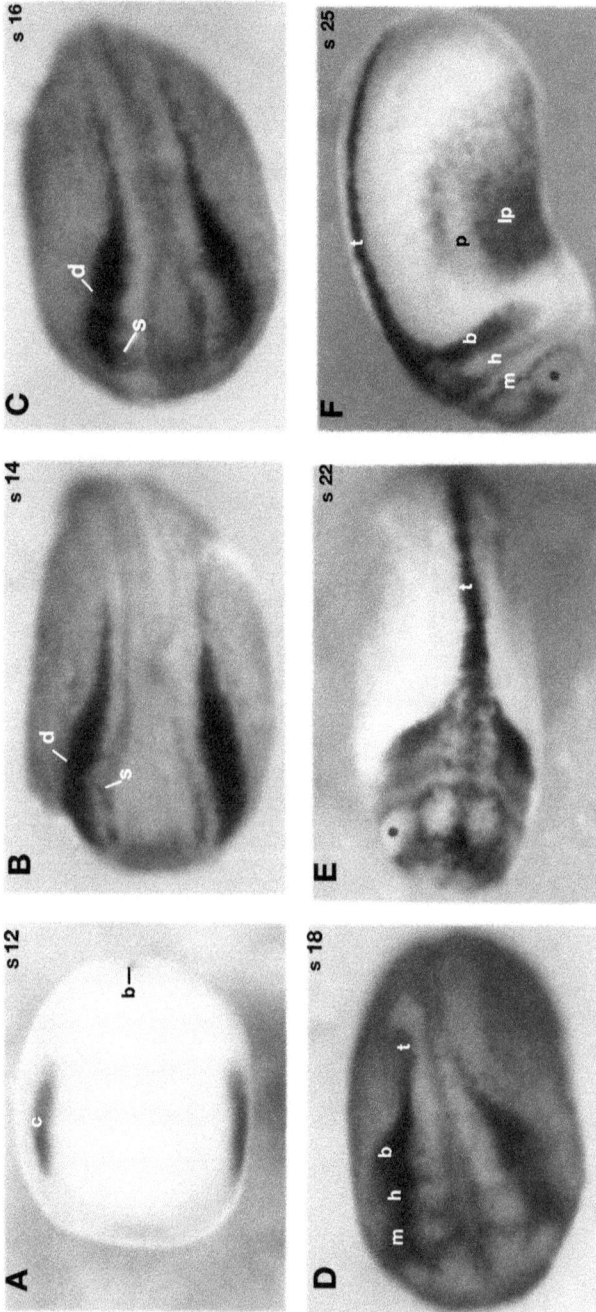

Figure 2.17. (*See Color Plate 4*) Expression of *Xslu* in neural folds and neural crest tissue. (Reprinted, with permission, from Mayor et al. 1995 [© Company of Biologists].) (*A*) Stage-12 embryo; (*B*) stage-14 embryo; (*C*) stage-16 embryo; (*D*) stage-18 embryo; (*E*) stage-22 embryo; (*F*) stage-25 embryo. (*A–E*) Dorsal views; (*F*) side view.

Figure 2.18. Pattern of ectodermal placodes at the neural plate stage. (AHP) Adenohypophyseal placode; (AP) auditory (or otic) placode; (LP) lens placode; (CGA) cement gland anlage; (OP) olfactory placode; (NC) neural crest; (NPL) neural plate; (PPT) primitive placodal thickening. (Reprinted, with permission, from Hausen and Riebesell 1991; after Knouff 1935.)

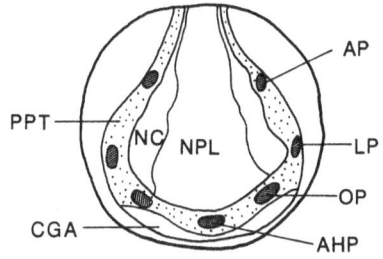

to suspend from the surface of the water; it degenerates in later life. Several genes are expressed in the cement gland anlage from early neural stages onward (Figure 2.20), such as *XCG13* (Jamrich and Sato 1989) *XCG*, and *XAG* (Sive et al. 1989); *XAG* is shown in Figure 2.21.

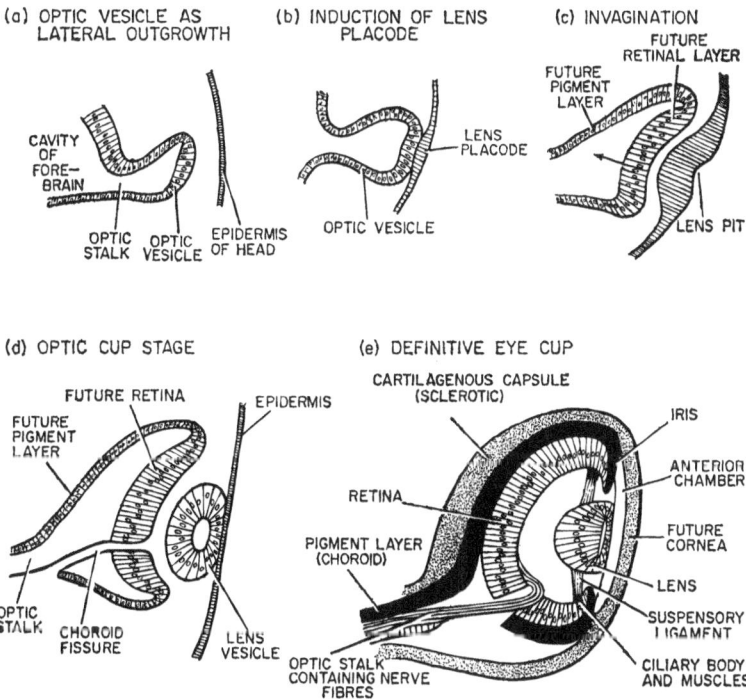

Figure 2.19. Frog eye development. The progression from the neurula stage through tadpole stages illustrates the formation of the lens from surface ectoderm and retinal layers from the optic vesicle. (Reprinted, with permission, from Deuchar 1966.)

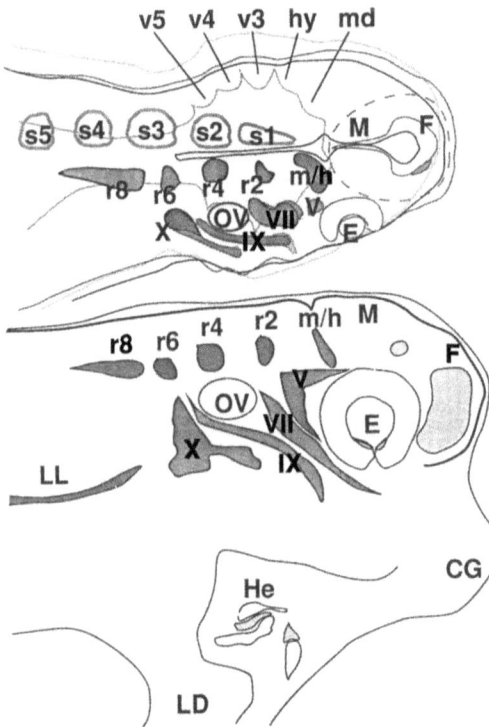

Figure 2.20. Summary of tanabin expression in stage-36 embryos; dorsal and lateral views. (F) Forebrain; (M) midbrain; (m/h) midbrain-hindbrain boundary; (r2-r8) rhombomeres 2–8. Cranial ganglia and nerves: V, VII, IX, and X (E) eye; (OV) otic vesicle; (s1-s5) somites; (H) heart; (LD) liver diverticulum; (md) mandibular arch; (hy) hyoid arch; (v3-v5) viseral arches. (Reprinted,with permission, from Hemmati-Brivanlou et al. 1992 [© Cell Press].)

MORPHOLOGICAL LANDMARKS OF THE NEURAL TUBE

Relatively little is known about the anatomy of the embryonic *Xenopus* brain. It is only recently, with the discovery of genes that are expressed in the early neural tube in a regionalized manner, that interest in early neuroanatomy has been revived. Most classical neurobiology texts are limited to descriptions of the anatomy of the adult brain (see Figure 2.22). In this section, some basic landmarks of the neural tube that are identifiable from early embryonic stages are described. A diagram illustrating some of the major features of later embryonic brain morphology in urodele embryos is shown in Figure 2.23 (taken from Slack and Tannahill 1992). The hindbrain is relatively easy to identify because of its triangular shape and its thin roof (late stage 20 onward). Posteriorly, the hindbrain tapers into the spinal cord,

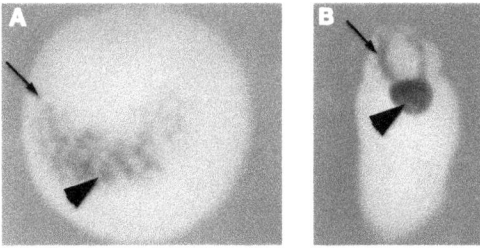

Figure 2.21. (*See Color Plate 5*) Expression of cement gland marker XAG. Anterior views of early neurula (*A*) and tailbud (*B*) stage embryos. This gene is expressed in the cement gland (arrowheads) and adjacent hatching gland (arrows). (Reprinted, with permission, from Sive and Bradley 1996; [© Wiley-Liss].)

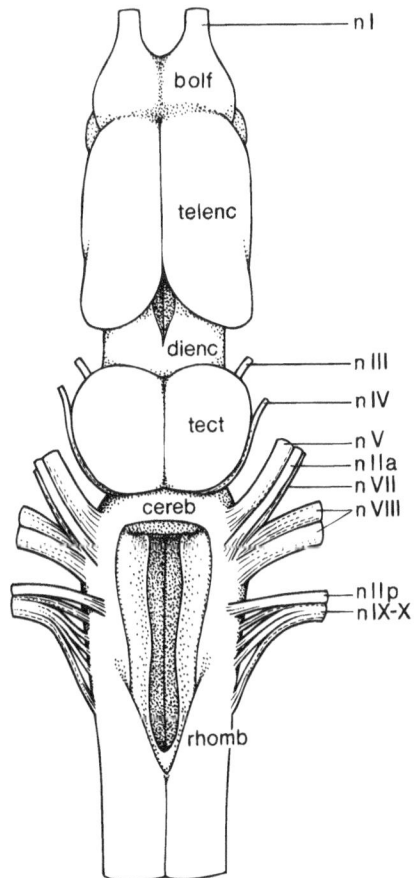

Figure 2.22. Diagram of adult *Xenopus* brain with cranial nerves. The nII, optic nerve, is not shown. (bolf) Olfactory bulb; (cereb) cerebellum; (dienc) diencephalon; (rhomb) rhombencephalon; (tect) tectum; (telenc) telencephalon; (nI) olfactory nerve; (nIII) oculomotor nerve; (nIV) trochlear nerve; (nV) trigeminal nerve; (nVII) facial nerve; (nVIII) acoustic nerve; (nIX-X) glossopharyngeal and vagus nerves. (Reprinted, with permission, from Nikundiwe and Niuwenhuys 1983 [© Wiley-Liss].)

A

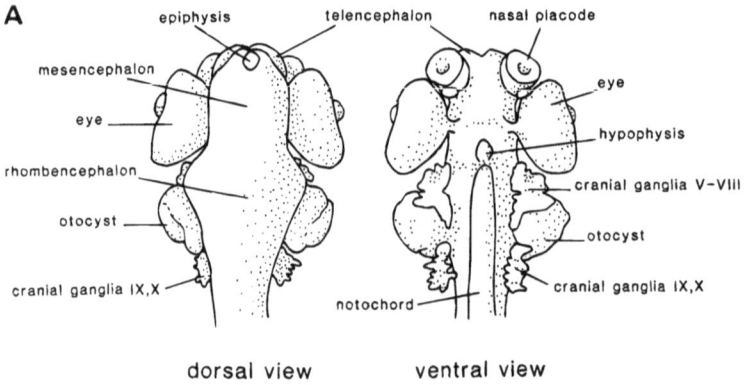

dorsal view   ventral view

B

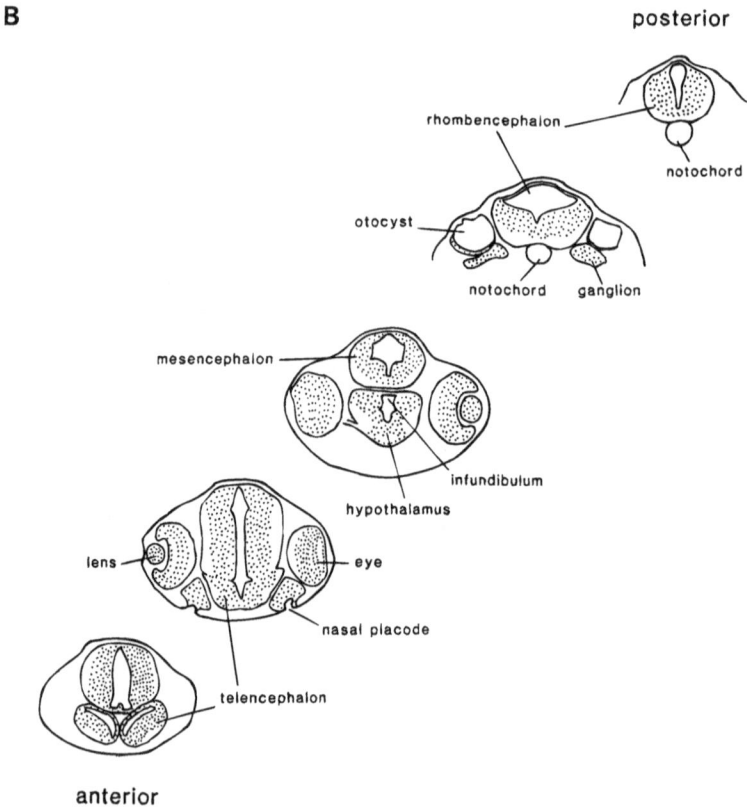

Figure 2.23. Structures in the central nervous system of a urodele embryo. (*A*) Views of the whole brain; (*B*) transverse sections through the brain. (Reprinted, with permission, from Slack and Tannahill 1992 [© Company of Biologists].)

which is very long and thin. Anteriorly, the hindbrain is separated from the midbrain by a prominent constriction, the isthmus. The hindbrain shows a series of seven to eight transient bulges, which are called rhombomeres and are thought to be segmental units of neuronal development (Lumsden 1990). Most of the work on hindbrain segmentation has been done in chickens and zebrafish, and it is unclear how much of the research can be applied to *Xenopus* (Hartenstein 1993). Nevertheless, rhombomeres can be seen in sectioned or whole-mount immunostained tadpoles, at about stage 43, perhaps even earlier (e.g., see tanabin expression, Figure 2.20). As in other species, the roots of some cranial nerves show a two-segment relationship with the rhombomeres. Thus, the root of the trigeminal (Vth) root is located in rhombomere 2 (r2), that of the facial (VIIth) in r4, and that of the glossopharyngeal (IXth) in r6 (the full range of cranial nerves is shown in Figure 2.15).

Gene expression can be used to identify some prospective structures in this area of the brain at a much earlier stage than would be possible using morphology. For example, the *Engrailed-2 (En-2)* gene is routinely used as a marker for the midbrain-hindbrain boundary (Bolce et al. 1992), and *Krox-20* (Bradley et al. 1993) is used as a marker for rhombomeres 3 and 5 (Figure 2.24). Generally speaking, however, one should be cautious in using gene expression as a stable cell marker, since expression may be modified between early and late stages.

The otocyst, the prospective ear, provides another landmark that is easy to identify under the dissecting microscope or with Nomarski optics. The otocyst develops from the otic placode and, at stage 27, appears as a small vesicle, adjacent to the middle of the hindbrain. By stage 43, the center of the otocyst lies opposite rhombomere 5.

The midbrain develops dorsally to the posterior aspect of the cephalic flexure and is the primary visual center in frogs; the dorsal part of the midbrain forms the tectum where optic fibers from the eyes synapse. The anterior border of the midbrain is difficult to identify accurately at early stages (stages 20–30). By analogy to zebrafish and chicken embryos, the tract of the posterior commissure classically defines the dorsal aspect of midbrain-forebrain boundary and can be identified by staining with any antibody that recognizes axons such as the anti-acetylated tubulin or anti-neurofilament antibody (Fidgor et al. 1993; Macdonald et al. 1994).

A number of landmarks can be used to delineate regions of the forebrain. The epiphysis, or pineal gland, is a small evagination that develops on top of the diencephalon and can be identified by staining with HNK-1 antibody (Hartenstein 1993) and *Xlim-3* (Taira et al. 1993) expression, starting at stage 24. The eyes provide a useful, easily identifiable landmark of the dien-

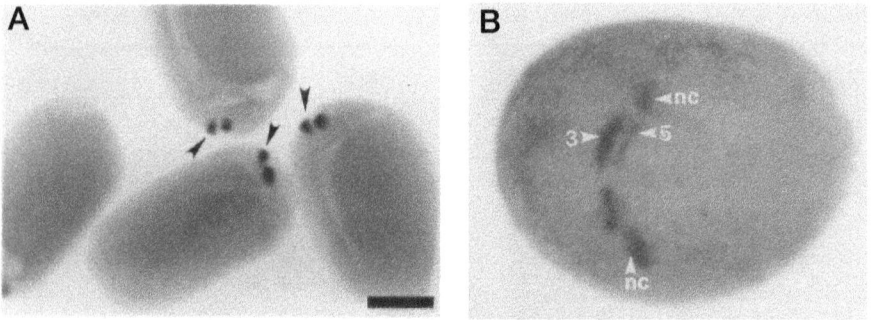

Figure 2.24. (*See Color Plate 6*) Early gene expression demarcates future regional properties in the neural plate. (*A*) Engrailed expression is restricted to the boundary between midbrain and hindbrain. (Reprinted, with permission, from Bolce et al. 1992 [© Company of
Biologists].) (*B*) Krox-20 expression is found in rhombomeres 3 and 5 of the hindbrain and
adjacent cells destined to be part of the neural crest. (Reprinted, with permission, from
Bradley et al. 1993 [© Elsevier Science].)

cephalon; eye development is summarized in Figure 2.19. Around stage 22,
the future eyes develop as protrusions from the diencephalic wall, the optic
vesicles. Where the optic vesicles contact surface ectoderm, the lens placode
and subsequently the lens vesicle will form. The optic vesicles invaginate to
form double-layered optic cups, connected with the ventral diencephalon by
long thin tubular structures, the optic stalks. The outer layer of each optic
cup forms the pigment epithelium and the inner layer forms the neural retina. Axons of the optic nerve grow from the retina along the optic stalks, and
enter the diencephalon in the chasmiatic ridge. The position where the optic
stalks are connected to the brain, the optic recess, defines the ventral aspect
of the telencephalic-diencephalic border. The dorsal aspect of this boundary
is indicated by a slight depression at mid-20 stages. The telencephalon
shows the greatest growth relative to other parts of the nervous system. The
cerebral hemispheres begin to evaginate around stage 46 and the part of the
telencephalon that receives input from the olfactory neurons forms the
paired olfactory bulbs.

ADULT MORPHOLOGY

Adult morphology and physiology are thoroughly described by Deuchar
(1975). Figure 2.25 shows the key morphological features of the adult brain.
Illustrations of the viscera of *Xenopus* and a diagram of the urinogenital system of both males and females are shown in Figure 2.25.

A.

r. auricle
ventricle

l. auricle
heart
liver (deflected)
gall bladder
pancreas
spleen
stomach
duodenum
intestine
rectum
l. lung
bladder

B.

inferior vena cava
kidney
fat body
testis

bladder
Wolffian duct

C.

l. ovary — deflected to r.
l. fat body — deflected to r.

l. oviducal funnel
l. oviduct, highly coiled
inferior vena cava
l. kidney — deflected to r.

r. ovisac
l. ovisac

bladder — deflected to r.

cloaca

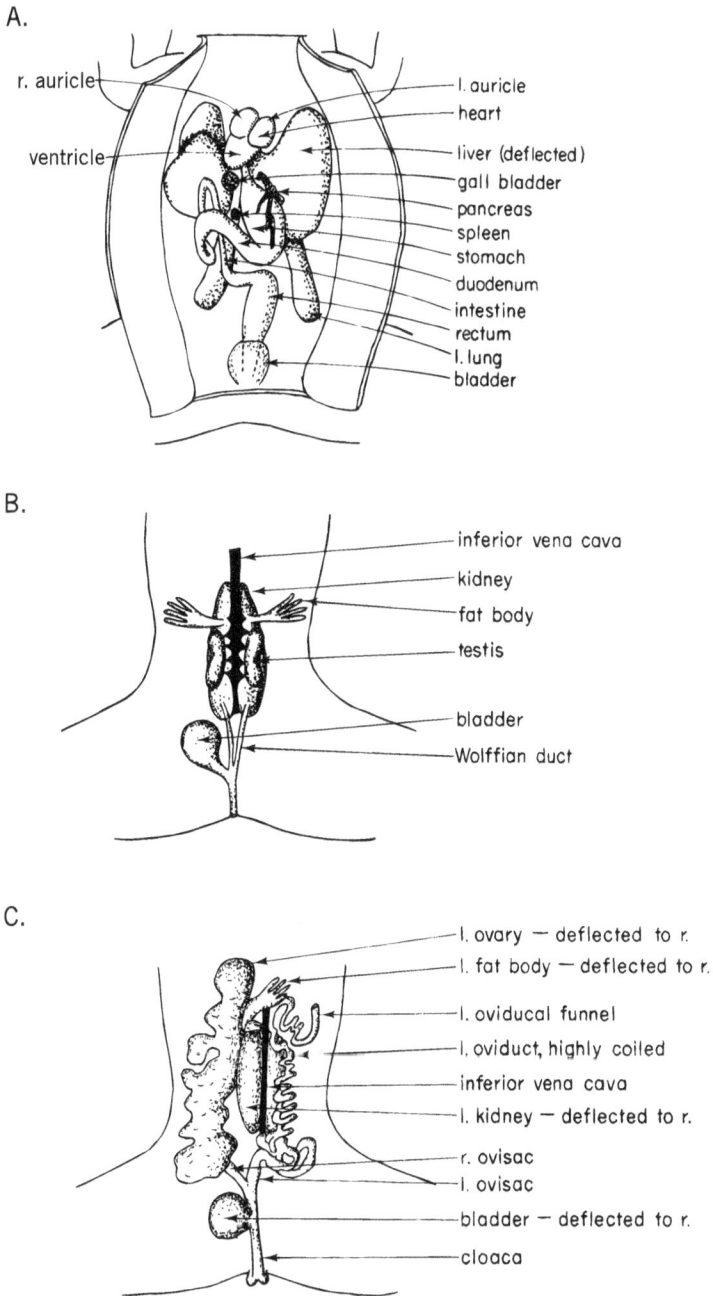

Figure 2.25. Adult *Xenopus* tissues. (*A*) Ventral view of viscera; (*B*) ventral view of male urinogenital system; (*C*) ventral view of female urinogenital system. (Reprinted, with permission, from Deuchar 1975; © John Wiley & Sons.)

REFERENCES

Baker C.V. and Bronner-Fraser M. 1997. The origins of the neural crest. Part I: Embryonic induction. *Mech. Dev.* **69:** 3–11.

Bolce M.E., Hemmati-Brivanlou A., Kushner P.D., and Harland R.M. 1992. Ventral ectoderm of *Xenopus* forms neural tissue, including hindbrain, in response to activin. *Development* **115:** 681–688.

Balinsky B.I. 1981. *An introduction to embryology*, Fifth edition. Saunders, Philadelphia, College Publishing.

Bradley L.C., Snape A., Bhatt S., and Wilkinson D.G. 1993. The structure and expression of the *Xenopus Krox-20* gene: Conserved and divergent patterns of expression in rhombomeres and neural crest. *Mech. Dev.* **40:** 73–84.

Carroll T.J. and Vize P.D. 1996. Wilms' tumor suppressor gene is involved in the development of disparate kidney forms: Evidence from expression in the *Xenopus* pronephros. *Dev. Dynam.* **206:** 131–138.

Chitnis A., Henrique D., Lewis J., Ish-Horowicz D., and Kintner C. 1995. Primary neurogenesis in *Xenopus* embryos regulated by a homologue of the *Drosophila* neurogenic gene *Delta. Nature* **375:** 761–766.

Cho K.Y., Blumberg B., Steinbeisser H., and DeRobertis E.M. 1991. Molecular nature of Spemann's Organizer: The role of the *Xenopus* homeobox gene *goosecoid. Cell* **67:** 1111–1120.

Cleaver O.B., Patterson K.D., and Krieg P.A. 1996. Overexpression of the *tinman*-related genes *XNkx-2.5* and *XNkx-2.3* in *Xenopus* embryos results in myocardial hyperplasia. *Development* **122:** 3549–3556.

Deuchar E. 1966. *Biochemical aspects of amphibian development*. Methuen London and John Wiley & Sons, New York.

———— 1975. Xenopus: *The South African clawed frog*. John Wiley London & Sons, New York.

Eagleson G.W. and Harris W.A. 1990. Mapping of the presumptive brain regions in the neural plate of *Xenopus laevis. J. Neurobiol.* **21:** 427–440.

Eagleson G., Ferreiro B., and Harris W.A. 1995. Fate of the anterior neural ridge and the morphogenesis of the *Xenopus* forebrain. *J. Neurobiol.* **28:** 146–158.

Elul T., Koehl M.A., and Keller R. 1997. Cellular mechanism underlying neural convergent extension in *Xenopus laevis* embryos. *Dev. Biol.* **191:** 243–58

Espeseth A., Johnson E., and Kintner C. 1995. *Xenopus* F-cadherin, a novel member of the cadherin family of cell adhesion molecules, is expressed at boundaries in the neural tube. *Mol. Cell Neurosci.* **6:** 199–211

Evans S.M., Yan W., Murillo M.P., Ponce J., and Papalopulu N. 1995. *Tinman*, a *Drosophila* homeobox gene required for heart and visceral mesoderm specification, may be represented by a family of genes in vertebrates: *XNkx-2.3* a second vertebrate homologue of tinman. *Development* **121:** 3889–3899.

Fernandez A.S., Pieau C., Reperant J., Boncinelli E., and Wassef M. 1998. Expression of the *Emx-1* and *Dlx-1* homoeobox genes defines three molecularly distinct domains in the telencephalon of mouse, chick, turtle and frog embryos—Implications for the evolution of telencephalic subdivisions in amniotes. *Development* **125:** 2099–2111.

Figdor M.C. and Stern C.D. 1993. Segmental organization of embryonic diencephalon. *Nature* **363:** 630–640.

Hartenstein V. 1993. Early pattern of neuronal differentiation in the *Xenopus* embryonic brainstem and spinal cord. *J. Comp. Neurol.* **328:** 213–231.

Hausen P. and Riebesell M. 1991. *The early development of* Xenopus laevis. Springer-Verlag, Berlin.

Hemmati-Brivanlou A. and Melton D. 1997. Vertebrate nerual induction. *Annu. Rev. Neurosci.* **20:** 43–60.

Hemmati-Brivanlou A., Mann R.W., and Harland R.M. 1992. A protein expressed in the growth cones of embryonic vertebrate neurons defines a new class of intermediate filament protein. *Neuron* **9:** 417–428

Hopwood N.D., Pluck A., and Gurdon J.B. 1989. A *Xenopus* mRNA related to *Drosophila* twist is expressed in response to induction in the mesoderm and the neural crest. *Cell* **59:** 893–903.

Jamrich M. and Sato S. 1989. Differential gene expression in the anterior neural plate during gastrulation of *Xenopus laevis*. *Development* **105:** 779–786.

Keller R.E. 1986. The cellular basis of amphibian gastrulation. In *Developmental biology: A comprehensive synthesis.* Volume 2: *The cellular basis of morphogenesis* (ed. Browder L.), pp. 241–327. Plenum Press, New York.

———1991. Early embryonic development of *Xenopus laevis*. *Methods Cell Biol.* **36:** 61–113.

Keller R., Shih J., and Sater A.K. 1992a. The cellular basis of the convergence and extension of the *Xenopus* neural plate. *Dev. Dynam.* **193:** 199–217.

Keller R., Shih J., Sater A.K., and Moreno C. 1992b. Planar induction of convergence and extension of the neural plate by the organizer in *Xenopus*. *Dev. Dynam.* **193:** 218–234.

Knouff R.A. 1935. The developmental pattern of ectodermal placodes in *Rana pipiens*. *J. Comp. Neurol.* **62:** 17–71.

Lumsden A.G.S. 1990. The development and significance of hindbrain segmentation. *Semin. Dev. Biol.* **1:** 117–125.

Macdonald R., Xu Q., Barth K.A., Mikkola J., Holder N., Fjose A., Krauss S., and Wilson S.W. 1994. Regulatory gene expression boundaries domarcate sites of neuronal differentiation in the embryonic zebrafish forebrain. *Neuron* **13:** 1039–1053.

Mathers P.H., Grinberg A., Mahon K.A., and Jamrich M. 1997. The Rx homeobox gene is essential for vertebrate eye development. *Nature* **387:** 603–607.

Mathers P.H., Miller A., Doniach T., Dirksen M.L., and Jamrich M. 1995. Initiation of anterior head-specific gene expression in uncommitted ectoderm of *Xenopus laevis* by ammonium chloride. *Dev. Biol.* **171:** 641–654.

Mayor R., Morgan R., and Sargent M.G. 1995. Induction of the prospective neural crest of *Xenopus*. *Development* **121:** 767–777.

Nelson O.E. 1953. *Comparative embryology of the vertebrates*. Blakiston, New York.

Nieuwkoop P.D. and J. Faber. 1994. *Normal table of* Xenopus laevis. Garland Publishing, New York.

Nikundiwe A.M. and Niuwenhuys R. 1983. The cell masses in the brainstem of the South African clawed frog *Xenopus laevis. J. Comp. Neurol.* **213:** 199–219.

Rugh R. 1951. *The frog. Its reproduction and development.* Blakiston, Philadelphia.

Sadaghiani B. and Thiebaud C.H. 1987. Neural crest development in the *Xenopus laevis* embryo, studied by interspecific transplantation and scanning electron microscopy. *Dev. Biol.* **124:** 91–110.

Saha M.S., Miles R.R., and Grainger R.M. 1997. Dorsal-ventral patterning during neural induction in *Xenopus*: Assessment of spinal cord regionalization with xHB9, a marker for the motor neuron region. *Dev. Biol.* **187:** 209–223.

Schroeder T.E. 1970. Neurulation in *Xenopus laevis*. An analysis and model based upon light and electron microscopy. *J. Embryol. Exp. Morphol.* **23:** 427–462.

Sive H. and Bradley L. 1996. A sticky problem: The *Xenopus* cement gland as a paradigm for anteroposterior patterning. *Dev. Dynam.* **205:** 265–280.

Sive H.L., Hattori K., and Weintraub H. 1989. Progressive determination during formation of the anteroposterior axis in *Xenopus laevis. Cell* **58:** 171–180.

Slack J.M.W. and Tannahill D. 1992. Mechanism of anteroposterior axis specification in vertebrates: Lessons from the amphibians. *Development* **114:** 285–302.

Smith J.L. and Schoenwolf G.C. 1997. Neurulation: Coming to closure. *Trends Neurosci.* **20:** 510–517.

Smith W.C. and Harland R.M. 1992. Expression cloning of noggin, a new dorsalizing factor localized to the Spemann organizer in *Xenopus* embryos. *Cell* **70:** 829–840

Taira M., Hayes W.P., Otani H., and Dawid I.B. 1993. Expression of LIM class homeobox gene Xlim-3 in *Xenopus* development is limited to neural and neuroendocrine tissues. *Dev. Biol.* **159:** 245–256.

Zimmerman K., Shih J., Bars J., Collazo A., and Anderson D.J. 1993. XASH-3, A novel *Xenopus* achaete-scute homolog, provides an early marker of planar neural induction and position along the mediolateral axis of the neural plate. *Development* **119:** 221–232.

The video series "Manipulating the Early Embryo of Xenopus laevis*" presents extensive illustrations of the external and internal morphological changes during gastrulation. In addition, a number of operations on bastula, gastrula, and neurula stage embryos illustrate features described in this chapter.*

CHAPTER 3

Manipulating Gene Expression in *Xenopus*

Discussed in this chapter are the various approaches that have been used to manipulate gene expression in *Xenopus*. It has not yet been possible to target mutations to particular genes in *Xenopus*, nor to work with a large collection of mutants. However, with the recent advances in molecular genetics, it is now possible to increase expression or activity of a particular gene and to design interventions that block the activity of a gene. Increased activity (a gain-of-function approach) can be useful to elucidate the function of a particular protein during development. Overexpression of a gene to elevate its activity or use of dominant mutant copies causes specific reproducible developmental changes. Such changes can indicate the normal function of the gene. Alternatively, by excess activity, a protein can make an obvious difference in the expression of downstream genes, thereby indicating the biochemical activity of the gene product.

Injection of cleavage-stage embryos with messenger RNA has been the most popular way to express foreign genes in *Xenopus*. This straightforward procedure allows high levels of expression to be obtained (Krieg and Melton 1987b). Injection of mRNAs into cleavage-stage embryos results in widespread and high-level gene expression; however, injected mRNAs are generally translated immediately after their introduction, resulting in expression inconsistent with the normal timing of gene activity and giving potentially misleading results. Therefore, to analyze gene function during development, it is more useful to express foreign genes in a temporally controlled manner, particularly when examining events of the gastrula or neurula stages. In some instances, gene expression can be obtained from injected plasmid DNA, although expression tends to be nonuniform. Alternative methods of obtaining foreign gene expression include the use of transgenic embryos, where heterologous DNA is integrated stably into the genome, and hormone-inducible proteins. Both of these approaches have been used successfully to control the timing of gene activity.

Since experimental intervention must be related to events that occur in the animal during normal development, it is very important to know when and where foreign gene activity may be altering the natural series of events. Targeting molecules to particular regions of the embryo can be especially informative, and explants can be taken from embryos to isolate and study particular developmental programs.

Loss-of-function approaches rely on mutant molecules that interfere with a pathway in a dominant-negative fashion (Herskowitz 1987). The means of expressing dominant negatives are similar to those used for gain-of-function assays. Dominant-negative approaches are particularly valuable, since they work by interfering with a normal cellular component and therefore provide strong evidence that a particular component is important in development. In contrast, gain-of-function approaches only demonstrate that a particular component could, in principle, work in a certain way.

Many of these manipulations involve the embryo, but it is important not to lose sight of the usefulness of frog oocytes as a "spherical test tube." Oocytes can be used as vessels for protein expression or to study signal transduction events. In the past, they have frequently offered a useful first step in understanding basic cell physiology, prior to the hard work of fractionation, purification, and reconstitution of biochemical components. Although most uses of oocytes are beyond the scope of this chapter, the use of oocytes for making secreted proteins is reviewed here.

GAIN OF FUNCTION

The injection of functional DNA, mRNA, or protein can lead to striking effects on embryo development. However, these effects are often difficult to interpret, and a variety of experiments may be required to determine the primary effect of increased gene expression. Genes are usually experimentally expressed at a very high level, and the investigator should be aware of the potential criticism that an unphysiological manipulation is likely to lead to a nonspecific effect on gene expression or cell physiology. Although this criticism is obvious, the potential benefits of this kind of approach are also clear. Development in vertebrates is frequently controlled by multiple overlapping and redundant signals, which may be impossible to analyze using a loss-of-function approach. Instead, insights may be gained by increasing the expression of a particular gene. Even though a protein may usually work as a single component in a cocktail of gene products, the one overexpressed gene product may still interact with or stimulate a simple cellular response. Thus, even though overexpressed proteins might not mimic the exact normal function of the gene, the resulting phenotypes can be extremely informative to putative developmental control mechanisms.

ASSAYS

Scoring Whole Embryos and Explants

Analysis of perturbed development includes inspection of the overall anatomy of the embryo and a deeper level of analysis using molecular markers or histology. Overexpression of some genes can lead to an obvious change in morphology, which can be scored simply. As an example, injection of the ventral side of fertilized eggs with certain mRNAs may result in induction of a complete axis, including a head. If the axis induction is partial, then more detailed analysis is required at either a histological or molecular level. Ectopic axes that are incomplete can appear as bumps on the flank of the embryo and are not necessarily overtly different from a malformation of normal tissue. They should be scored for the kinds of tissues that might be expected in the axis, such as notochord, muscle, and neural tube (see, e.g., Steinbeisser et al. 1993). This is done most easily in whole-mount analysis, where many embryos can be processed simultaneously for immunohistochemistry or in situ hybridization (see Chapters 12 and 13). Analysis is often assisted by using a lineage tracer (such as *lacZ* mRNA) with the injected mRNA (see Chapter 9). The tracer shows whether the mRNA injection causes a particular pathway of differentiation or whether the injected cells become a source of secondary signals, recruiting uninjected cells into the ectopic structure (Smith and Harland 1991; Sokol et al. 1991). Even if the altered phenotype is subtle, lineage tracing can show that cells have changed their morphogenetic behavior considerably. For example, *goosecoid* or *Smad2* mRNA injection causes dorsal mesodermal cells to adopt the migrating properties of the early involuting endomesoderm (Niehrs et al. 1993; Baker and Harland 1996).

Molecular markers for different tissues or regional identities are now standard, having largely replaced histological analyses (see Chapter 14). Histology requires more experience than molecular marker techniques, and the results can be ambiguous. Histology is very useful in whole embryos, where different tissues are arranged in stereotyped ways, but it can be more difficult in explants. For example, morphology of neural development can be interpreted confidently with sufficient experience of histology (Nieuwkoop and Koster 1995), but expression of molecular markers provides a more objective identification of tissues. The problems of using histology are quite clear in some cases; e.g., if animal cap cells are induced by some signal to become endodermal cells, the resulting tissue would lack one of the main histological criteria for endoderm, namely, large cells, and would go unrecognized. It is also difficult to display convincing histological results in print. In contrast, interpretation of molecular marker expression does not require great skill, although with increasingly sensitive techniques, it still requires the usual caution. Two general approaches are taken to analyze molecular marker expression. The first is to extract RNA from explants and assay for expression of particular genes (see Appendix 3), and the second approach involves staining of explants (see below).

RNA Assays

Following RNA extraction, tissue-specific gene expression can be measured using Northern blotting, S1 nuclease protection, RNase protection, or reverse transcription–polymerase chain reaction (RT-PCR). All of these techniques provide valid approaches, with RT-PCR being the most sensitive.

Staining of Explants

Staining of embryo explants for molecular markers has been in use for some time, first with antibody staining of sections (Slack et al. 1984), then with antibody staining of whole-mount explants (Hemmati-Brivanlou et al. 1990), and finally with in situ hybridization analysis of explants (Bolce et al. 1992) (for a description of these methods, see Chapters 12 and 13). Staining of explants reveals not only the identity of the tissue, but also its arrangement and extent within the explant. Although not as quantitative as direct RNA methods, explant staining provides a different and useful kind of qualitative information about the reproducibility of an effect from explant to explant. Whereas analysis of RNA by Northern blot or RT-PCR may display robust expression of a gene after some manipulation, explant analysis might show that just a very few explants express a large amount of transcript. One example is the examination of explants treated with noggin (a neural inducer) for expression of midbrain or hindbrain markers. Whole RNA analysis might show that the marker En-2 is expressed, but explant analysis would show that only one or two explants from a large batch express En-2; these explants are invariably slightly larger than most and probably are contaminated with some kind of mesoderm (T.M. Lamb and R.M. Harland, unpubl.). When concluding that a particular manipulation induces some kind of gene expression, it is very important that the induction be reproducible and not due to an unanticipated combination of poorly controlled signals.

Interpretation of Molecular Markers

The sensitivity of new techniques has led to new problems in interpreting expression of molecular markers. It is almost an article of faith that expression of a molecular marker (or preferably a group of molecular markers) provides a diagnostic test for the presence of a particular tissue. However, explants are inevitably somewhat artificial, and both the level and pattern of gene expression may be very different from those found in the intact embryo. Two examples demonstrate the uncertainty of an interpretation. The first is that a combination of fibroblast growth factor (FGF) and noggin, applied to animal cap explants, leads to expression of various markers of anterior-posterior position in the neural tube (Lamb and Harland 1995). A simple interpretation would be that the explants have mimicked the development of the normal neural tube. However, analysis of the spatial expression of the markers leads to a different interpretation. Although expression of some of the markers, such as HoxB9 and Otx2, is restricted to patches or bands of tissue (as in the normal neural tube), expression of others, such as En-2, and sometimes Krox20, is diffuse. The experiment has therefore induced a new kind of tissue that does not exist in vivo, where cells are probably coexpressing En-2 and Krox20, and perhaps even other markers indicative of anterior-posterior position. Interpreting these details is not straightforward.

A second example illustrates the usefulness of comparing the intensity of gene expression on a per cell basis with the intensity of expression in embryos. For example, expression of activin-like signals or bone morphogenetic protein (BMP) antagonists in animal caps leads to expression of endodermal markers (Jones et al. 1993; Gamer and Wright 1995; Henry et al. 1996; Sasai et al. 1996). However, although activin or Vg1 leads to intense staining of cells, comparable to that seen in the normal endoderm, the BMP antagonists lead to diffuse and low-level staining. In this case, the molecular marker may indicate either that endoderm is formed or that there is some poorly understood and leaky gene expression. In conclusion, both whole RNA assays and explant analyses have their advantages, and since they provide complementary information, it is best to use both approaches.

TEMPORAL AND SPATIAL CONTROL OF GENE EXPRESSION

Addition of Protein to Explants

The most straightforward way to control the point at which cells are exposed to a secreted signaling protein is to add known amounts of protein directly to explants. This circumvents any potential problems with low-level, but biologically significant, expression from injected promoter constructs. It also allows the investigator to choose the precise time at which to introduce the protein, rather than relying on available promoters whose temporal activity might be unsuitable. An example of the value of this temporal control is demonstrated by the measurement of cells' competence to respond to growth factors such as activin (Green et al. 1990). In Green's experiments, animal cap explants were taken from embryos at successively later stages of development and incubated in a medium containing activin protein. It was found that cells taken from explants at later stages had dramatically declined in their ability to respond to activin. This decline correlated with a decline in competence to respond to mesoderm induction from vegetal cells.

Similarly, addition of noggin protein to explants was used to show that the induction of muscle in ventral mesodermal explants by noggin differed in mechanism from the induction of muscle by activin. Activin was only able to induce muscle actin gene expression at the blastula stage, whereas noggin could induce muscle at both blastula and gastrula stages (Smith et al. 1993). Such experiments have allowed rigorous measurements of the competence of cells to respond to a signal, measurements that would be difficult using any other method.

Use of Explants from Different Regions of the Embryo

It is often informative to examine how signaling molecules interact with the signaling already going on in the embryo. The normal fate map of the embryo (see Chapter 9) can be exploited to study the effects of proteins on particular embryonic tissues or inductive events. In broad terms, manipulations can be targeted to particular germ layers by using the animal cap (prospective ectoderm), marginal zone (prospective mesoderm and endoderm), or vegetal mass (prospective endoderm). In the case of ectoderm manipulation, the animal cap is used, since it has ectodermal fate both when left in the context of the embryo and when explanted. When left in place, most of the animal cap forms epidermis, although some becomes the cement gland and anterior neural plate. When isolated and cultured in simple buffered saline (e.g., MMR or equivalent, see Appendix 1), the animal cap forms epidermis exclusively (although the cells in this case form a poorly organized tissue known as "atypical epidermis"). However, the animal cap is sensitive to inducing signals and can respond to mesoderm inducers at the blastula stage (Smith 1987) or neural inducers at late blastula and gastrula stages (Lamb et al. 1993). It therefore provides a valuable assay for studying the effects of inducers.

In contrast to the animal cap, explants of the equatorial region of the blastula have already been exposed to mesoderm-inducing signals, and their repertoire of potential responses is more limited. Mesodermal explants respond to patterning signals such as BMPs and Wnts and can be used to assay dorsalizing or ventralizing signals (Dale et al. 1992; Jones et al. 1992; Christian and Moon 1993; Smith et al. 1993; Leyns et al. 1997; Wang et al. 1997).

As the embryo develops further, and organogenesis begins, parts of the embryo can be removed and tested for sensitivity to soluble proteins. However, the attractiveness of *Xenopus* as an experimental animal at these stages is then rivaled by other vertebrates, such as the chick and the rat, where embryos grow considerably, making dissection more straightforward than with the constant-sized *Xenopus* embryo. In addition, explanted tissue from chicks and rodent embryos develops well in enriched, although often serum-free, culture medium. Early embryonic explants from *Xenopus* require only buffered saline to develop well, but later explants, which may be running out of yolk stores, are likely to require an enriched medium for long-term culture. Nevertheless, *Xenopus* can still be used for these manipulations. Phenomena such as lens induction have been well studied with explants of *Xenopus* embryos (Saha et al. 1992), and as the molecular mechanisms of lens induction become clearer, *Xenopus* explants will be useful to study the effects of candidate inducers.

Addition of Protein at Various Times after Explant Preparation

Protein can be added to fresh explants and to explants that have been aged in vitro. If left in the context of the embryo, cells continue to be exposed to normal signals. Therefore, it is often useful to isolate the cells from the strong signaling centers of the embryo. Animal caps can be removed from early-stage embryos and aged in vitro to the stage at which the protein is added. This kind of manipulation is particularly valuable when examining neural inducing signals, since responsive ectoderm becomes progressively modified and restricted in its response to inducers when left in place. In the case of noggin or FGF protein addition, in vitro aging has been used to examine neural inducing effects during gastrula stages, without complications arising from the presence of induced mesoderm in the explant (Lamb and Harland 1995). Of course, an obvious criticism of such experiments is that the explant may not resemble any normal target that remains in the context of the embryo. However, to examine the mechanisms of embryonic development, it is often necessary to take a simplified approach, rather than a more holistic but experimentally intractable approach.

Injection of Synthetic mRNA into Embryos

Injection of mRNA is the most commonly used manipulation of *Xenopus* early development. Injected mRNA directs early synthesis of a large amount of protein. Although this can be a strength, in some cases, early activity may preclude analysis of a different later-acting activity of the same protein (e.g., the effects of Wnt signals at early stages of development are opposite to their effects at late stages (Christian and Moon 1993).

Choice of Vector

A number of plasmids containing bacteriophage promoters have been used for in vitro mRNA synthesis. Most such transcripts are translated in vivo to some extent, but little systematic work has been aimed at optimizing this translation. It has long been known that natural (in-vivo-transcribed) mRNAs, such as globin, which are normally stable and efficiently translated, are also stable in embryos after injection (Gurdon et al. 1973). The two determinants that contribute most to messenger stability are the presence of a cap structure and a poly(A) tail (Colman and Drummond 1986; Krieg and Melton 1987b; Harland and Misher 1988). Synthetic mRNAs can be capped during transcription, by including a cap analog in the transcription reaction (see Appendix 3). It is more difficult to direct synthesis of a long poly(A) tail, since polynucleotide tracts are not stable in plasmids. Nevertheless, short 3 poly(A) and poly(C) tracts have been included in a popular expression vector, pSP64T (Krieg and Melton 1984), and they do increase the stability of mRNA in polysomes. This leads to more efficient translation in oocytes over the long term when compared to transcripts obtained from vectors lacking these signals (Galili et al. 1988). pSP64T is generally thought to increase translatability of most cloned cDNAs. This is certainly true in comparison to pBluescript vectors (Stratagene) which appear to be quite poisonous to translation. However, in comparison to simple plasmids containing the SP6 promoter, such as the pGEM vectors, the advantages of pSP64T are less obvious (B.D. Brown and R.M. Harland, cited in Vize et al. 1991). The poisonous effect of some plasmids may be due, at least in part, to 5 -untranslated sequences. Indeed, a short stretch of G residues at the 5 end can be particularly poisonous to translation (Kuo et al. 1996). There is anecdotal evidence that the globin sequences present in pSP64T and pRN3 can rescue translatability, but this translational context is not guaranteed to increase translation above the level provided by a simple SP6 vector (Vize et al. 1991; Lemaire et al. 1995). In cases where a cDNA contains a particularly poor initiation site, translation efficiency can be increased by cloning the initiation codon in frame with the globin initiation codon in pSP64Xβm, an SP6 plasmid that contains a *Xenopus* β-globin cDNA (Cho et al. 1991b). Since most globin cDNAs contain an *Nco*I site (CCATGG) at their initiator methionine, amplification of a cDNA with a 5 primer containing such a sequence facilitates cloning into this site.

More recently, it was recognized that inclusion of a polyadenylation signal, as opposed to a stretch of poly(A), leads to large increases in translation relative to transcripts that lack a polyadenylation signal. CS plasmids (Rupp et al. 1994; Turner and Weintraub 1994) contain convenient polylinkers, an SV40 polyadenylation signal, and an unstructured 5'-untranslated region (secondary structure at the 5 end of transcripts, even in the untranslated region, can inhibit translation). In vitro transcripts from CS plasmids are cleaved and polyadenylated in vivo, leading to increased stability and enhanced translation. Experiments using noggin as an expressed sequence show that in such a context, these plasmids lead to far greater biological effects than the simple SP6 plasmids, and to SP6 plasmids containing the β-globin cDNA (D. Hsu et al., unpubl.).

Synthesis of mRNA

Synthesis of mRNAs is discussed in Appendix 3. Both commercial kits and home-assembled reagents work quite well. Variation in yields can arise for different reasons. Poor yields are usually due to impure template, but occasionally, sequences that cause premature termination of transcription by phage polymerases are encountered. However, such sequences do not block transcription completely, and their effects can be alleviated by lowering the incubation temperature of the transcription reaction and maintaining a high concentration of nucleotide precursors (Krieg and Melton 1987b).

Following transcription, and prior to embryo injection, it is essential to remove the cap analog, since it is a competitive inhibitor of cap-dependent translation and is quite toxic to embryos. Purification schemes involving spin columns or LiCl precipitation remove the cap analog effectively and may also remove other toxic contaminants. Well-purified mRNA can be injected at quite high doses into embryos, and as long as the volume is small, embryos injected with 5 ng of mRNA should develop quite well. Doses up to 18 ng have been reported (Ryan et al. 1996), although in the hands of most investigators, such doses would be nonspecifically lethal. Animal caps usually tolerate higher doses of mRNA than do whole embryos. (For details of microinjection, see Chapter 8.)

Controls for mRNA Injection

Although the usefulness of mRNA injection is clear, the interpretation of results requires appropriate controls. In particular, investigators must guard against drawing firm conclusions from results that suggest the inhibition of physiological processes. Such inhibition can result all too easily from the introduction of toxic quantities of mRNA or protein. Control experiments range from injection of a relatively neutral mRNA, such as globin or β-galactosidase, to injection of an mRNA that encodes a nonfunctional polypeptide that is almost identical to the test protein, with similar subcellular localizations and stabilities. Even these "neutral" mRNAs can cause artifactual results when injected at high doses. An example of a well-controlled experiment is provided by the comparison of a dominant-negative FGF receptor with a point mutant, of the same protein, that is inactive. Like its wild-type counterpart, the control protein (HAVØ) is translated in vivo and inserted into the plasma membrane. The mutant protein has a stability similar to that of the wild-type polypeptide, but the mutant FGF receptor does not bind FGF. It is completely nonfunctional and neutral (Amaya et al. 1991, 1993). This control therefore rules out uninteresting effects that might be obtained by overexpression of some other mutant transmembrane protein. Such effects might include the nonspecific saturation of the secretory system, which might prevent secretion of unrelated proteins with important signaling properties.

Many experiments, particularly those using DNA-binding proteins, may be difficult to control, since a large excess of any DNA-binding protein is likely to interfere with nuclear processes. In such cases, the best control is probably to inject transcripts from a cDNA encoding a class of protein similar to the test protein, but which has no effects or different effects (Cho et al. 1991b). Transcripts that delete the DNA-binding domain are likely to encode the proteins that are cytoplasmic or unstable, and therefore only control for RNA toxicity (Harvey and Melton 1988; Cho et al. 1991b). Point mutations that affect the DNA-binding specificity may also provide a useful control. However, if the protein is active as a dimer, such mutations can turn the protein into a dominant negative. A point mutant that acts as a dominant negative can be considered an optimal control, since wild-type and mutant mRNAs may have reciprocal effects (Mead et al. 1996).

A built-in control for microinjection experiments is often provided by lineage tracing of the injected cells. In the worst case, injected cells may die and slough off the embryo, leading to the deletion of structures. Persistence of lineage tracer in the embryo can demonstrate how well cells survive. If the labeled cells contribute to a recognizable differentiated organ, it is reasonable to argue that they are viable, and not moribund (Niehrs and De Robertis 1991; Niehrs et al. 1993). However, even if the injected cells survive, they are not necessarily healthy. They may be impaired in their ability to contribute to normal development or morphogenesis. An effect is unlikely to be artifactual if the injected cells are shown to (1) contribute to some unexpected structure, which cannot be explained by their normal fate map; (2) express a gene that they would normally not express; or (3) be active in a morphogenetic process in which they would not normally participate.

Site of Injection

The fate map of *Xenopus* has been determined for both the early blastula and the gastrula stages (Keller 1975, 1976; Dale and Slack 1987; Moody 1987; Chapter 9). It can be exploited to direct the introduction of mRNAs into particular regions of the embryo and to assess changes in cell fate. However, it is still difficult to target mRNAs to a single tissue type, since there is always some mixing of cell and tissue types; e.g., somites and spinal cord are scattered over a large area in the fate map.

The most common use of the fate map is to inject mRNAs into either dorsal or ventral blastomeres at the four- to eight-cell stage. Cortical rotation leads to a lighter pigmentation of the dorsal side, and in many batches of embryos, this is sufficiently clear to differentiate the dorsal and ventral sides. A fixed dorsoventral axis can be imposed on the embryo by tipping and staining (see Protocol 7.5). Although this approach is laborious, it leads to a reliable assignment of dorsal and ventral sides. Dorsal and ventral injections have been used to demonstrate the differential effects of several genes, including the neutral effects of *Xwnt8* mRNA on the dorsal side and the potent axis-inducing effects of the same molecule on the ventral side (Sokol et al. 1991). A second example is provided by the neutral effects of BMP mRNAs on the ventral side, although BMPs suppress axis development on the dorsal side (Dale et al. 1992; Jones et al. 1992).

Different tiers of blastomeres can also be injected, exploiting the good fate maps available for the 32-cell stage embryo. This kind of injection (e.g., into vegetal tier blastomeres) has been used to address the question of whether injected signaling molecules can mimic the Nieuwkoop center (the early-acting vegetal dorsalizing center that can induce axis development without contributing to axial structures). Such a test could be combined with analysis of the effective time for activity of a molecule, thus providing a useful operational test for Nieuwkoop center activity (Smith and Harland 1991). Such injections must be used in combination with a lineage tracer (see below) to determine whether the fate of the injected cell has been altered.

Use of Lineage Tracers with Injected mRNA

Fate maps have been constructed for early developmental stages by injecting lineage tracers into particular blastomeres. These lineage tracers (e.g., fluorescein dextran, horseradish peroxidase, and colloidal gold) diffuse rapidly through the cell and label all the cell's progeny. They also have the important property that they do not alter cell fate and so act as neutral reporters of developmental potential.

Lineage tracers are injected along with mRNA and used to label the progeny of an injected cell, thus easing the interpretation of results. Although the tracer is generally found in each of the injected cell's progeny, the same cannot be assumed for the injected mRNA. mRNA is slow to diffuse from the site of injection, and at cell division, the distribution can be so uneven as to leave one daughter cell with the entire injection load and the other devoid of foreign material. To trace the postinjection path of the mRNA more accurately, lineage tracers such as LacZ mRNA (Kintner 1988; Vize et al. 1991) or green fluorescent protein (GFP) mRNA (Tannahill et al. 1995; Zernicka-Goetz et al. 1996) should be coinjected with the test mRNA. Although injection of a lineage-tracer mRNA is useful, it has been known to compete for translation with the coinjected test RNA (C. Kintner, pers. comm.). LacZ can be visualized in fixed specimens by staining with histochemical substrates or antibodies. In combination with an independent stain, the lineage tracer can test the effects of injection on expression of a particular marker (Amaya et al. 1993) or demonstrate a radical alteration of cell fate (Hemmati-Brivanlou et al. 1994). Different colored tracers (e.g., magenta-gal and red-gal) are used to enhance contrast between the lineage tracer and other stains such as the blue NBT/BCIP (see Chapter 13) stain from in situ hybridization (Knecht and Harland 1997; Pan and Rubin 1997). GFP localization can be monitored in live embryos and to a lesser extent in fixed embryos (see Chapter 9).

Even though an mRNA lineage tracer may appear to be a reliable marker for diffusion of an injected test RNA, diffusion rates vary with the tracer used; e.g., the FGF receptor mRNA diffuses much more slowly than *lacZ* mRNA (T.M. Lamb and R.M. Harland, unpubl.), presumably because signal sequences on nascent protein attach the mRNA to the endoplasmic reticulum (Colman and Drummond 1986). Conversely, LacZ staining can be misleading since enzymatic activity depends on the formation of a tetramer, giving a nonlinear assessment of where RNA exists. In these cases, the ideal solution is to stain directly for the test protein using specific antibodies (Amaya et al. 1993). If specific antibodies are not available, the test protein can be tagged with an epitope, such as a *myc* tag, for which antibodies are commercially available (McMahon and Moon 1989). This approach has been useful for relatively stable proteins, but many biologically active secreted molecules are present at levels that cannot be detected using antibodies. Coinjection of a tracer mRNA in these cases still provides the best currently available way of monitoring the distribution of the test mRNA. An obvious way of tracing an unstable or rare test protein that is secreted is to coinject a tracer mRNA that encodes an easily detectable transmembrane protein. This approach has not been used extensively, although transmembrane reporters are becoming available (Siegel and Isacoff 1997; E. Amaya, unpubl.).

Finally, the stability of polyadenylated transcripts, such as those made using CS vectors, allows injected mRNA to be traced directly using in situ hybridization, at least up to tailbud stages. However, effective use of this method requires the injection of more than 100 pg of mRNA, and even then, in situ hybridization may still fail to detect cells where low levels of RNA may have been degraded.

In conclusion, a variety of methods for lineage tracing are available. The information obtained by lineage tracing provides valuable data that complement other methods of analysis. Lineage tracing is considered an essential part of any set of experiments.

Number of Injections

Because of the restricted diffusion of mRNA, it is useful to inject the embryo more than once. In some cases, it is absolutely necessary to achieve widespread expression of an injected mRNA to see the most dramatic effects. For example, with a dominant-negative activin receptor, strong effects are most clearly seen when the entire marginal zone expresses the dominant negative (Hemmati-Brivanlou and Melton 1992). Since slow diffusion of mRNA prevents a single injection from filling the embryo, it is necessary in these cases to inject at several sites around the equator of the embryo. Even when this is done, it is likely that only the marginal zone will be effectively filled with mRNA. The animal and vegetal portions of the embryo will probably remain unaffected.

Timing of Injection

In principle, it is possible to assess the temporal activity of a test molecule by varying the time of injection. For example, *Xwnt8* mRNA only mimics the Nieuwkoop center when injected into early blastomeres. Injecting later blastomeres does not induce an axis, even though a *lacZ* tracer demonstrates expression of the injected mRNA (Smith and Harland 1991).

For some purposes, it is useful to inject late-stage cells, although this requires more specialized equipment. As an example, lineage tracing has been accomplished, in both superficial and deep cells of the neural plate, to assess whether the fates of neural epithelium cells are affected by their position (Hartenstein 1989). Late-stage injection of lipophilic lineage tracers has also been used to monitor the fate of neural crest cells (Collazo et al. 1993).

The main reason to vary the time of injection is to restrict expression of a molecule. As the embryo cleaves, the potential fates of particular blastomeres become more restricted. Therefore, the injection can be correlated with the fate map, as discussed above.

In practice, assessing the temporal activity of a protein is a difficult technique and therefore rather inefficient. This is partly because the blastomeres become smaller as cleavage progresses, and therefore several cells must be injected to achieve an equivalent dose of mRNA to an early injection into a large cell. Injection of later blastomeres also becomes more difficult using a conventional injection setup that is designed for rapid injections of large embryos. Alternative methods for later activation of injected molecules, using expression from DNA or activation of inducible proteins, are generally more useful (see Chapter 8).

Use of Promoters to Control Timing and Location of Gene Expression

A large quantity of work on promoters has been done in *Xenopus*, but very little has been published. The ease of injection of *Xenopus* embryos would seem to make an ideal case for using specific promoters to control timing or location of gene expression. However, problems arise because injected DNA is not efficiently incorporated into the genome and, for unknown reasons, promoters on episomes are poorly regulated (Krieg and Melton 1987a). When assayed in the tadpole, injected DNAs are expressed in a mosaic fashion, with a few "jackpot" cells expressing large amounts of transcript while most cells express none (Sargent and Mathers 1991; Vize et al. 1991; Kroll and Amaya 1996). Investigators can now be confident that the reason for the mosaic expression is because the DNA is episomal; transgenic frogs express the same genes from a stable position within the genomic DNA in a uniform manner and restrict their expression to the appropriate tissues for the injected promoter (Kroll and Amaya 1996).

Despite the problems with injected DNA, there have been several instances where it has been extremely useful to control the timing of injected gene expression. An early and striking example came from the use of DNA to express *Xwnt8*. Expression of *Xwnt8* after the mid-blastula stage has an effect opposite to that of the injection of mRNA (Christian and Moon 1993). Whereas protein made from mRNA has a potent dorsalizing activity, mimicking the Nieuwkoop center discussed above, later expression has a ventralizing activity. To accomplish later expression, a cytoskeletal actin promoter is used. This promoter is activated around/close to the onset of gastrulation. During gastrulation, the mosaic effect is less severe than in the later-stage embryo. Furthermore, the problems of mosaicism are alleviated by the limited diffusibility of the Xwnt8 protein; although not all cells express the protein, most cells in the region of the injected DNA are likely exposed to the protein by diffusion.

DNA injections are subject to much more variation in expression than RNA injections. In particular, there are strong threshold effects, and replication of injected DNA may result in a nonlinear amount of transcript compared to injected DNA; replication efficiency can also be affected by linearizing the DNA prior to injection (for discussion, see Harland and Misher 1988; Cambridge et al. 1997). DNA is quite toxic to embryos, with 50 pg being near the upper limit for injection. When combined with the relative inefficiency of transcription of injected DNA, as compared to transcription of endogenous DNA, levels of expression are lower than can be obtained with mRNA injection. More positively, it could be argued that levels of expression from injected DNA might be closer to the normal physiological levels, so that any striking result might be more readily interpreted. As a counterargument, strong promoters may be expressed in a few cells at such high levels that they have toxic effects in those cells. Still, DNA injections can be very useful in timing gene expression, and particularly so when the gene product is diffusible, so that mosaic expression is not a severe problem. Alternatively, the approach is valuable when the presence of gene product can be monitored by immunohistochemistry.

Spatially Restricted Promoters and Transgenic Animals

It would be very useful to target gene expression to a particular tissue. However, to date, tissue-specific promoters have not been functional when injected as plasmids. Although much of the expression may appear in the appropriate tissue, it is leaky and is invariably found elsewhere. This problem, combined with the severe mosaicism of expression, indicates that plasmid-encoded tissue-specific promoters are not useful for the manipulation of gene expression in *Xenopus*.

Despite these problems, the use of tissue-specific promoters is expected to gain ground, now that effective transgenesis procedures have been established (see Chapter 11). Already the effects of expressing a dominant-negative FGF receptor in muscle have been examined using this approach. These experiments used a transgenic cardiac actin promoter to drive expression of the dominant-negative FGF receptor cDNA (Kroll and Amaya 1996). As this technique becomes more widely used, many other promoters, both from *Xenopus* and from other vertebrates, will be tested, providing a repertoire of

Table 3.1 *Published Fusion Proteins*

Protein	Type of protein	Receptor fused	System	References
DNA-binding proteins				
GAL4, GAL4-VP16	zinc finger	ER, ER*	tissue culture	Braselmann et al. (1993); Louvion et al. (1993)
GATA-1,2,3	zinc finger	ER	tissue culture	Briegel et al. (1993); Mattioni et al. (1994)
HNF4	nuclear receptor/zinc finger	ER	tissue culture	Drewes et al. (1994)
Adenovirus E1A	zinc finger	ER,GR	tissue culture	Picard et al. (1988; Becker et al. (1989); Mattioni et al. (1994)
Fos family	leucine zipper	ER	tissue culture	Superti-Fugara et al. (1991); Schuermann et al. (1993); Mattioni et al. (1994)
Jun family	leucine zipper	ER	tissue culture	Mattioni et al. (1994); Francis et al. (1995); Fialka et al. (1996)
GCN4	leucine zipper	MR	yeast	Fankhauser et al. (1994); Mattioni et al. (1994)
C/EBP	CCAAT BP	ER, GR	tissue culture	Umek et al. (1991); Mattioni et al. (1994)
Rel	NFκB family	ER	tissue culture	Boehmelt et al. (1992); Mattioni et al. (1994)
Myc	HLH	ER,ER*, GR	tissue culture	Eilers et al. (1989); Mattioni et al. (1994); Littlewood et al. (1995); Tikhonenko et al. (1995)
MyoD	HLH	GR, ER, TR	tissue culture, Xenopus	Hollenberg et al. (1993); Kolm and Sive (1995b); Mattioni et al. (1994)
HNF3	forkhead domain	ER	tissue culture	Drewes et al. (1994)
Otx-2	homeodomain	GR	Xenopus	Gammill and Sive (1997)
LFB1/HNF1	homeodomain	ER	tissue culture	Drewes et al. (1994)
Brachyury	T box	GR	Xenopus	Tada et al. (1997)
Myb	DNA binding	ER	tissue culture	Burk and Klempnauer (1991); Mattioni et al. (1994)
p53	DNA binding	ER*	tissue culture	Mattioni et al. (1994); Vater et al. (1996)

Epstein-Barr virus:				
EBNA1; EBNA2	DNA binding	ER	tissue culture	Middleton and Sugden (1992); Kempkes et al. (1995)
RNA-binding proteins				
HIV Rev		GR	tissue culture	Hope et al. (1990); Mattioni et al. (1994)
HTLV-1 Rex	RNA binding	ER	tissue culture	Rehberger et al. (1997)
Cell cycle regulators				
Adenovirus E1A		ER	tissue culture	Picard et al. (1988); Mattioni et al. (1994)
p53		ER*	tissue culture	Mattioni et al. (1994); Vater et al. (1996)
Kinases				
c-Abl	tyrosine kinase	ER	tissue culture	Jackson et al. (1993); Mattioni et al. (1994)
Raf-1	serine/threonine kinase	ER	tissue culture	Samuels et al. (1993); Mattioni et al. (1994)
STE11	MAPKKK	ER	tissue culture	Mattioni et al. (1994)
Enzymes and other				
Methyl transferase	ER,GR	ER	tissue culture	Ishibashi et al. (1994)
FLP recombinase			tissue culture	Logie and Stewart (1995)
Cre recombinase	ER, ER*, PR*		tissue culture, mice	Metzger et al. (1995); Kellendonk et al. (1996); Zhang et al. (1996)
LexA repressor	GR, TR		tissue culture	Godowski et al. (1988)
Transmembrane proteins				
Fas	TNFR family	ER, RAR, EcR	tissue culture, mice	Takebayashi et al. (1996)

*Mutant form.

tissue- or location-specific expression patterns. Since the transgenic promoters are only expressed zygotically, the technique may have a limited impact on studying very early developmental decisions. However, the technique is expected to make the study of later developmental stages much more tractable in *Xenopus*. *Xenopus* has not been widely used for studying organogenesis, because it is difficult to interpret the effects of early mRNA injection on a process that will occur many hours later, and which relies on a cascade of interactions. Most such experiments have been done in culture using chick or rodent tissues. In chicks, gene expression is manipulated through use of viruses, whereas in mice, the tissues can be taken from transgenic or knockout animals (Hogan et al. 1994). The use of transgenic *Xenopus* will offer new opportunities to understand later developmental events.

Hormone-inducible Fusion Proteins

A steroid hormone-inducible system can be used to produce high levels of temporally controlled foreign gene expression. The system uses fusions between the hormone-binding domain of steroid receptors and a heterologous protein (for review, see Mattioni et al. 1994). In the absence of hormone, the fusion protein is held in an inactive state, presumably due to complex formation between the hormone-binding domain and the heat shock protein, hsp90 (Scherrer et al. 1993). Addition of hormone causes a conformational change, dissociating hsp90 and resulting in the rapid activation of the fusion protein (Tsai and O'Malley 1994). There are several advantages to this system: (1) In vitro transcripts can be injected at cleavage stages to produce widely distributed high levels of protein; (2) addition of the hormone-binding domain can stabilize the fusion protein, relative to the wild-type protein (Kolm 1997; Tada et al. 1997), allowing activation through late neurula/early tailbud stages (Gammill and Sive 1997); (3) steroid hormones are small lipophilic molecules that are readily taken up by the embryo; activation of the protein can be achieved by incubation of the embryos in standard culture medium containing the hormone; and (4) hormone addition rapidly activates the protein, such that increases in the levels of downstream targets can be seen in 2 hours (Gammill and Sive 1997).

What Types of Protein Can Be Rendered Inducible?

A wide variety of protein-hormone-binding domain fusions have been reported, including both DNA- and RNA-binding proteins, kinases, and other enzymes (see Table 3.1). To ensure that the addition of the ligand-binding domain does not alter protein function, the function of the native protein in question must be known a priori. A knowledge of the domain structure of the protein is also useful to ensure that insertion of the ligand-binding domain does not disrupt any important functional domain.

Which Receptor Ligand-binding Domain Should Be Used?

Hormone-binding domains from both the steroid and thyroid hormone families of receptors can be used to regulate protein function. Two points should be borne in mind when choosing a hormone-binding domain.

1. The ligand used should not be present in the embryo during the stages under examination. One possibility is to use ligand-binding domains whose ligands are not present in the embryo and that can therefore be controlled by addition of ligand to the culture medium. The activity of the glucocorticoid receptor (GR) ligand-binding domain is very tightly regulated in *Xenopus* embryos (Kolm and Sive 1995), even though there is a GR mRNA present in the embryo as early as gastrula stages (Gao et al. 1994). A potential disadvantage of this domain is that it has a relatively low affinity for ligand, and therefore, high levels of hormone are required in the medium (see Table 3.2) (Mattioni et al. 1994). In contrast, the estrogen receptor (ER) hormone-binding domain is somewhat leaky when used in whole embryos, but it is regulated quite tightly in animal caps (Kolm and Sive 1995). Preliminary data suggest that the thyroid hormone receptor (TR) hormone-binding domain may also be useful during gastrula and neurula stages (Gammill and Sive 1997). Other potential ligand-binding domains include the mineralocorticoid receptor (MR) and the androgen receptor (AR).

    Alternatively, several mutant hormone-binding domains are now available that are capable of binding synthetic hormones, rather than the normal endogenous ligand (see Table 3.2). They include a mutant ER (ER*) that renders the ligand tamoxifen-regulated and a mutant progesterone receptor (PR*) that renders the ligand RU486-regulated. However, Tada et al. (1977) have reported that expression of ER* is also leaky in *Xenopus* embryos.

    In addition, tissue culture data suggest that the *Drosophila* ecdysone receptor (EcR) hormone-binding domain may be used to make muristerone-inducible proteins (Christopherson et al. 1992; No et al. 1996). However, a drawback of this system is that muristerone inducibility appears to be dependent on the coexpression of a heterodimeric partner, Ultraspiracle (USP) or retinoid X receptor (RXR) (No et al. 1996; Takebayashi et al. 1996).

2. The second point to consider when choosing a receptor ligand-binding domain is that addition of the hormone alone should not affect gene expression or development. Many hormones fit this criterion for *Xenopus* gastrula or neurula embryos, including dexamethasone (Gao et al. 1994; Kolm and Sive 1995), β-estradiol (Baker and Tata 1990; Kolm and Sive 1995), tri-iodothyronine (Old et al. 1992; Smith et al. 1994), and muristerone (P.J. Kolm and M. Patel, unpubl.).

Where Should the Ligand-binding Domain be Inserted?

For maximal control of regulation, the hormone-binding domain should be inserted relatively close to the functional domain to be regulated (Mercola et al. 1988). Caution must be used when inserting a hormone-binding domain (which is normally present at the carboxyl terminus) at the amino terminus. This can lead to initiation from an internal AUG, producing proteins lacking the regulatory domain (Mattioni et al. 1994). It is also critical to insert the hormone-binding domain in a region of the protein where the structure of functional domains will not be disrupted. Inserting the hormone-binding domain into several different regions and testing the resulting molecules for native protein activity are advisable.

Is Hormone-induced Activity Reversible?

In some cases, removal of hormone from the culture medium can reverse the activity of hormone-binding domain fusion proteins (Jackson et al. 1993; Mattioni et al. 1994; Spitkovsky et al. 1994). However, these assays were done in tissue culture over a period of days. It is not clear whether simple removal of ligand can reverse protein activity in the short time period necessary to make this useful during *Xenopus* development. As an alternative, it may be possible to reverse hormone-induced activity by adding hormone-binding domain antagonists to the medium (Boehmelt et al. 1992).

Does the Trans-*activation Function of the Hormone-binding Domain Interfere with Function of the Fusion Protein?*

The ligand-binding domain of most nuclear receptors contains a *trans*-activation function, termed TAF-2. This can cause unexpected levels of gene expression when the hormone-binding domain is attached to a weak *trans*-activator (Schuermann et al. 1993).

Table 3.2. *Steroid Receptors and Ligands*

Receptor	Amino acid positions[a]	Ligand	Concentration[b]	References
GR	~512 to ~777 (human)	dexamethasone	10 μM (X)	Mattioni et al. (1994)
ER	282–595 (human)	17b-estradiol	10 nM (tc) to 1 μM (X)	Mattioni et al. (1994)
ER*	282–595 (human)	4-hydroxy tamoxifen	200 nM	Littlewood et al. (1995)
PR	~670–933 (human)	progesterone	100 nM – 1 μM	Mattioni et al. (1994)
PR*	671–891 (human)	RU486	1–100 nM	Kellendonk et al. (1996)
MR	685–981 (rat)	aldosterone	10 nM	Mattioni et al. (1994)
AR	~650–918 (human)	5a-dihydrotestosterone	100 nM	Mattioni et al. (1994)
Tra	164–410 (rat)	tri-iodothyronine	100 nm (tc, X)	Hollenberg et al. (1993)
EcR[c]	329–878 (*Drosophila*)	muristerone A	100 nM to 10 μM	Christopherson et al. (1992); No et al. (1996)
RARα	176–462 (human)	retinoic acid	0.1–10 μM	Takebayashi et al. (1996)

[a] Listed are the approximate amino acid positions for the ligand-binding domain from the species noted. Since these proteins are highly conserved, the corresponding hormone-binding domain sequences from other species can also be used.

[b] Concentrations listed are optional for tissue culture cells unless otherwise noted. (X) Used in *Xenopus*; (tc) used in tissue culture.

[c] May require coexpression of a heterodimeric partner such as RXR or OSP (No et al. 1996; Takebayashi et al. 1996).

The *trans*-activation domain can also cause problems in experiments designed to investigate the activity of a transcriptional repressor. The activation function of TAF-2 and the repressor domain of the transcription factor being tested will compete, and thus the experiment may yield ambiguous results. Repression function is increasingly being experimentally added to a DNA-binding domain to convert a transcriptional activator into a transcriptional repressor. An example of such a domain is the repressor region of the *Drosophila* engrailed protein (EnR) (Jaynes and O'Farrell 1991; Conlon et al. 1996; Fan and Sokol 1997). However, when such a DNA-binding domain/engrailed repressor chimera is fused to a hormone-inducible domain, the effect of the engrailed repressor may be masked by the TAF-2 activation domain. In the case of TAF-2, this problem can be prevented by using mutant forms of GR or ER that inactivate TAF-2 (Mattioni et al. 1994). Indeed, tamoxifen-inducible ER, which does not have TAF-2 function, effectively regulates the activity of a Myb-EnR fusion protein (Lyon and Watson 1995).

Future Additions to the Repertoire of Gene Manipulations

A separate approach for controlling gene activation is the use of photoactivatable, caged transcription factors that can be coinjected with a plasmid containing transcription-factor-binding sites. In principle, this approach allows the precise activation of genes in a subset of cells (Cambridge et al. 1997).

A variety of imaginative techniques for producing predictable gene expression have been developed in several animal systems. Since promoters in transgenic *Xenopus* embryos are regulated in a predictable way, it is hoped that some of the new techniques will be transferable to *Xenopus*. In *Drosophila*, one of the most powerful techniques involves the use of fly lines expressing a GAL4 transcriptional activator in particular patterns. The spatial activity of the transcriptional activator can be assessed in a background where *lacZ* is expressed in a GAL4-dependent manner. Such lines can be crossed to other test lines containing different genes under the control of the GAL4-binding sites. The consequences of localized expression of the gene under study can then be assessed. The long generation time of *Xenopus laevis* precludes the generation of many lines of this type, but the technique could be adapted to first-generation transgenics, where the various GAL-expressing, reporter, or test constructs are introduced into sperm nuclei and analyzed in the first generation of transgenic *Xenopus* embryos.

The predictability of transgenic promoters may lead to the development of more sophisticated techniques using inducible promoters, whether hormone- or heat-shock-inducible, in *Xenopus*. Such developments ultimately rely on the routine production of transgenic frogs throughout the *Xenopus* research community.

LOSS OF FUNCTION

The most extensive loss-of-function analyses have been carried out with dominant-negative proteins, i.e., proteins that interfere with the function of a wild-type protein. For this approach to work, the target protein must form a complex with other like or unlike subunits. A mutant form of the protein is engineered such that it still interacts with the subunits but does not form a functional complex. Specific examples illustrate how a mutant protein might poison the wild-type function. To facilitate interpretation of loss-of-function analyses, it is important to ensure that the results of the manipulation are not due to toxic effects. Two kinds of control experiments are generally done to ensure that the dominant-negative manipulation is having a specific effect: (1) It should be possible to overcome the effects of the dominant negative by titrating in the wild-type gene product and (2) it should be possible to bypass the effects of the dominant negative by activating the pathway downstream or in parallel to the one being inhibited.

Since the dominant-negative approach involves overexpressing an active domain of a protein, it tends to interfere with the function of not only the protein of interest, but related family members as well. This is a drawback because it means that the dominant-negative manipulation is not as precise as a genetic mutation. However, there is a silver lining to this cloud: A dominant negative will interfere with functionally redundant family members, often revealing a phenotype that could not be revealed by single mutations.

Dominant-Negative Receptors

The first striking use of dominant-negative constructs came from interference with the function of peptide growth factor receptors. Invariably, the ligand activates a receptor by causing dimerization or clustering of receptor subunits. In the case of the FGF receptor (Amaya et al. 1991), dimerization of subunits causes the tyrosine kinase intracellular domains of the receptor to mutually phosphorylate one another. In this case, the dominant negative was made by truncation of the wild-type receptor and introduced by injection of synthetic mRNA. The dominant-negative receptor is expressed in excess so that, upon ligand binding, wild-type subunits dimerize with truncated subunits. The truncated subunit cannot phosphorylate the wild type, and the wild-type subunit has no substrate to phosphorylate. Since no phosphorylation occurs, signal transduction is blocked.

This use of a dominant-negative FGF receptor also illustrates a number of controls required to verify the specificity of the dominant negative. An oocyte assay for receptor activity was used to test what ratio of dominant negative to wild type was needed to extinguish activity. The oocyte assay, as well as embryo assays, was used to show that a modest amount of wild-type receptor could rescue the effects of the dominant-negative subunits. This is an important class of control and rules out a number of trivial possibilities. For example, if the entire secretory pathway were to be saturated with dominant-negative subunits, then signal transduction might be blocked by prevention of important transduction components from reaching the membrane. In addition to the rescue control, the stability of receptor subunits was measured to ensure that there were no unexpected alterations in the number of wild-type or mutant receptors. To further model the activity of the dominant-negative subunits, reconstruction experiments were carried out in animal caps, where the effects of exogenous FGF could be measured. Such experiments showed that the dominant negative could block FGF-mediated induction. With all of these controls, the effect that the dominant-negative FGF receptor had on the embryo was easier to interpret and more compelling. The effect on embryos was also reversible when the wild-type receptor was injected along with the dominant negative. However, even after all the elegant controls for the specificity of action of the dominant-negative FGF receptor, Amaya et al. (1991) still could not rule out the possibility that they were interfering with a family of FGF receptors in early embryos, or even a larger family of receptor tyrosine kinases. Ultimately, what one learns from dominant negatives is not so much the role of specific molecules, like FGFR1, but the role of a process, namely, FGF signaling.

These experiments illustrate that a dominant negative is easier to interpret if a detailed understanding of the mechanism is available. However, even if the mechanism is not known in detail, dominant negatives can still be made and tested. In the case of the activin type II receptor, it was assumed that the receptor dimerized and that a similar truncation to the dominant-negative FGF receptor might work (Hemmati-Brivanlou and Melton 1992). Injections into animal caps were used to show that the dominant negative blocked induction by exogenous activin, and subsequent effects on embryos could be interpreted with some confidence. However, it has gradually become clear that the effects of a truncated activin type II receptor are quite pleiotropic, inhibiting most, if not all, transforming growth factor-β (TGF-β) receptor activities (Wilson and Hemmati-Brivanlou 1995). Even though the truncated activin receptor inhibits TGF-β signaling, it does show specificity; thus, activation of a different signal transduction pathway, the FGF pathway, could still induce mesoderm in the presence of the dominant-negative activin receptor. The latter control illustrates the power of bypassing the dominant-negative protein to demonstrate specificity.

Another class of dominant-negative receptor is one in which the extracellular domain is expressed and is able to bind ligand, thus preventing the ligand from accessing receptor. In the case of the activin type II receptor, the extracellular domain is not as pleiotropic as the truncated receptor (Dyson and Gurdon 1997). However, the interpretation of such results still relies on knowing the binding specificity of the extracellular domain. To some extent, this can be tested, and indeed the extracellular domain of the activin receptor does block activin signaling but not Xnr2 or BMP4 signaling. However, it was not ruled out that the effects of the receptor could be due to blockage of BMP7, which is also known to bind the activin receptor, or to the blockage of yet unknown ligands. Natural examples of such secreted domains of receptors exist and can be used as dominant negatives. For example, *frzb* encodes a secreted polypeptide that is similar to the extracellular domain of the Frizzled class of Wnt receptors. Ectopic expression of *frzb* has been shown to block the effects of added wnt ligands (Leyns et al. 1997; Wang et al. 1997).

As knowledge of the TGF-β signaling pathways has increased, it has become evident that type II receptors may interact with numerous transducing (type I) receptors, especially when the receptors are overexpressed. In practice, dominant-negative type I receptors have proven to be more specific in their effects, inhibiting a subset of pathways (Graff et al. 1994; Suzuki et al. 1994; Chang et al. 1997). This illustrates that as knowledge of the biochemistry of a signal transduction pathway increases, it will become possible to design more specific tools to interfere with specific functions.

Dominant-Negative Ligands

In cases where a ligand is dimeric, it may be possible to design dominant negatives that interfere with processing and secretion. The TGF-β family provides a useful example because, after dimerization, the prodomain of the ligand is protcolytically removed from the mature ligand. However, this approach is only useful in cases where there are no maternal stores of protein (Wittbrodt and Rosa 1994). The technique was used on the BMPs and contributed to a large body of evidence from the use of dominant-negative BMP receptors, dominant-negative ligands, and BMP antagonists that BMP signaling is crucial for ventral mesodermal and epidermal development (Hawley et al. 1995).

It has proved quite difficult to design dominant-negative ligands that act as receptor antagonists, i.e., bind to receptor and prevent its action by excluding wild-type ligands. Several natural examples of receptor antagonists exist (Hannum et al. 1990), and some have been constructed artificially by mutating the ligand sequence (Wittbrodt and Rosa 1994). However, the effects of mutation are difficult to predict, and it is far easier to destroy the activity of a ligand altogether than to turn it into a receptor antagonist. Nevertheless, there have been some examples of success. A truncated form of *delta* appears to act as a dominant-negative ligand for the Notch receptor (Chitnis et al. 1995). Although the mechanism is unclear, the activity was inferred because the effects of the truncated delta ligand were opposite to those of overexpressed or activated Notch receptors (Chitnis et al. 1995). These results with a truncated delta ligand illustrate the value of testing several mutant ligands for possible dominant-negative effects.

In some cases, a dominant negative can be modeled on a mutant in a different organism. For example, a temperature-sensitive allele of wingless in *Drosophila* has a phenotype as a heterozygote, suggesting that the mutant protein may interfere with wild-type function. When the mutation was engineered into a *Xenopus* Wnt, the mutant form was indeed shown to be an effective dominant negative, suppressing the effects of added *wnt* mRNA (Hoppler et al. 1996).

Signal Transduction Components

Wherever an interaction of proteins occurs in a pathway, there is in principle a way to interfere with function. Expression of individual protein domains may allow titration of a signal transduction component, either upstream or downstream in the pathway, and thus prevent signaling. For example, a version of the Raf-1 kinase, lacking the kinase domain, acts as a dominant-negative, presumably because it binds to activated *ras* and competes with wild-type *raf-1* (MacNicol et al. 1993). Kinases with mutated ATP-binding pockets are often dominant negatives. Presumably, these may titrate upstream or downstream components. By binding to upstream components, the mutated kinase may block further signal transduction. The GSK3 dominant negative is an example that clearly works, even though it is not known whether it is titrating upstream or downstream components. The GSK3 kinase is normally constitutively active, but it is down-regulated by activation of Wnt signaling. The dominant-negative GSK3 mimics this down-regulation. The production of an ectopic axis, when this dominant-negative kinase is expressed on the ventral side of embryos, implies an important role for the normal constitutive kinase in maintaining ventral identities (Dominguez et al. 1995; He et al. 1995; Pierce and Kimelman 1995).

Dominant-Negative Transcription Factors

Where transcription factors dimerize, there is also an opportunity to make dominant negatives. Such "negatively complementing" mutants have been known for a long time (see, e.g., Kelley and Yanofsky 1985). Normally, the affinity of a dimeric DNA-binding protein for its target is the product of the individual binding affinities, so that elimination of DNA binding by one of the subunits reduces the overall affinity to negligible levels. Recently, transcription factors have been mutated in their DNA-binding domains to provide a dominant-negative function in *Xenopus* (Mead et al. 1996; Wettstein et al. 1997). This approach is powerful, but it suffers from the same complications of interpretation as other dominant negatives, in that all potential partners of a transcription factor may be inhibited.

Another example of a dominant negative that works in *Xenopus* is where interactions with transcriptional coactivators may be prevented. In TCF3, the amino-terminal domain interacts with free β-catenin, and the complex provides the combination of DNA binding and transcriptional activation. Removal of the β-catenin interaction domain causes TCF3 to become a dominant negative (Molenaar et al. 1996). Presumably, it binds to target sites, such that the *trans*-activation function of β-catenin cannot be supplied.

Transcription Factors with Neomorphic Activities

Finally, transcription factors can be modified to reverse their normal activity. This has exploited the understanding of domains that appear to activate or actively repress transcription. Transcriptional activators can be turned into repressors by adding a strong repressor domain from another protein, e.g., engrailed from *Drosophila* (Jaynes and O'Farrell 1991). Conversely, repressors can be turned into activators by fusing a strong transcriptional activator domain from a protein such as VP16 from cytomegalovirus. However, such modifications do not lead to a dominant negative function in the usual sense. A dominant-negative protein interferes with the function of wild-type protein. Therefore, a result from a "true" dominant-negative experiment can be taken to mean that the wild-type protein is required for a particular activity. The fusion of engrailed or VP16 domains to a transcription factor will lead to a new kind of activity that cannot necessarily be interpreted in the same way. Examples include the fusion of the engrailed domain to the DNA-binding domains of Xbrachyury, eomesodermin, and Siamois (Conlon et al. 1996; Ryan et al. 1996; Fan and Sokol 1997; Kessler 1997). Although all of these fusions produce striking effects on development, it is not completely clear how the experiments should be interpreted. Presumably, when a transcription factor is overexpressed, it binds to its target sites, as well as to other sites that may resemble its normal targets. The transcriptional repressing function of the chimeric protein could easily block the activities of any proteins that bind such a class of DNA sites. The fusion protein may even have an effect in tissues where the normal gene is not expressed, clearly illustrating that it is not simply blocking the wild-type gene activity. In the case of the Xbrachyury/engrailed fusion, the phenotypes were reassuringly similar to the known mutant phenotypes of brachyury mutations in mice and zebrafish (Conlon et al. 1996). However, if a novel phenotype is encountered, does that necessarily mean that this mimics the absence of the normal protein? As an example, a Siamois/engrailed fusion blocks the formation of dorsal structures in *Xenopus*. This result does not rule out the possibility that Siamois is specifically required for organizer formation and function, but it does not prove it either. More likely, the result indicates that DNA-binding sites occupied by Siamois and similar proteins must be bound by transcriptional activators in order to perform organizer functions. The result is interesting, but the experiment is not as rigorous as finding a phenotype resulting from mutation of a single gene.

In conclusion, it is worth reemphasizing the caveats attached to dominant-negative experiments. Controls showing that the effect can be rescued by restoring wild-type function or bypassing the function are essential. Conclusions about specificity should be drawn with caution; until all of the proteins with which a dominant negative may interact are known, the specificity of the effect cannot be confirmed.

Injection of Antibodies

Antibodies are highly specific reagents that interact with target proteins. If these antibodies can identifiably block the function of a protein, then they may provide valuable tools for analyzing the requirement for that protein. Antibodies that are injected intracellularly into *Xenopus* embryos are remarkably stable, although the disulfide bonds become reduced in the intracellular environment.

In practice, there are only a few cases where antibody injection has been used to interfere with specific gene products (Warner et al. 1984; Wright et al. 1989; Ruiz i Altaba et al. 1991; Kume et al. 1997). It is laborious to generate a high-titer antibody that is concentrated enough for injection, and it can be difficult to avoid nonspecific effects of the protein injection. The most convincing controls use the same antibody as that used in the experiment but blocked with excess antigen, instead of preimmune antibodies or commercial antibodies that have been purified separately. Partly because of the general toxic effects of concentrated antibodies, antibody experiments have been used much more extensively in vitro, rather than in injection experiments. In vitro, antibodies bound to beads can be used to deplete a protein from an extract to study whether the protein is necessary, and as a control, the protein can then be reintroduced to a depleted extract, thus demonstrating that only this component was removed.

Antisense

The use of antisense has produced extremely variable results in different *Xenopus* embryos and different cell types within the embryos. *Xenopus* has not proven to be the most facile example, in contrast to nematodes, where injection of either sense or antisense RNA can phenocopy the mutant phenotypes of a high proportion of maternally expressed genes.

A noted *Xenopus* researcher has suggested that there are three potential outcomes from an antisense experiment. Either it does nothing, it does something unexpected and unexplainable, or it does exactly what one wants. Only in the latter case are the results rushed into press.

The nonspecific results that can be obtained by injecting antisense RNAs indicate that particular caution must be taken in the design of control experiments. The controls should be of the same kind as those in the dominant-negative case, i.e., the mutant phenotype should be rescued by adding more of the sense RNA, and if possible, the inhibition caused by the antisense RNA should be bypassed by use of a downstream component. A demonstrated reduction in target protein levels should also be documented. When the controls are done well, the results are convincing.

Ablation of mRNA in Oocytes

The most compelling cases of successful antisense experiments come from oligodeoxynucleotide (ODN) injection into oocytes. ODNs are short molecules of single-stranded DNA that hybridize to their targets. Once hybridization has occurred, ribonuclease H degrades the RNA portion of the heteroduplex, thus destroying the target RNA and hence the gene's activity. When the oocytes have recovered from the injection procedure and the large dose of ODN necessary to effect complete message degradation, the oocytes are reimplanted into a primed female. The oocytes complete maturation in vivo and are then fertilized as normal. This is a complex procedure. It is not easy to design an ODN that degrades a target; usually, several possibilities must be tested to find one that works well. For optimal results, chemically modified ODNs must be used. Short DNA molecules are rapidly degraded, and when injected at sufficient doses to be effective, they become nonspecifically toxic to further development. Although chemically modified oligonucleotides may be more toxic than unmodified DNA, the greater stability means that a smaller and ultimately less toxic dose can be used. Oocytes must be of the highest quality to survive these manipulations and complete normal development.

Despite these problems, there have been spectacular successes using this technique. Depletion of maternal β-catenin mRNA using ODNs resulted in loss of axial structures in the resulting embryos (Heasman et al. 1994). The embryos were still capable of responding to signals, because they could be rescued by β-catenin mRNA injection (using a truncated β-catenin that did not contain the ODN hybridization site).

Unfortunately, other experiments have proven to be the opposite of what may have been hoped, i.e., that an mRNA is not required for early development (Shuttleworth et al. 1988; Kloc et al. 1989; Minshull et al. 1991). Indeed, there may be only a few cases where new translation is required from maternal mRNA for early development; in most cases, adequate pools of protein have been accumulated during oogenesis.

Effects of Antisense RNA in Embryos

An increasing number of antisense RNA experiments have been successful in embryos. If the determinants that have made the technique effective in particular cases, and not effective in others, become understood, then the technique will become more useful. Currently, results from antisense experiments are fraught with inconsistencies and are extremely difficult to interpret. The most convincing cases have shown that the antisense injection did indeed result in a reduction of the amount of native protein synthesized (Lallier et al. 1996; Lombardo and Slack 1997). In another case, the effects of antisense mRNA on BMP4 were consistent with other manipulations that reduce BMP4 signaling (Steinbeisser et al. 1995). Antisense experiments are meaningless without a control showing that the phenotype can be rescued by addition of sense mRNA.

Antisense RNA Encoded by DNA

The isolated examples where antisense DNA constructs have been shown to affect late developmental events are likely to be artifacts, especially now that the mosaicism of RNA expression from injected plasmids is appreciated. However, with the more uniform expression that can be achieved from transgenic frogs, it may now be possible to revisit this issue.

Special Antisense RNA: Hammerhead

A unique case has exploited hammerhead structures in an antisense RNA to inhibit accumulation of histone H1. The native mRNA was reduced in abundance by this manipulation, but the amount of protein synthesized was affected more severely. It was suggested that both degradation and hybrid arrest of translation may have occurred in response to the antisense injection (Bouvet et al. 1994; Steinbach et al. 1997). Although the use of hammerhead structures might provide a more general approach to making antisense methods more effective, it has not yet proved straightforward to apply to mRNAs other than that encoding histone H1 (L. Snider et al., pers. comm.).

USE OF OOCYTES IN ASSAYING GENE FUNCTIONS: MAKING SECRETED PROTEIN

Oocytes provide a valuable translation system, where injected mRNAs are correctly translated and their products processed and transported to the appropriate subcellular compartment (for review, see Smith et al. 1991). This makes oocytes particularly valuable for studies of secreted proteins. Oocytes will tolerate injection of large amounts of mRNA (see Chapter 8), which can effectively compete with the endogenous mRNA. An additional advantage of this system is that secreted proteins, if correctly processed, are biologically active. Oocyte injection provided one of the first means of expressing interferon mRNA, and thus facilitated the cloning of interferon cDNA.

Synthetic mRNAs encoding secreted proteins are efficiently translated in the oocyte and can provide a valuable first step in synthesizing proteins for bioassays. Oocytes of course limit the scale of expression because of their size, but large numbers can be injected. Although oocytes are very active in translation, synthesizing several nanograms of protein per hour (Smith et al. 1991), it is not possible to subvert all this capacity for the production of secreted proteins from injected mRNA. The practical amounts of secreted protein obtained are on the order of 1 ng per oocyte per day .

Oocytes can also be efficiently radioactively labeled, and if the injection is done cleanly, and oocyte health maintained, then the secreted proteins are isotopically almost pure (see, e.g., Smith et al. 1993). Oocytes have also been used in expression cloning experiments where animal caps are applied to injected oocytes, and respond to the proteins secreted from them (Lustig and Kirschner 1995).

REFERENCES

Amaya E., Musci T.J., and Kirschner M.W. 1991. Expression of a dominant negative mutant of the FGF receptor disrupts mesoderm formation in *Xenopus* embryos. *Cell* **66:** 257–270.

Amaya E., Stein P.A., Musci T.J., and Kirschner M.W.1993. FGF signalling in the early specification of mesoderm in *Xenopus*. *Development* **118:** 477–487.

Baker B. and Tata J. 1990. Accumulation of proto-oncogene c-erb-A related transcripts during *Xenopus* development: Association with early acquisition of response to thyroid hormone and estrogen. *EMBO J.* **9:** 879–885.

Baker J.C. and Harland R.M. 1996. A novel mesoderm inducer, Madr2, functions in the activin signal transduction pathway. *Genes Dev.* **10:** 1880–1889.

Becker D., Hollenberg S., and Ricciardi R. 1989. Fusion of adenovirus E1A to the glucocorticoid receptor by high-resolution deletion cloning creates a hormonally inducible viral transactivator. *Mol. Cell. Biol.* **9:** 3878–3887.

Boehmelt G., Walker A., Kabrun N., Mellitzer G., Beug H., Zenke M., and Enrietto P.J. 1992. Hormone-regulated v-rel estrogen receptor fusion protein: Reversible induction of cell transformation and cellular gene expression. *EMBO J.* **11:** 4641–4652.

Bolce M.E., Hemmati-Brivanlou A., Kushner P.D., and Harland R.M. 1992. Ventral ectoderm of *Xenopus* forms neural tissue, including hindbrain, in response to activin. *Development* **115:** 681–688.

Bouvet P., Dimitrov S., and Wolffe A.P. 1994. Specific regulation of *Xenopus* chromosomal 5S rRNA gene transcription in vivo by histone H1. *Genes Dev.* **8:** 1147–1159.

Braselmann S., Graninger P., and Busslinger M. 1993. A selective transcriptional induction system for mammalian cells based on Gal4-estrogen receptor fusion proteins. *Proc. Natl. Acad. Sci.* **90:** 1657–1661.

Briegel K., Lim K.C., Plank C., Beug H., Engel J.D., and Zenke M. 1993. Ectopic expression of a conditional GATA-2/estrogen receptor chimera arrests erythroid differentiation in a hormone-dependent manner. *Genes Dev.* **7:** 1097–1109.

Burk O. and Klempnauer K. 1991. Estrogen-dependent alterations in differentiation state of myeloid cells caused by a v-myb/estrogen receptor fusion protein. *EMBO J.* **10:** 3713–3719.

Cambridge S.B., Davis R.L., and Minden J.S. 1997. *Drosophila* mitotic domain boundaries as cell fate boundaries. *Science* **277:** 825–828.

Chang C., Wilson P., Mathews L.S., and Hemmati-Brivanlou A. 1997. A *Xenopus* type I activin receptor mediates mesodermal but not neural specification during embryogenesis. *Development* **124:** 827–837.

Chitnis A., Henrique D., Lewis J., Ish-Horowicz D., and Kintner C. 1995. Primary neurogenesis in *Xenopus* embryos regulated by a homologue of the *Drosophila* neurogenic gene Delta. *Nature* **375:** 761–766.

Cho K.W., Blumberg B., Steinbeisser H., and De Robertis E.M. 1991a. Molecular nature of Spemann's organizer: The role of the *Xenopus* homeobox gene goosecoid. *Cell* **67:** 1111–1120.

Cho K.W., Morita E.A., Wright C.V., and De Robertis E.M. 1991b. Overexpression of a homeodomain protein confers axis-forming activity to uncommitted *Xenopus* embryonic cells. *Cell* **65:** 55–64.

Christian J.L. and Moon R.T. 1993. Interactions between Xwnt-8 and Spemann organizer signaling pathways generate dorsoventral pattern in the embryonic mesoderm of *Xenopus. Genes Dev.* **7:** 13–28.

Christopherson K.S., Mark M.R., Bajaj V., and Godowski P.J. 1992. Ecdysteroid-dependent regulation of genes in mammalian cells by a *Drosophila* ecdysone receptor and chimeric transactivators. *Proc. Natl. Acad. Sci.* **89:** 6314–6318.

Collazo A., Bronner-Fraser M., and Fraser S.E. 1993. Vital dye labelling of *Xenopus laevis* trunk neural crest reveals multipotency and novel pathways of migration. *Development* **118:** 363–376.

Colman A. and Drummond D. 1986. The stability and movement of mRNA in *Xenopus* oocytes and embryos. *J. Embryol. Exp. Morphol.* (suppl.) **97:** 197–209.

Conlon F., Sedgwick S., Weston K., and Smith J. 1996. Inhibition of Xbra transcription activation causes defects in mesodermal patterning and reveals autoregulation of Xbra in dorsal mesoderm. *Development* **122:** 2427–2435.

Dale L. and Slack J.M. 1987. Fate map for the 32-cell stage of *Xenopus* laevis. *Development* **99:** 527–551.

Dale L., Howes G., Price B.M., and Smith J.C. 1992. Bone morphogenetic protein 4: A ventralizing factor in early *Xenopus* development. *Development* **115:** 573–585.

Dominguez I., Itoh K., and Sokol S.Y. 1995. Role of glycogen synthase kinase 3β as a negative regulator of dorsoventral axis formation in *Xenopus* embryos. *Proc. Natl. Acad. Sci.* **92:** 8498–8502.

Drewes T., Clairmont A., Klein-Hitpass L., and Ryffel G.U. 1994. Estrogen-inducible derivatives of hepatocyte nuclear factor-4, hepatocyte nuclear factor-3 and liver factor B1 are differently affected by pure and partial antiestrogens. *Eur. J. Biochem.* **225:** 441–448.

Dyson S. and Gurdon J.B. 1997. Activin signalling has a necessary function in *Xenopus* early development. *Curr. Biol.* **7:** 81–84.

Eilers M., Picard D., Yamamoto K., and Bishop J. 1989. Chimaeras of Myc oncoprotein and steriod receptors cause hormone-dependent transformation of cells. *Nature* **340:** 66–68.

Fan M. and Sokol S. 1997. A role for Siamois in Spemann organizer formation. *Development* **124:** 2581–2589.

Fankhauser C.P., Briand P.A., and Picard D. 1994. The hormone binding domain of the mineralocorticoid receptor can regulate heterologous activities in *cis*. *Biochem. Biophys. Res. Commun.* **200:** 195–201.

Fialka I., Schwarz H., Reichmann E., Oft M., Busslinger M., and Beug H. 1996. The estrogen-dependent c-JunER protein causes a reversible loss of mammary epithelial cell polarity involving a destabilization of adherens junctions. *J. Cell Biol.* **132:** 1115–1132.

Francis M.K., Phinney D.G., and Ryder K. 1995. Analysis of the hormone-dependent regulation of a JunD-estrogen receptor chimera. *J. Biol. Chem.* **270:** 11502–11513.

Galili G., Kawata E.E., Smith L.D., and Larkins B.A. 1988. Role of the 3 -poly(A) sequence in translational regulation of mRNAs in *Xenopus laevis* oocytes. *J. Biol. Chem.* **263:** 5764–5770.

Gamer L.W. and Wright C.V. 1995. Autonomous endodermal determination in *Xenopus:* Regulation of expression of the pancreatic gene XlHbox 8. *Dev. Biol.* **171:** 240–251.

Gammill L. and Sive H. 1997. Identification of *otx2* target genes and restrictions in ectodermal competence during *Xenopus* cement gland formation. *Development* **124:** 471–481.

Gao X., Stegman B., Lanser P., Koster J., and Destrée O. 1994. GR transcripts are localized during early *Xenopus laevis* embryogenesis and overexpression of GR inhibits differentiation after dexamethasone treatment. *Biochem. Biophys. Res. Comm.* **199:** 734–741.

Godowski P.J., Picard D., and Yamamoto K.R. 1988. Signal transduction and transcriptional regulation by glucocorticoid receptor-LexA fusion proteins. *Science* **241:** 812–816.

Graff J.M., Thies R.S., Song J.J., Celeste A.J., and Melton D.A. 1994. Studies with a *Xenopus* BMP receptor suggest that ventral mesoderm-inducing signals override dorsal signals in vivo. *Cell* **79:** 169–179.

Green J.B., Howes G., Symes K., Cooke J., and Smith J.C. 1990. The biological effects of XTC-MIF: Quantitative comparison with *Xenopus* bFGF. *Development* **108:** 173–183.

Gurdon J.B., Lingrel J.B., and Marbaix G. 1973. Message stability in injected frog oocytes: Long life of mammalian α and β globin messages. *J. Mol. Biol.* **80:** 539–551.

Hannum C.H., Wilcox C.J., Arend W.P., Joslin F.G., Dripps D.J., Heimdal P.L., Armes L.G., Sommer A., Eisenberg S.P., and Thompson R.C. 1990. Interleukin-1 receptor antagonist activity of a human interleukin-1 inhibitor. *Nature* **343:** 336–340.

Harland R. and Misher L. 1988. Stability of RNA in developing *Xenopus* embryos and identification of a destabilizing sequence in TFIIIA messenger RNA.

Development **102:** 837–852.

Hartenstein V. 1989. Early neurogenesis in *Xenopus*: The spatio-temporal pattern of proliferation and cell lineages in the embryonic spinal cord. *Neuron* **3:** 399–411.

Harvey R.P. and Melton D.A. 1988. Microinjection of synthetic Xhox-1A homeobox mRNA disrupts somite formation in developing *Xenopus* embryos. *Cell* **53:** 687–697.

Hawley S.H., Wunnenberg-Stapleton K., Hashimoto C., Laurent M.N., Watabe T., Blumberg B.W., and Cho K.W. 1995. Disruption of BMP signals in embryonic *Xenopus* ectoderm leads to direct neural induction. *Genes Dev.* **9:** 2923–2935.

He X., Saint-Jennet J.-P., Woodgett J.R., Varmus H.E., and Dawid I.B. 1995. Glycogen synthase kinase-3 and dorsoventral patterning in *Xenopus* embryos. *Nature* **374:** 617–622.

Heasman J., Crawford A., Goldstone K., Garner-Hamrick P., Gumbiner B., McCrea P., Kintner C., Noro C.Y., and Wylie C. 1994. Overexpression of cadherins and underexpression of β-catenin inhibit dorsal mesoderm induction in early *Xenopus* embryos. *Cell* **79:** 791–803.

Hemmati-Brivanlou A., Stewart R.M., and Harland R.M. 1990. Region-specific neural induction of an engrailed protein by anterior notochord in *Xenopus*. *Science* **250:** 800–802.

Hemmati-Brivanlou A. and Melton D.A. 1992. A truncated activin receptor inhibits mesoderm induction and formation of axial structures in *Xenopus* embryos. *Nature* **359:** 609–614.

Hemmati-Brivanlou A., Kelly O.G., and Melton D.A. 1994. Follistatin, an antagonist of activin, is expressed in the Spemann organizer and displays direct neuralizing activity. *Cell* **77:** 283–295.

Henry G.L., Brivanlou I.H., Kessler D.S., Hemmati-Brivanlou A., and Melton D.A. 1996. TGF-β signals and a pattern in *Xenopus laevis* endodermal development. *Development* **122:** 1007–1015.

Herskowitz I. 1987. Functional inactivation of genes by dominant negative mutations. *Nature* **329:** 219–222.

Hogan B., Beddington R., Costantini F., and Lacy E. 1994. *Manipulating the mouse embryo*, 2nd edition. Cold Spring Harbor Laboratory Press, Cold Spring Harbor, New York

Hollenberg S., Cheng P., and Weintraub H. 1993. Use of a conditional MyoD transcription factor in studies of MyoD trans-activation and muscle determination. *Proc. Natl. Acad. Sci.* **90:** 8028–8032.

Hope T.J., Huang X.J., McDonald D., and Parslow T.G. 1990. Steroid-receptor fusion of the human immunodeficiency virus type 1 Rev transactivator: Mapping cryptic functions of the arginine-rich motif. *Proc. Natl. Acad. Sci.* **87:** 7787–7791.

Hoppler S., Brown J.D., and Moon R.T. 1996. Expression of a dominant-negative Wnt blocks induction of MyoD in *Xenopus* embryos. *Genes Dev.* **10:** 2805–2817.

Ishibashi T., Nakabeppu Y., and Sekiguchi M. 1994. Artificial control of nuclear translocation of DNA repair methyltransferase. *J. Biol. Chem.* **269:** 7645–7650.

Jackson P., Baltimore D., and Picard D. 1993. Hormone-conditional transformation by fusion proteins of c-abl and its transforming variants. *EMBO J.* **12:** 2808–2819.

Jaynes J. and O'Farrell P. 1991. Active repression of transcription by the engrailed homeodomain protein. *EMBO J.* **10:** 1427–1433.

Jones C.M., Lyons K.M., Lapan P.M., Wright C.V., and Hogan B.L. 1992. DVR-4 (bone morphogenetic protein-4) as a posterior-ventralizing factor in *Xenopus* mesoderm induction. *Development* **115:** 639–647.

Jones E.A., Abel M.H., and Woodland H.R. 1993. The possible role of mesodermal growth factors in the formation of endoderm in *Xenopus laevis. Roux's Arch. Dev. Biol.* **202:** 233–239.

Kellendonk C., Tronche F., Monaghan A., Angrand P., Stewart F., and Schütz G. 1996. Regulation of Cre recombinase activity by the synthetic steroid RU486. *Nucleic Acids Res.* **24:** 1404–1411.

Keller R.E. 1975. Vital dye mapping of the gastrula and neurula of *Xenopus laevis.* I. Prospective areas and morphogenetic movements of the superficial layer. *Dev. Biol.* **42:** 222–241.

———. 1976. Vital dye mapping of the gastrula and neurula of *Xenopus laevis.* II. Prospective areas and morphogenetic movements of the deep layer. *Dev. Biol.* **51:** 118–137.

Kelley R.L. and Yanofsky C. 1985. Mutational studies with the trp repressor of *Escherichia coli* support the helix-turn-helix model of repressor recognition of operator DNA. *Proc. Natl. Acad. Sci.* **82:** 483–487.

Kempkes B., Pawlita M., Zimber-Strobl U., Eissner G., Laux G., and Bornkamm G.W. 1995. Epstein-Barr virus nuclear antigen 2-estrogen receptor fusion proteins transactivate viral and cellular genes and interact with RBP-κ in a conditional fashion. *Virology* **214:** 675–679.

Kessler D. S. 1997. Siamois is required for formation of Spemann's organizer. *Proc. Natl. Acad. Sci.* **94:** 13017–13022.

Kintner C. 1988. Effects of altered expression of the neural cell adhesion molecule, N-CAM, on early neural development in *Xenopus* embryos. *Neuron* **1:** 545–555.

Kloc M., Miller M., Carrasco A.E., Eastman E., and Etkin L. 1989. The maternal store of the xlgv7 mRNA in full-grown oocytes is not required for normal development in *Xenopus. Development* **107:** 899–907.

Knecht A.K. and Harland R.M. 1997. Mechanisms of dorsal-ventral patterning in noggin-induced neural tissue. *Development* **124:** 2477–2488.

Kolm P. 1997. *Patterning of the posterior neurectoderm by labial-like Hox genes and retinoids.* Massachusetts Institute of Technology, Cambridge.

Kolm P.J. and Sive H.L. 1995. Efficient hormone-inducible protein function in *Xenopus laevis. Dev. Biol.* **171:** 267–272.

Krieg P.A. and Melton D.A. 1984. Functional messenger RNAs are produced by SP6 in vitro transcription of cloned cDNAs. *Nucleic Acids Res.* **12:** 7057–7070.

———. 1987a. An enhancer responsible for activating transcription at the midblastula transition in *Xenopus* development. *Proc. Natl. Acad. Sci.* **84:** 2331–2335.

———. 1987b. In vitro RNA synthesis with SP6 RNA polymerase. *Methods Enzymol.*

155: 397–415.

Kroll K.L. and Amaya E. 1996. Transgenic *Xenopus* embryos from sperm nuclear transplantations reveal FGF signaling requirements during gastrulation. *Development* **122:** 3173–3183.

Kume S., Muto A., Inoue T., Suga K., Okano H., and Mikoshiba K. 1997. Role of inositol 1,4,5-trisphosphate receptor in ventral signaling in *Xenopus* embryos. *Science* **278:** 1940–1943

Kuo J.S., Veale R., Maxwell B., and Sive H. 1996. Translational inhibition by 5 -polycytidine tracts in *Xenopus* embryos and in vitro. *Gene* **176:** 17-21.

Lallier T.E., Whittaker C.A., and DeSimone D.W. 1996. Integrin α 6 expression is required for early nervous system development in *Xenopus laevis. Development* **122:** 2539–2554.

Lamb T.M. and Harland R.M. 1995. Fibroblast growth factor is a direct neural inducer, which combined with noggin generates anterior-posterior neural pattern. *Development* **121**: 3627–3636.

Lamb T.M., Knecht A.K., Smith W.C., Stachel S.E., Economides A.N., Stahl N., Yancopolous G.D., and Harland R.M. 1993. Neural induction by the secreted polypeptide noggin. *Science* **262:** 713–718.

Lemaire P., Garrett N., and Gurdon J.B. 1995. Expression cloning of Siamois, a *Xenopus* homeobox gene expressed in dorsal-vegetal cells of blastulae and able to induce a complete secondary axis. *Cell* **81:** 85–94.

Leyns L., Bouwmeester T., Kim S.-H., Piccolo S., and De Robertis E.M. 1997. Frzb-1 is a secreted antagonist of Wnt signaling expressed in the Spemann organizer. *Cell* **88:** 747–756.

Littlewood T., Hancock D., Daniwlian P., Parker M., and Evan G. 1995. A modified oestrogen receptor ligand-binding domain as an improved switch for the regulation of heterologous proteins. *Nucleic Acids Res.* **23:** 1686–1690.

Logie C. and Stewart A.F. 1995. Ligand-regulated site-specific recombination. *Proc. Natl. Acad. Sci.* **92:** 5940–5944.

Lombardo A. and Slack J.M. 1997. Inhibition of eFGF expression in *Xenopus* embryos by antisense mRNA. *Dev. Dyn.* **208:** 162–169.

Louvion J.F., Havaux-Copf B., and Picard D. 1993. Fusion of GAL4-VP16 to a steroid-binding domain provides a tool for gratuitous induction of galactose-responsive genes in yeast. *Gene* **131:** 129–134.

Lustig K.D. and Kirschner M.W. 1995. Use of an oocyte expression assay to reconstitute inductive signaling. *Proc. Natl. Acad. Sci.* **92:** 6234–6238.

Lyon J.J. and Watson R.J. 1995. Conditional inhibition of erythroid differentiation by c-Myb/oestrogen receptor fusion proteins. *Differentiation* **59:** 171–178.

MacNicol A.M., Muslin A.J., and Williams L.T. 1993. Raf-1 kinase is essential for early *Xenopus* development and mediates the induction of mesoderm by FGF. *Cell* **73:** 571–583.

Mattioni T., Louvion J., and Picard D. 1994. Regulation of protein activities by fusion to steriod binding domains. *Methods Cell Biol.* **43:** 335–352.

McMahon A.P. and Moon R.T. 1989. Ectopic expression of the proto-oncogene int-1 in *Xenopus* embryos leads to a duplication of the embryonic axis. *Cell* **58:** 1075–1084.

Mead P.E., Brivanlou I.H., Kelley C.M., and Zon L.I. 1996. BMP-4-responsive regulation of dorsal-ventral patterning by the homeobox protein Mix.1. *Nature* **382:** 357–360.

Mercola M., Melton D.A., and Stiles C.D. 1988. Platelet-derived growth factor A chain is maternally encoded in *Xenopus* embryos. *Science* **241:** 1223–1225.

Metzger D., Clifford J., Chiba H., and Chambon P. 1995. Conditional site-specific recombination in mammalian cells using a ligand-dependent chimeric Cre recombinase. *Proc. Natl. Acad. Sci.* **92:** 6991–6995.

Middleton T. and Sugden B. 1992. A chimera of EBNA1 and the estrogen receptor activates transcription but not replication. *J. Virol.* **66:** 1795–1798.

Minshull J., Murray A., Colman A., and Hunt T. 1991. *Xenopus* oocyte maturation does not require new cyclin synthesis. *J. Cell Biol.* **114:** 767–772.

Molenaar M., van de Wetering M., Oosterwegel M., Peterson-Maduro J., Godsave S., Korinek V., Roose J., Destree O., and Clevers H. 1996. XTcf-3 transcription factor mediates β-catenin-induced axis formation in *Xenopus* embryos. *Cell* **86:** 391–399.

Moody S.A. 1987. Fates of the blastomeres of the 32-cell-stage *Xenopus* embryo. *Dev. Biol.* **122:** 300–319.

Niehrs C. and De Robertis E.M. 1991. Ectopic expression of a homeobox gene changes cell fate in *Xenopus* embryos in a position-specific manner. *EMBO J.* **10:** 3621–3629.

Niehrs C., Keller R., Cho K.W., and De Robertis E.M. 1993. The homeobox gene goosecoid controls cell migration in *Xenopus* embryos. *Cell* **72:** 491–503.

Nieuwkoop P.D. and Koster K. 1995. Vertical versus planar induction in early amphibian development. *Dev. Growth Differ.* **37:** 653–668.

No D., Yao T.P., and Evans R.M. 1996. Ecdysone-inducible gene expression in mammalian cells and transgenic mice. *Proc. Natl. Acad. Sci.* **93:** 3346–3351.

Old R., Jones E., Sweeney G., and Smith D. 1992. Precocious synthesis of a thyroid homone receptor in *Xenopus* embyos causes hormone-dependent developmental abnormalities. *Roux's Arch. Dev. Biol.* **201:** 312–321.

Pan D. and Rubin G.M. 1997. Kuzbanian controls proteolytic processing of Notch and mediates lateral inhibition during *Drosophila* and vertebrate neurogenesis. *Cell* **90:** 271–280.

Picard D., Salser S., and Yamamoto K. 1988. A movable and regulable inactivation function within the steroid binding domain of the glucocoticoid receptor. *Cell* **54:** 1073–1080.

Pierce S.B. and Kimelman D. 1995. Regulation of Spemann organizer formation by the intracellular kinase Xgsk-3. *Development* **121:** 755–765.

Rehberger S., Gounari F., DucDodon M., Chlichlia K., Gazzolo L., Schirrmacher V., and Khazaie K. 1997. The activation domain of a hormone inducible HTLV-1 Rex protein determines colocalization with the nuclear pore. *Exp. Cell Res.* **233:** 363–371.

Ruiz I Altaba A., Choi T., and Melton D.A. 1991. Expression of the Xhox3 home-obox protein in *Xenopus* embryos: Blocking its early function suggests the requirement of Xhox3 for normal posterior development. *Dev. Growth Differ.* **33:** 651–669.

Rupp R.A., Snider L., and Weintraub H. 1994. *Xenopus* embryos regulate the nuclear localization of XMyoD. *Genes Dev.* **8:** 1311–1323.

Ryan K., Garrett N., Mitchell A., and Gurdon J.B. 1996. Eomesodermin, a key early gene in *Xenopus* mesoderm differentiation. *Cell* **87:** 989–1000.

Saha M.S., Servetnick M., and Grainger R.M. 1992. Vertebrate eye development. *Curr. Opin. Genet. Dev.* **2:** 582–588.

Samuels M.L. Weber M.J., Bishop J.M., and McMahon M. 1993. Conditional trans-formation of cells and rapid activation of the mitogen-activated protein kinase cascade by an estradiol-dependent human raf-1 protein kinase. *Mol. Cell. Biol.* **13:** 6241–6252.

Sargent T.D. and Mathers P.H. 1991. Analysis of class II gene regulation. *Methods Cell Biol.* **36:** 347–365.

Sasai Y., Lu B., Piccolo S., and De Robertis E.M. 1996. Endoderm induction by the organizer secreted factors chordin and noggin in *Xenopus* animal caps. *EMBO J.* **15:** 4547–4555.

Scherrer L.C., Picard D., Massa E., Harmon J.M., Simons S., Jr., Yamamoto K.R., and Pratt W.B. 1993. Evidence that the hormone binding domain of steroid recep-tors confers hormonal control on chimeric proteins by determining their hor-mone-regulated binding to heat-shock protein 90. *Biochemistry* **32:** 5381–5386.

Schuermann M., Hennig G., and Muller R. 1993. Transcriptional activation and transformation by chimaeric Fos-estrogen receptor proteins: Altered properties as a consequence of gene fusion. *Oncogene* **8:** 2781–2790.

Shuttleworth J., Matthews G., Dale L., Baker C., and Colman A. 1988. Antisense oligodeoxyribonucleotide-directed cleavage of maternal mRNA in *Xenopus* oocytes and embryos. *Gene* **72:** 267–275.

Siegel M.S. and Isacoff E.Y. 1997. A genetically encoded optical probe of membrane voltage. *Neuron* **19:** 735–741.

Slack J.M., Dale L., and Smith J.C. 1984. Analysis of embryonic induction by using cell lineage markers. *Philos. Trans. R. Soc. Lond. B* **307:** 331–336.

Smith D., Mason C., Jones E., and Old R. 1994. Expression of a dominant negative retinoic acid receptor g in *Xenopus* embryos leads to partial resistance to retinoic acid. *Roux's Arch. Dev. Biol.* **203:** 254–265.

Smith J.C. 1987. A mesoderm inducing factor is produced by a *Xenopus* cell line. *Development* **99:** 3–14.

Smith L.D., Xu W.L., and Varnold R.L. 1991. Oogenesis and oocyte isolation. *Methods Cell Biol.* **36:** 45–60.

Smith W.C. and Harland R.M. 1991. Injected Xwnt-8 RNA acts early in *Xenopus* embryos to promote formation of a vegetal dorsalizing center. *Cell* **67:** 753–765.

Smith W.C., Knecht A.K., Wu M., and Harland R.M. 1993. Secreted noggin protein mimics the Spemann organizer in dorsalizing *Xenopus* mesoderm. *Nature* **361:**

547–549.

Sokol S., Christian J.L., Moon R.T., and Melton D.A. 1991. Injected Wnt RNA induces a complete body axis in *Xenopus* embryos. *Cell* **67**: 741–752.

Spitkovsky D., Steiner P., Lukas J., Lees E., Pagano M., Schulze A., Joswig S., Picard D., Tommasino M., Eilers M. et al. 1994. Modulation of cyclin gene expression by adenovirus E1A in a cell line with E1A-dependent conditional proliferation. *J. Virol.* **68**: 2206–2214.

Steinbach O.C., Wolffe A.P., and Rupp R.A. 1997. Somatic linker histones cause loss of mesodermal competence in *Xenopus. Nature* **389**: 395–399.

Steinbeisser H., De Robertis E.M., Ku M., Kessler D.S., and Melton D.A. 1993. *Xenopus* axis formation: Induction of goosecoid by injected Xwnt-8 and activin mRNAs. *Development* **118**: 499–507.

Steinbeisser H., Fainsod A., Niehrs C., Sasai Y., and De Robertis E.M. 1995. The role of gsc and BMP-4 in dorsal-ventral patterning of the marginal zone in *Xenopus:* A loss-of-function study using antisense RNA. *EMBO J.* **14**: 5230–5243.

Superti-Fugara G., Bergers G., Picard D., and Busslinger M. 1991. Hormone-dependent transcriptional regulation and cellular transformation by Fos-steroid receptor fustion proteins. *Proc. Natl. Acad. Sci.* **88**: 5114–5118.

Suzuki A., Thies R.S., Yamaji N., Song J.J., Wozney J.M., Murakami K., and Ueno N. 1994. A truncated bone morphogenetic protein receptor affects dorsal-ventral patterning in the early *Xenopus* embryo. *Proc. Natl. Acad. Sci.* **91**: 10255–10259.

Tada M., O'Reilly M., and Smith J. 1997. Analysis of competence and of Brachyury autoinduction by use of hormone-inducible Xbra. *Development* **124**: 2225–2234.

Takebayashi H., Oida H., Fujisawa K., Yamaguchi M., Hikida T., Fukumoto M., Narumiya S., and Kakizuka A. 1996. Hormone-induced apoptosis by Fas-nuclear receptor fusion proteins: Novel biological tools for controlling apoptosis in vivo. *Cancer Res.* **56**: 4164–4170.

Tannahill D., Bray S., and Harris W.A. 1995. A *Drosophila* E(spl) gene is "neurogenic" in *Xenopus*: A green fluorescent protein study. *Dev. Biol.* **168**: 694–697.

Tikhonenko A.T., Black D.J., and Linial M.L. 1995. v-Myc is invariably required to sustain rapid proliferation of infected cells but in stable cell lines becomes dispensable for other traits of the transformed phenotype. *Oncogene* **11**: 1499–508.

Tsai M. and O'Malley B. 1994. Molecular mechanisms of action of steroid/thyroid receptor superfamily members. *Annu. Rev. Biochem.* **63**: 451–486.

Turner D.L. and Weintraub H. 1994. Expression of achaete-scute homolog 3 in *Xenopus* embryos converts ectodermal cells to a neural fate. *Genes Dev.* **8**: 1434–1447.

Umek R.M., Friedman A.D., and McKnight S.L. 1991. CCAAT-enhancer binding protein: A component of a differentiation switch. *Science* **251**: 288–292.

Vater C.A., Bartle L.M., Dionne C.A., Littlewood T.D., and Goldmacher V.S. 1996. Induction of apoptosis by tamoxifen-activation of a p53-estrogen receptor fusion protein expressed in E1A and T24 H-ras transformed p53-/- mouse embryo fibroblasts. *Oncogene* **13**: 739–748.

Vize P.D., Melton D.A., Hemmati-Brivanlou A., and Harland R.M. 1991. Assays for gene function in developing *Xenopus* embryos. *Methods Cell Biol.* **36:** 367–387.

Wang S., Krinks M., Lin K., Luyten F.P., and Moos M., Jr. 1997. Frzb, a secreted protein expressed in the Spemann organizer, binds and inhibits Wnt-8. *Cell* **88:** 757–766.

Warner A.E., Guthrie S.C., and Gilula N.B. 1984. Antibodies to gap-junctional protein selectively disrupt junctional communication in the early amphibian embryo. *Nature* **311:** 127–131.

Wettstein D.A., Turner D.L., and Kintner C. 1997. The *Xenopus* homolog of *Drosophila* Suppressor of Hairless mediates Notch signaling during primary neurogenesis. *Development* **124:** 693–702.

Wilson P.A. and Hemmati-Brivanlou A. 1995. Induction of epidermis and inhibition of neural fate by Bmp-4. *Nature* **376:** 331–333.

Wittbrodt J. and Rosa F.M. 1994. Disruption of mesoderm and axis formation in fish by ectopic expression of activin variants: The role of maternal activin. *Genes Dev.* **8:** 1448–1462.

Wright C.V., Cho K.W., Hardwicke J., Collins R.H., and De Roberts E.M. 1989. Interference with function of a homeobox gene in *Xenopus* embryos produces malformations of the anterior spinal cord. *Cell* **59:** 81–93.

Zernicka-Goetz M., Pines J., Ryan K., Siemering K.R., Haseloff J., Evans M.J., and Gurdon J.B. 1996. An indelible lineage marker for *Xenopus* using a mutated green fluorescent protein. *Development* **122:** 3719–3724.

CHAPTER 4

Equipment for Embryo Experiments

Described in this chapter are the various pieces of equipment that can be used to manipulate *Xenopus* embryos. Although embryos develop over a wide range of temperatures, from 14°C to 23°C, optimal and predictable development requires a more restricted and controlled environment. A microscope is essential for manipulating fairly small embryos and for documenting the effects on development. The embryos are manipulated under the microscope by dissection or injection, and a variety of tools must be made or purchased for these purposes. There are many variations in types of equipment and this chapter provides some suggestions.

VARIABLE TEMPERATURE INCUBATORS

Xenopus embryos develop normally within a temperature range of 14–20°C, but their rate of development varies greatly with temperature (see Appendix 2). Variable temperature incubators provide a convenient way of controlling development rates and are used to produce embryos at a number of different developmental stages simultaneously. However, incubators are expensive. A commonly used but expensive incubator is the HiLo temperature incubator from LabLine. An inexpensive alternative is a small refrigerator in which the thermostat is hooked up to a Honeywell temperature regulator.

MICROSCOPES

A good-quality dissecting microscope that provides at least 50x magnification is essential for both microinjection and microdissection of embryos. *Xenopus* embryos are relatively large (until neurula, about 1 mm in diameter) and thus the microscope does not need to be mounted on a vibration-free surface. However, a large flat base on the microscope is useful so that the user's hands can rest comfortably during dissection. A phototube with a "beam splitter" is a good addition to a microscope. Some phototubes redirect the light from one eyepiece to the camera, which is a disadvantage when the phototube is used for video monitoring. Many companies (e.g., Zeiss, Nikon, and Wild) make good-quality microscopes, but it is well worth testing the apparatus before purchasing. Dissection can be tiring with the wrong equipment.

MICROINJECTORS, MICROMANIPULATORS, AND NEEDLE PULLERS

The tools essential to most procedures in embryo manipulation are microinjectors, micromanipulators, and needle pullers. Microinjectors vary from inexpensive hand-cranked devices through models such as the Drummond Nanoject Variable, the Picospritzer, and the Narishige Medical Systems. Less expensive models are acceptable if resources are limited, but they must be capable of delivering specified volumes accurately (between 10 and 50 nl for most applications).

A micromanipulator is required to hold the glass capillary needles with which embryos are injected. Narishige makes some excellent basic models, as does Brinkman. Needles must be held firmly at an angle of approximately 45° for injection. A micromanipulator that is bolted to a microscope and cannot be angled is not suitable.

A needle puller is an obvious requirement for microinjection work. Again, a wide range of devices are available. Sutter Instruments supplies an excellent model, but inexpensive versions, such as those made by Narishige, are adequate. Homemade pullers can be used, but needle tip size tends to be less uniform than that obtained with a commercial puller. Although these machines are expensive, they allow a large number of needles to be pulled in a short time and so the cost can be shared by several laboratories. Needle bevelers are used to grind needle tips to a sharp but smooth angle. This is not usually necessary for microinjection, since needle tips can be broken to a suitably sharp angle with forceps.

NEEDLES

Recommended for microinjection and nuclear transplantation are glass capillaries such as the 30-μl Drummond micropipette needles (Fisher). These are approximately 8 mm long and 1 mm in diameter, and needles can be pulled either by hand over a Bunsen burner or by one of the commercial needle pullers described above. Specialized microinjection capillaries are also available (e.g., 1 mm × 10 cm glass thin-wall capillaries from World Precision Instruments). These capillaries contain a glass filament that helps the injection sample to run to the end of the pipette.

For microinjection, the needle tip must have steep shoulders followed by a gradual taper approximately 50–75 mm in length. Break the tip with forceps to a diameter of approximately 10 μm. The gradual taper allows the tip to be broken repeatedly to adjust the orifice or to reopen a clogged tip. The taper should not be so long that it bends instead of penetrating the embryo cleanly. Needles with narrow tips must be back-filled, using an automatic pipettor fitted with a long narrow tip (of the kind used for loading sequencing gels, e.g., Bio-Rad Seque/Pro). If the needle is front-filled by suction, the tip must be somewhat wider.

For nuclear transplantation, needles must have long, gently sloping tips, which are produced by pulling the same micropipette twice (as shown in Figure 11.1 in Chapter 11). The initial pull should draw the pipette to 10–15 cm in length. Transplantation needles should have a taper of 200–400 μm and a beveled tip of 60–75 mm in diameter. Size measurements are made using the ocular micrometer (eyepiece reticle) of a dissecting microscope, which in turn can be calibrated for different magnifications using a stage micrometer.

Needles used for injecting sperm nuclei should be siliconized before use (this prevents shearing of the nuclei during injection). Siliconizing is achieved by forcing a small volume (10–50 μl) of a liquid siliconizing agent (e.g., Sigmacote, Sigma) through the needle. The easiest way to do this is to attach a 1-cm length of 0.78-mm internal diameter Tygon tubing (Fisher) to a yellow micropipette tip on an automatic micropipetter. Use the micropipetter to draw the siliconizing agent into the tubing before attaching it to the blunt end of the needle. Depress the micropipetter plunger until a few drops of liquid emerge from the tip of the needle. Rinse needles before use by repeating this process several times with sterile water. Transplantation needles can be siliconized for 10 minutes to several months ahead of time.

For more details on the preparation of needles for nuclear transplantation, see Chapter 11.

DISSECTION AND MICROMANIPULATION TOOLS

For excellent advice on dissection and micromanipulation tools, see Hamburger (1960). All tools must be kept clean and rinsed with 70% ethanol to keep them sterile.

Forceps

Stainless-steel forceps, such as the Dumont 5 or 5A (Fine Science Tools), are adequate for membrane removal and most micromanipulations. They do not rust easily, but they cannot be sharpened to as fine a point as carbon-steel forceps. To sharpen the forcep points, use a fine stone or fine emery paper, observing the process under the dissecting microscope. Use only slight pressure and keep the points together so that the sharpened points meet cleanly at their tips. Slight bends can be introduced or corrected by pressing the tips against the microscope stage.

Label the forceps for different uses, good ones being reserved for fine work and old battered ones (which can be bent and filed back to a blunt end) for all noncritical purposes. Needle-nose pliers or coarse forceps can be used to make adjustments to the angles of fine forceps, but any excessive bend will break the tips.

Hair Loops

Loops of human hair mounted in beeswax at the tip of a cut-off pasteur pipette are used for steadying and moving embryos around during micromanipulations. It is useful to have loops in a variety of sizes. Small loops are resilient and useful for fine dissections, whereas larger loops are good for sorting and pushing embryos around. Hair loops can be prepared easily in the laboratory as follows:

1. Heat a long-stem pasteur pipette below the shoulders in a small Bunsen burner flame and pull to about 15 cm. Then make a second and finer pull to approximately 25 cm on the same pipette. Break the end after scoring with a diamond pencil. The opening should be big enough to thread the hair easily (it may be necessary to use the dissecting microscope and forceps to view and hold the hair). Careful flame polishing of the severed glass tip helps prevent the hair from being damaged, but this is not essential.

2. Select a human hair longer than the pipette and thread it through the smaller opening of the pipette until the end emerges from the larger opening. Then push the free end of the hair far enough into the small opening so that it does not pop out. Tighten the loop by pulling on the first end (see Figure 4.1).

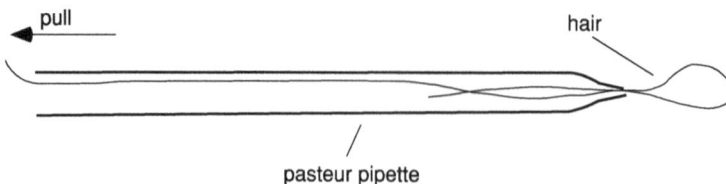

Figure 4.1. Hair loop.

3. Scrape a little beeswax into the end of another pasteur pipette and warm it over a flame until molten. Apply the waxed tip to the small opening of the hair loop. Wax is drawn by capillary action into the tip. Make final adjustments to the size of the loop by pulling on the hair, and then remelt the wax with a warmed pipette tip. The set wax holds the loop in place. This can also be achieved by dipping the hair loop into molten beeswax (melted on a hotplate). After removing the loop from the wax, blot excess wax from the loop by touching the loop to a hot Kimwipe or tissue fragment (preheat the tissue on a warm hotplate or piece of metal that has been prewarmed in the Bunsen burner).

Eyebrow Knives

Xenopus embryos are sufficiently soft to be cut with human eyebrow hairs, which offer a combination of resilience and sharpness that is difficult to match. Eyebrow knives are made in the same way as hair loops, but more easily. Eyebrow (or eyelash) hairs vary enormously from person to person in their quality as cutting tools. Try different people for sources of hair. The ideal hair is strong but not brittle. Simply insert the root of the hair far enough into the drawn-out pipette so that it leaves the desired amount of curvature/resilience (2–5 mm). Set in place with beeswax as described above.

Eyebrow knives are not used to saw down into the embryo as are conventional knives; instead, they are pushed or threaded through the tissue and flicked upward through it (see Chapter 10). They can also be used to trim isolated tissue that has been removed from the embryo. Trimming is done by pressing the eyebrow against the solid base of the dissecting dish (see Chapter 10 and Video Guide).

Fine Tungsten Needles

Tungsten needles and glass knives are useful for manipulating and dissecting embryos of more advanced developmental stages (neurula and beyond) when the embryonic tissues become tougher and somewhat resistant to eyebrow knives. Tungsten needles with 1-mm tips can be purchased from Fine Science Tools as an alternative to eyebrow knives. In contrast to eyebrow knives, tungsten needles become permanently bent quite easily. To avoid this, dissect the embryos on an agarose bed.

Tungsten needles can be made quite easily. Mount short pieces of fine tungsten wire (California Fine Wire, size 0.001) in pipette tips that have been prepared as described above for hair loops. Cut the wire to a length of 10 mm and insert it into the drawn-out tip, which is then fused carefully around the wire by gentle heating in the edge of a flame. Overheating makes the wire brittle. Alternatively, tungsten wire holders can be purchased from Fine Science Tools or Carolina Biological.

Glass Needles

Glass needles are useful for dissecting older embryos. They are made very simply from microinjection needles or hand-pulled capillaries. The needles are mounted in holders used for tungsten needles (available from Fine Science Tools or Carolina Biological). The fine needle tips should be broken off to provide a tool sturdy enough for dissection. Because this method leaves the tips open, the needles are difficult to sterilize and thus should be discarded after use.

PHOTOGRAPHY

Cameras

The most convenient way to take high-quality photographs of subjects under the microscope is to use a purpose-built, high-quality microscope camera. However, good results can be obtained with a single-lens-reflex (SLR) camera. The SLR camera can perform double duty for making photographic slides when fitted with a macro lens and used on a copystand.

Digital photography is rapidly improving and is considered as an alternative to conventional photography. It has the advantage of convenience, since results can be viewed immediately and downloaded as computer files using desktop publishing software. However, high-quality digital cameras are expensive, and even they cannot compete with 35-mm film for image quality. Conventional slides and prints can be converted into digital format using relatively inexpensive scanners in conjunction with imaging software.

SLR cameras are commonly used on dissecting microscopes. A plain focusing screen is optimal, since focusing screens that use a prism require higher light levels and are thus difficult to use. Since it is extremely difficult to focus low-power pictures, it is essential to magnify the image to be photographed. Magnifiers are available as accessories to some cameras, but a focusing telescope (e.g., Zeiss 3x12B, part 522012) can be applied to the viewfinder of a camera, or the eyepiece of a microscope, to enlarge the image and ensure that it is in focus with the photographic reticle.

The lines on the focusing screen and the image must be in sharp focus simultaneously, which can be achieved by focusing on the screen by adjusting the telescope, and then focusing on the specimen using the microscope focus. In general, eyes quickly accommodate at low magnification, so it is important to keep checking that the lines on the focusing screen, or focusing reticle, are sharp.

Film

It is important to choose the correct film for the purpose in hand. For non-fluorescent microscopy, where light is relatively plentiful, a good choice is Ektachrome 64 slide film (Kodak). It provides high resolution and can usually be processed quickly at a local photography store. Royal Gold 25 (Kodak) is a high-resolution print film that is also recommended for these purposes. If exposed approximately correctly, this film can be printed perfectly, avoiding the need to bracket exposures. Local 1-hour print shops usually do a reasonable job of printing, but they must be willing to adjust the computer exposure for each roll of film. For bright-field pictures of embryos, prints should be one stop lighter than normal; for dark-field pictures, they should be one stop darker. When exposures must be matched exactly, it is often worth the extra expense of custom printing.

For fluorescent photography, film resolution must be sacrificed in the interest of reasonable exposure times. Ektachrome 400 (Kodak) is a popular choice.

Helpful Hints

It is very easy to destroy a day's photography with a few simple errors. The following recommendations are designed to avoid some of the common pitfalls:

- When loading the film into a manual camera, it is vital to wind the film securely onto the take-up spool before closing the camera back. Once the slack is taken up, the rewinding crank should be seen to turn.

- Adjust the ASA setting on manual cameras according to the film used. Automatic cameras will set the ASA and ensure that the film is properly loaded.

- When taking the pictures, it is important to realize that the human eye makes considerable corrections to the image. Uneven illumination and slight imperfections due to dust particles may be barely noticeable to the naked eye but will show clearly on the film. Make a deliberate attempt to to check the background imperfections. Decide whether the field is going to be mostly light or mostly dark. For bright-field pictures of embryos, set the auto exposure at +1. For dark-field pictures, set at –1 or –2 (depending on how much bright embryo occupies the field). If the background is dark, it is particularly important to remove dust particles. With films such as Royal Gold 25 (Kodak), there is no need to bracket exposures since any slight misexposure can be corrected during printing. However, for slide film, it is crucial to bracket exposures.

- To make the background disappear, set a 90-mm petri dish on top of the stage instead of the normal black baseplate (see Figure 4.2). Use a clean new petri dish for specimens. Place the embryos on an agarose bed to push the base of the petri dish out of focus. Wash the embryos well to remove debris and fill the space between the top and bottom dish with water or 70% ethanol. The background (under the microscope) will now be very distant and out of focus. Use at least two light sources to maintain even illumination. Alternatively, place the petri dish on a colored background (electrical tape works well).

Compound Microscopes

Although the observations that applied to dissecting microscopes apply to compound microscopes, there are some aspects of photography that apply specifically to the compound microscope (see Figure. 4.3).

Always take the time to set up the correct Kohler illumination. The best resolution and image quality apply when the optical elements are aligned and all focused on the specimen. First, focus on the specimen. Then close down the field diaphragm at the base of the microscope, and use the condenser focusing knob to bring the edge of the field diaphragm into sharp focus.

For low-power photographs of embryos, it is necessary to use a low-numerical-aperture condenser lens. This allows the whole field to be evenly illuminated and the light to be focused on the specimen. The alternative, i.e.,

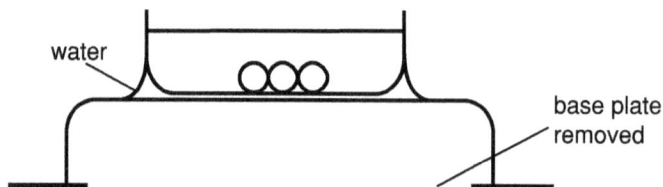

Figure 4.2. Dark-field photography of embryos.

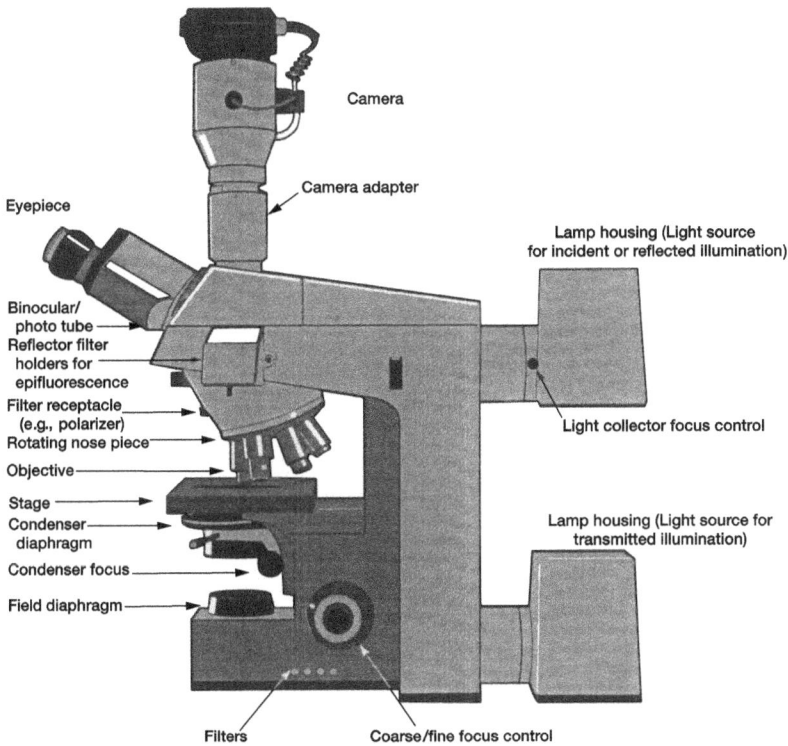

Figure 4.3. The light microscope. Basic components of the light microscope are arranged for transmitted and incident illumination. (Reprinted, with permission, from Spector et al. 1998.)

defocusing a higher-numerical-aperture condenser lens, yields a much lower-resolution image. The field diaphragm should then be moved to the edge of the field and centered using the screws on the condenser.

Once the image is in focus, adjust the condenser diaphragm. A closed-down condenser provides a larger depth of field, but significantly worse resolution in the image. As a rule of thumb, the condenser should be closed until there is a barely discernible effect on the brightness of the image. If cleared embryos look incompletely cleared, an excessively closed condenser diaphragm may be the problem.

Filters are often left in place on compound microscopes. Make sure that the chosen filter set is engaged. The presence of a daylight filter can easily go unnoticed. In addition, if filters are not properly seated, they can project into the light path and cause uneven illumination. Check to ensure that all filters and stops are properly engaged.

Making Slides

An SLR camera, equipped with a macro lens and a copy stand, can be used to take presentation slides. The general guidelines are the same as those for low-power photography. Use a slow film with high resolution (Ektachrome 64), set the f-stop to the minimum to maximize the depth of field (e.g., f22), and use the telescope to magnify and focus the image. For exposures, use the automatic setting on the camera, but correct the exposure to achieve the right shade of background. It is best to bracket around +2 stops if the picture is light (e.g., a line drawing) and around –2 stops for pictures in which the background is dark. Again, it is easy to ignore uneven illumination and glare, so turn off the overhead lights and look critically for evenness of illumination.

REFERENCES

Hamburger V. 1960. *A manual of experimental embryology*. University of Chicago Press, Illinois.

Spector D.L., Goldman R.D., and Leinwand L.A. 1998. *Cells: A laboratory manual*. Vol. 1 *Light microscopy and cell structure*. Cold Spring Harbor Laboratory Press, Cold Spring Harbor, New York.

The video series "Manipulating the Early Embryo of Xenopus laevis*" presents discussions and illustrations of needles used in embryo experiments and of dissection and micromanipulation tools.*

CHAPTER 5

Obtaining Embryos

Because of their large size and rapid developmental rate, *Xenopus* embryos are ideal subjects for the developmental embryologist. *Xenopus* embryos can be obtained by inducing ovulation and fertilizing the resulting eggs in vitro (see In Vitro Fertilization below) or alternatively, by "natural" matings (see Natural Matings below). A really good female lays many hundreds of eggs. However, because the actual number of eggs laid and the efficiency of fertilization are unpredictable, ovulation should be induced in more than one female, even when only a few eggs are required. Females can be induced to lay repeatedly for several years, but rest periods of 2–3 months are required between ovulations. The testes from one male contain sufficient sperm to fertilize several thousand eggs.

PROTOCOL 5.1

Handling *Xenopus* Adults

Handle the frogs with clean, soap-free hands or wear smooth latex gloves. Powdered or textured gloves must not be used since they abrade the animal's skin. Before handling a living animal, it may be helpful to practice holding a model frog.

1. Pick up the frog up by placing one hand across its back with a forefinger between the animal's hindlegs and wrapping the rest of the hand around the animal's middle (see Figure 5.1).

2. Use the other hand to cover the frog's eyes. This has a calming effect and will also physically restrain the animal. *Xenopus* is a quick mover and entirely aquatic. Escaped frogs dehydrate and die within a matter of hours of leaving the water, and so must be immediately captured and rinsed well in distilled water.

Figure 5.1. Handling *Xenopus* adults. Note position of forefinger between animal's hindlegs. (Courtesy of Mark Curtis.)

Inducing Ovulation

Ovulation is induced by injection of human chorionic gonadotropin (hCG) (available from many suppliers including Sigma and ProVet) into the dorsal lymph sac of a female frog (Figure 5.2). Newly purchased females and those that have not been induced to ovulate for more than 6 months should be primed before induction of full ovulation. Approximately 50 units of hCG should be injected into the dorsal lymph sac, at least 5 days before ovulation is induced. Higher doses of hCG may induce partial ovulation, which can cause subsequent ovulation to yield small numbers of poor quality eggs. If ovulation is induced less than 5 days after priming, egg laying commences very quickly (probably in the middle of the night when no one is there). Optimal quality eggs are laid 1–2 weeks after priming, after which egg quality declines. The priming effect wears off after about 1 month. For induction of full ovulation, 500–800 units of hCG, depending on frog size, should be injected into the dorsal lymph sac.

1. Wrap the frog firmly in wet paper towels, completely covering the eyes, but leaving the hind legs and lower abdomen exposed. With experience, this step can be omitted.

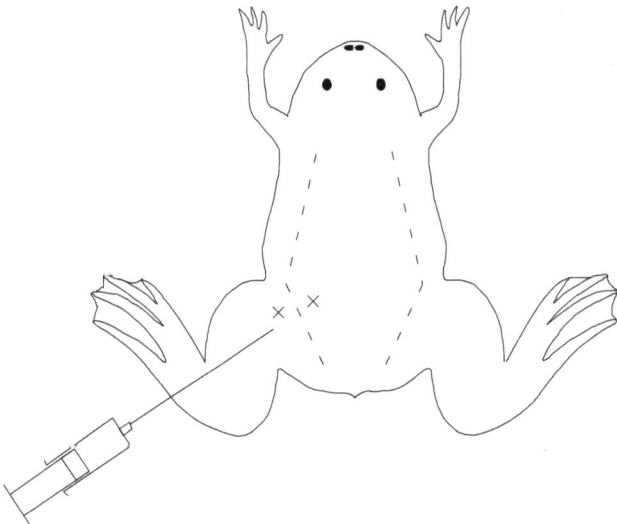

Figure 5.2. Injection of hCG into the dorsal lymph sac of a female frog. Crosses mark alternative insertion points for needle. See text for details. (Courtesy of Mark Curtis.)

2. Place the frog belly down on a clean smooth surface to avoid abrading the frog's skin. Plastic-coated paper towels (e.g., Benchkote, VWR) turned plastic side up work well.

3. Place the index and middle fingers over the frog's thighs so that if it starts to move during injection, pressure can be exerted on its thighs to prevent movement.

    Note: *In general, the frog will lie quietly, but before moving, it will start to breathe very quickly, thus giving the handler several seconds warning.*

4. Use a fine needle (26 gauge) attached to a 1-ml syringe and place the needle posteriorly, at the level of the hindlimb near the lateral line sense organs (see Figure 5.2). Penetrate the skin with a firm push, holding the syringe almost parallel to the back (penetration is easily recognized since the skin is quite loose and not attached to the underlying tissue). Pushing the needle straight down will penetrate the muscle beneath and cause bleeding (see Note in step 6.)

5. If the skin has been penetrated outside the lateral line sense organs, slip the needle laterally toward the dorsal midline, across the lateral line "stitch" marks to the dorsal lymph sac (see Figure 5.2). If initial penetration was made inside the lateral line, push the needle gently directly down into the lymph sac.

6. Penetrate the wall of the sac (some small resistance will be felt and then relief thereof) and inject the liquid. Wait five seconds and then slowly pull out the needle.

    Notes: *It is most important that the injected liquid does not run out of the frog when the needle is removed. No bleeding should occur. Bleeding indicates that either the needle went in too deep, possibly into muscle, or the skin was torn by sideways shearing of the needle. It is not necessary (and may even be detrimental) to swab skin with alcohol before injection. Frog skin contains naturally occurring antibacterial agents and its protective mucus may be damaged by alcohol.*

    Frogs kept at room temperature (~23°C) begin laying eggs about 9–10 hours after induction of ovulation, whereas frogs kept at 15°C begin egg laying approximately 14 hours after injection.

PROTOCOL 5.3

Isolating the Testes

1. Sacrifice a male frog by submerging it in 0.05% benzocaine brought to room temperature for 30 minutes to 1 hour before dissection (immersion of serveral hours does not damage sperm quality). When laid on its back, the frog should be completely limp and the heart should not be beating.

   Note: *Prepare fresh dilutions of benzocaine and store at 4°C. A 10% stock of benzocaine can be prepared in ethanol.*

2. Place the frog belly up on clean plastic-coated paper towels (e.g., Benchkote, VWR) turned plastic side up.

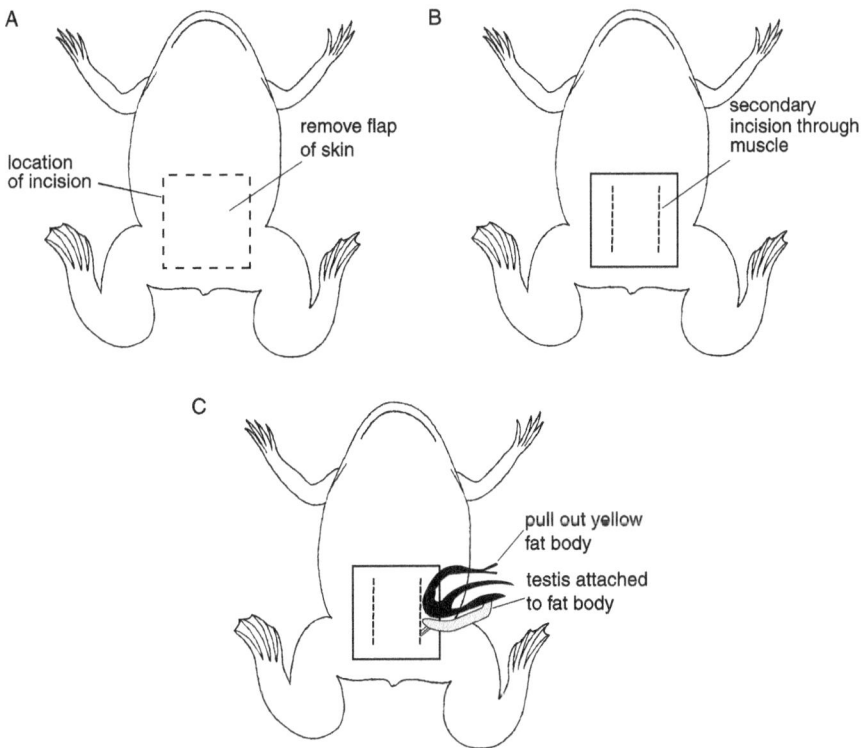

Figure 5.3. Isolation of testis. In the fresh male, the testis is attached to the fat body on either side. (*A*) Removal of skin flap; (*B*) incisions into muscle; (*C*) removal of fat body with attached testis.

3. Use a tissue or sharp forceps to pick up the loose skin on the belly and make a small cut with scissors. Then cut a large flap of skin open to expose the lower belly (Figure 5.3A).

4. Lift the abdominal muscles with forceps and cut to expose the viscera by making a slit on either side of the dorsal midline. Be careful not to cut across the midline as this will sever the large abdominal blood vessel that runs along the midline (Figure 5.3B).

5. Use blunt forceps to push aside the liver and pull out the yellowish fat body. The testes is pulled out with the fat body. Do not penetrate the intestine since release of intestinal contents contaminates the testis, and later the embryos, with bacteria (Figure 5.3C).

   Note: The testes lie at the base of the fat bodies and are easily recognized (see Figure 5.3). They are whitish, about 1 cm long, and covered with capillaries.

6. Use scissors or forceps to free each testis from the fat body and surrounding connective tissue.

7. To verify that the organs removed are the testes, crush a little on a microscope slide and view with phase contrast. The sperm are easy to identify by their fine helical shape.

8. Place the isolated testes in 80% serum (calf), 20% 1x modified Barth's saline (MBS) plus high salt (see Apendix 1) with antibiotic (0.05 mg/ml gentamycin can be used) and store at 4°C. Under these conditions, testes can be kept for at least 48 hours, after which sperm viability drops.

9. Wrap the frog carcass and freeze at –20°C. If it was not clear whether the animal was dead during dissection, freeze for at least 24 hours to ensure death before disposal.

COLLECTING EGGS

The stress of induction of ovulation and subsequent egg collection is the most common cause of disease in frogs (see Chapter 1). To minimize the risk of disease, keep females in 20 mM NaCl, supplemented with 5 µg/ml gentamycin during egg laying, and for 12 hours after (gentamycin is nephrotoxic to adult *Xenopus*; do not use in concentrations higher than 5 µg/ml). It is imperative to keep the frogs in isolation for 12–24 hours in *clean* water after egg collection and to monitor them for signs of illness. Change the water at least on a daily basis to ensure against septicemia and death.

High-quality eggs are laid individually rather than in tight clumps or strings. They should be round, firm, and not flaccid. White eggs are a sign of poor egg quality. Pigmented eggs should have a clearly defined animal hemisphere. Eggs should be of uniform size: The presence of small eggs may indicate that pieces of ovary have been torn and laid, which may lead to infection of the frog. The number of eggs laid is less important than the quality of the eggs. A hundred good-quality eggs are far preferable to thousands of poor quality eggs.

Manual Egg Collection

Prior to manual egg collection, female frogs should be lying quietly in their containers (4-liter plastic beakers with plastic mesh lids are suitable) in approximately 3 liters of 20 mM NaCl supplemented with 5 µg/ml of gentamycin. The pH of the water should be close to 7.0.

The idea is to mimic the actions of a male frog and to encourage the female to lay spontaneously. Do not physically squeeze the eggs out of the animal. Learn to use a firm but gentle touch. The eggs collect in a sac near the cloaca and simultaneous lateral and vertical pressure should expel them. If performed correctly, the females remain relaxed during the procedure, except when actually pushing the eggs out. After induction of ovulation, the cloaca becomes red and swollen and is probably rather sensitive, so avoid touching it when picking up the frog. As shown in Figure 5.4A, hold the frog with two hands, and gently but firmly massage the belly with one thumb over a sterile/clean petri dish (80 mm) containing 1x MBS plus high salt (see Appendix 1). After less than 1 minute, she will begin to lay. If she does not lay, repeat the massage with increased pressure. Eggs may be laid without movement from the frog or she may push them out in a couple of vigorous bursts. An agitated frog can be calmed by holding her against the handler's clean lab coat briefly. Once the animal is calm, egg collection can be resumed.

A

B

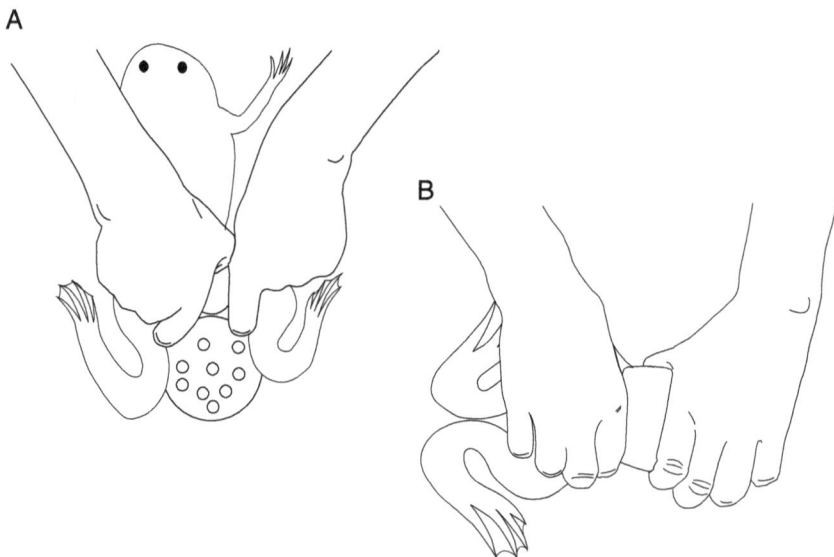

Figure 5.4. Manual egg collection. (*A*) Two-handed frog hold; eggs are collected in dish. (*B*) Hold frog down with left hand and stroke sides of the frog with right hand. Eggs are collected into plastic wrap or aluminum foil. For a small frog, the eyes can be covered with the right hand. The belly is massaged with the right thumb. If possible, do not touch the cloaca. (Courtesy of Danny deBruin.)

Egg collection should not be continued for more than 2–3 minutes total. Make collections every hour for the first 2 or 3 hours of laying and then less frequently as the day progresses. A maximum of four to six collections can be expected from a frog in 1 day.

An alternative method of manual egg collection is to place the frog belly down on clean aluminum foil or plastic wrap. The frog should be held immobilized by holding a hand over her, making sure the eyes are covered with one hand, and stroked firmly down either side with the fingers of the other hand, as shown in Figure 5.4B. Stroking the frog's sides also approximates amplexus movements. In this case, eggs are collected dry and then washed off the foil into a petri dish containing high-salt MBS, supplemented with 5 µg/ml of gentamycin.

Be sure not to not rub the back of the frog in attempting to collect eggs. Some eggs may be obtained, but there is very little subcutaneous fat on its back and the animal is very likely to bruise and may succumb to infection.

Egg Collection into High-salt Solution

After induction of ovulation, some handlers maintain the females in high-salt MBS (see Appendix 1) throughout the day and then collect the eggs after they have been laid in the water. This is thought to be less stressful for the females than manual collection, although the number of eggs obtained is generally lower. This method can work quite well, but the eggs must be removed from the container soon after laying for optimal fertilization (within ~15 minutes) and the 1x MBS plus high salt maintained at pH 7.0 (higher or lower pH irritates the frogs). When using this method, it is important to remember that a happy frog is a placid frog. High levels of activity indicate that the solution is irritating the frog's skin. If this occurs, the pH of the solution should be checked and/or the frog transferred to low-salt solution.

Females usually lay eggs for about 8 hours. Keep each laying in a separate dish and note the time of each on the dish. Use the earliest layings first as competence for fertilization decreases with age. Eggs can remain competent for fertilization for up to 12 hours after laying when kept in high-salt MBS at 15°C, although viability does drop during this time. The actual time of competence is dependent on the health of the female frog, and with only moderately healthy animals, fertilizations should be performed as soon as possible after laying. During the time of laying, females should be kept separately so that each can be observed independently.

IN VITRO FERTILIZATION

Eggs that will be fertilized immediately (within a few minutes of collection) can be collected into 1x MBS. Otherwise, maintain eggs in high-salt MBS. Just prior to fertilization, remove all buffer from the eggs with a pipette. Use forceps (cleaned with 70% ethanol) to tease a piece of testis apart and rub the tissue over the waiting eggs. It is important to touch every egg with the testis. Alternatively, crush part of the testis in high-salt MBS and mix the sperm "slurry" (~1/20 testis) with the eggs. With this approach, a pair of forceps is used to break up clumps of eggs and ensure that they are well dispersed across the dish. The eggs are tough at this point and will not break easily. After contact with sperm, flood the eggs with 0.1x MBS.

The first sign of fertilization (within a few minutes) is a contraction of the pigmented animal hemisphere to less than one half of the egg. A fertilized egg is more elastic and resistant to deformation than an unfertilized egg when squeezed gently with forceps. This is due to thickening of the vitelline membrane after fertilization. Approximately 20 minutes after fertilization after swelling of the perivitelline space (between the egg and vitelline membrane), the eggs rotate within the vitelline membrane so that the animal hemisphere faces upward. To assess egg quality, perform a test fertilization on a few eggs soon after laying. If fertilization efficiency is poor, test the eggs from a subsequent batch because egg quality can vary from one laying to the next. Good quality eggs should produce fertilization efficiencies of 80–100%.

NATURAL MATING

Males and females can be allowed to mate naturally (in a process called "amplexus") in containers fitted with a plastic mesh false-bottom through which the eggs can fall. Alternatively, eggs can be collected from the bottom of a container with a large-bore pipette. Square or rectangular containers large enough to allow frogs to move about should be used. Rat cages work well. Keep frogs in 20 mM NaCl during mating. Water depth should be approximately 6 inches to allow the frog to breathe. Induce ovulation by injecting females with 500–800 units of hCG as described above (In vitro Fertilization). Stimulate male sexual activity by injection of 50 units of hCG into the dorsal lymph sac (see Inducing Ovulation above), a few days before mating. If the male frogs are healthy, they should mate spontaneously; however, the females do not lay without stimulation. Place male and female frogs together. Natural matings are a convenient way to obtain many different stages of embryos at once, with minimal stress to the frogs.

In amplexus, the male frog grasps the female around the top of her hindlegs, using nuptual pads on his forelimbs. Male and female frogs periodically and simultaneously deposit sperm and eggs. Amplexus may continue for 12 hours or longer.

KEEPING TRACK OF FROGS

A frog can be induced to lay eggs repeatedly, but it must rest between ovulations. Rest periods of 2–3 months appear to be optimal. After natural matings, frogs can be used more frequently (6 weeks to 2 months). It is useful to keep accurate records of ovulations. For individual frogs, include in the records the dates of ovulation and the number of eggs laid, along with some indication of fertilization efficiency. A frog that repeatedly performs badly (three times) is unlikely to improve and should not be used again.

Most methods of marking individual frogs for identification seem cruel (e.g., acid-etched numbers) or impermanent (e.g., clipping toenails in specific patterns). Some laboratories implant subcutaneous microchips. These were originally designed for mice and are detected using a hand-held computer. The chips (~15 mm long and 1.5 mm in diameter) are relatively expensive, but they are simple to implant and can be reused. They are injected into the subcutaneous area, rather than into connective tissue, from which they tend to be difficult to retrieve. When placed within a few centimeters of the animal, the hand-held computer displays a 9-digit code, identifying the individual. The reader does not work if the animal is under water, and animals must be removed from the water for scanning. The chips are available from Avid Inc. (3179 Hamner Ave., Norco, California 91760) and are used routinely by Jonathan Slack's group at the University of Bath, from whom this procedure was obtained.

An alternative tracking strategy is to keep batches of previously ovulated frogs together in "good," "medium," or "poor" tanks, designated according to their previous performance. Within these categories, frogs are grouped according to the date of their last ovulation and are reassessed after each subsequent ovulation. The fewer the number of frogs grouped, the more precise the tracking can be. However, groups of up to 50 frogs can be usefully tracked. This method is somewhat imprecise, but it is very simple and adequate for the purposes of most laboratories.

REFERENCES

Deuchar E. 1975. Xenopus: *The South African clawed frog*. John Wiley & Sons, London.

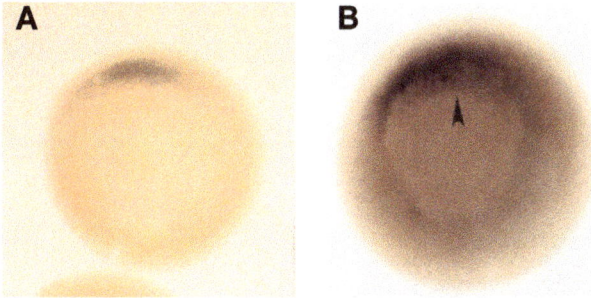

Color Plate 1. (*See Chapter 2, p. 17*) Genes expressed in dorsal mesoderm at the beginning of gastrulation. (*A*) Goosecoid (Reprinted, with permission, from Cho et al. 1991 [© Cell Press]); (*B*) Noggin (Reprinted, with permission, from Smith and Harland 1992 [© Cell Press]). Both images are views from the vegetal pole of the embryo.

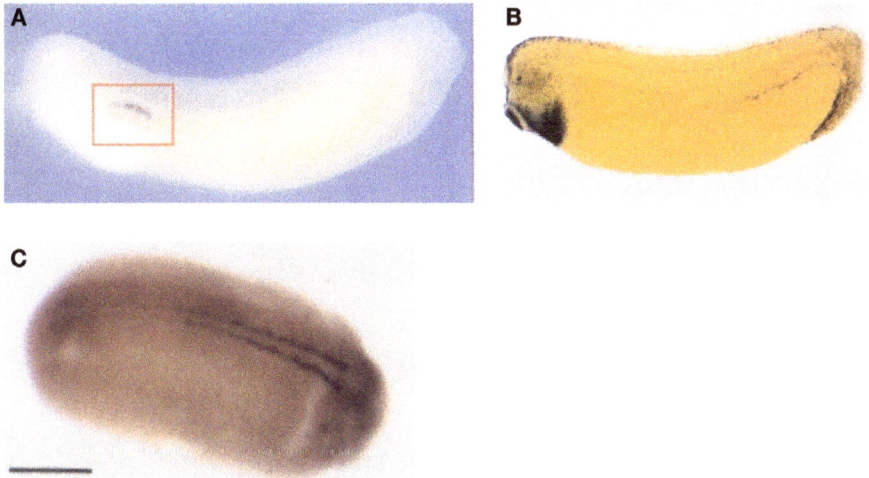

Color Plate 2. (*See Chapter 2, p. 24*) Regionalized gene expression at neural tube and tail-bud stages. (*A*) *xWT1* defines the pronephric region. (Reprinted, with permission, from Carroll and Vize 1996 [© Wiley-Liss].) (*B*) *XTin1* delineates the presumptive heart region. (Reprinted, with permission, from Cleaver et al. 1996 [© Company of Biologists].) (*C*) *xHB9* is expressed in the presumptive motor neuron area of the spinal cord. (Reprinted, with permission, from Saha et al. 1997.)

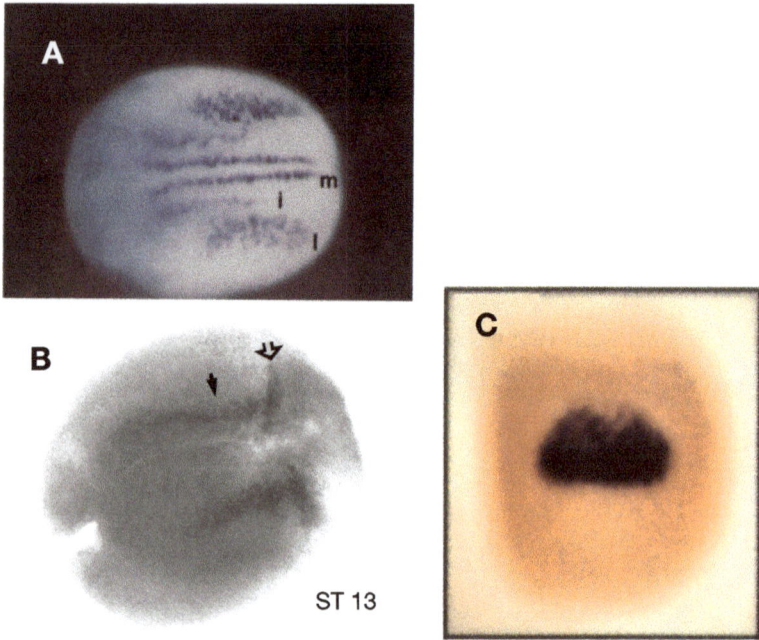

Color Plate 3. (*See Chapter 2, p. 26*) Genes expressed in particular domains of neural tissue during neurulation. (*A*) Expression of *N-tubulin* in stripes corresponding to the sites where primary neurons will form (dorsal view). (Reprinted, with permission, from Chitinis et al. 1995 [© Macmillian Magazines].) (*B*) Lateral domains of *Xash-3* expression demarcate the future sulcus limitans, the border between dorsal and ventral parts of the spinal cord (dorsal view). (Reprinted, with permission, from Zimmerman et al. 1993 [© Company of Biologists].) (*C*) *Rx* expression in the domain of the future retina (anterior view). (Reprinted, with permission, from Mathers et al. 1997 [© Macmillian Magazines].)

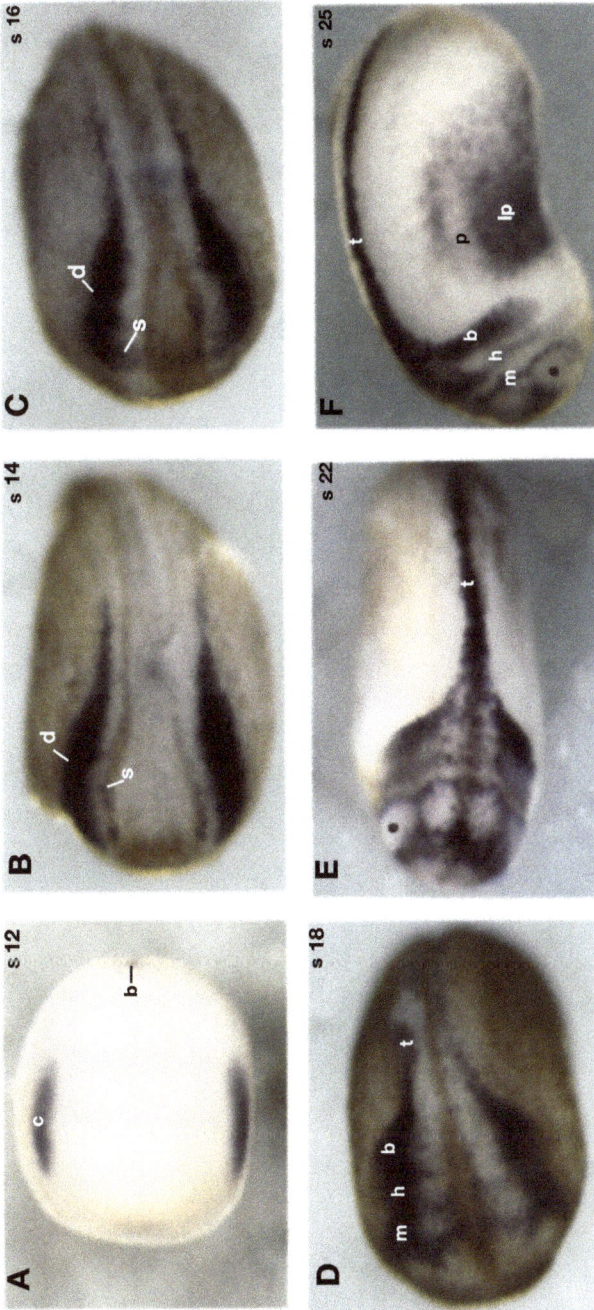

Color Plate 4. (*See Chapter 2, p. 28*) Expression of *Xslu* in neural folds and neural crest tissue. (Reprinted, with permission, from Mayor et al. 1995 [© Company of Biologists].) (*A*) Stage-12 embryo; (*B*) stage-14 embryo; (*C*) stage-16 embryo; (*D*) stage-18 embryo; (*E*) stage-22 embryo; (*F*) stage-25 embryo. (*A*–*E*) Dorsal views; (*F*) side view.

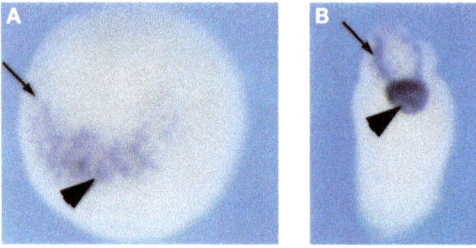

Color Plate 5. (*See Chapter 2, p. 31*) Expression of cement gland marker XAG. Anterior views of early neurula (*A*) and tailbud (*B*) stage embryos. This gene is expressed in the cement gland (arrowheads) and adjacent hatching gland (arrows). (Reprinted, with permission, from Sive and Bradley 1996 [© Wiley-Liss.])

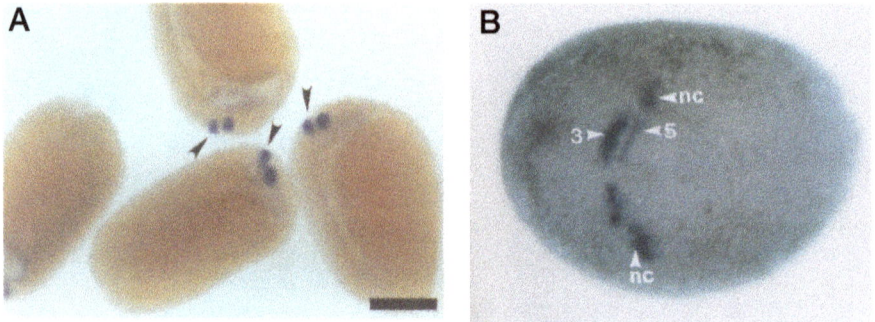

Color Plate 6. (*See Chapter 2, p. 34*) Early gene expression demarcates future regional properties in the neural plate. (*A*) Engrailed expression is restricted to the boundary between midbrain and hindbrain. (Reprinted, with permission, from Bolce et al. 1992 [© Company of Biologists].) (*B*) Krox-20 expression is found in rhombomeres 3 and 5 of the hindbrain and adjacent cells destined to be part of the neural crest. (Reprinted, with permission, from Bradley et al. 1993 [© Elsevier Science].)

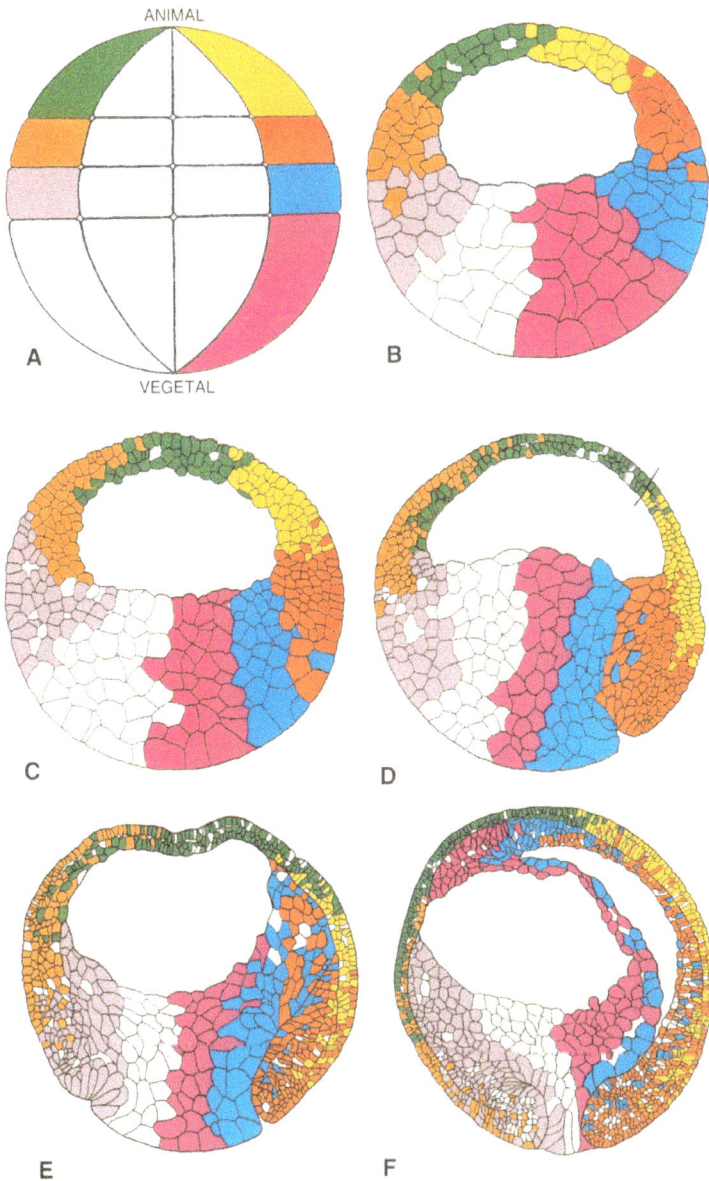

Color Plate 7. (*See Chapter 9, p. 151*) Summary diagrams illustrating the locations of the clones derived from the midline blastomeres of the 32-cell embryo (*A*) at stage 8 (*B*), stage 9 (*C*), stage 10 (*D*), stage 11 (*E*), and stage 12.5–13 (*F*). Data were derived from tissue sections in which two adjacent blastomere clones were labeled with different lineage dyes. Diagrams are oriented as in Figure 9.6, top. Clones are represented by the following colors: (*lilac*) C4; (*orange*) B4; (*green*) A4; (*yellow*) A1; (*red*) B1; (*blue*) C1; (*purple*) D1. (*A*, modified, with permission, from Bauer et al. 1994 [© company of biologists Ltd.]; *B–F*, modified, with permission, from Hausen and Riebesell 1991.)

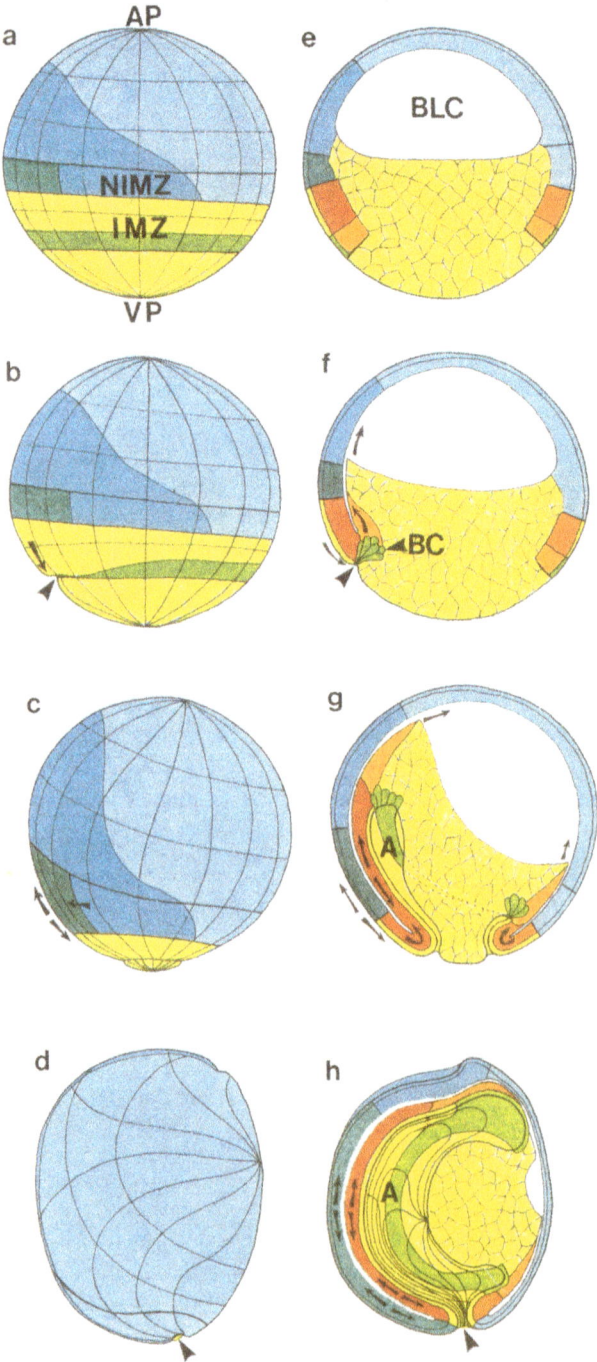

Color Plate 8. (*See facing page for legend.*)

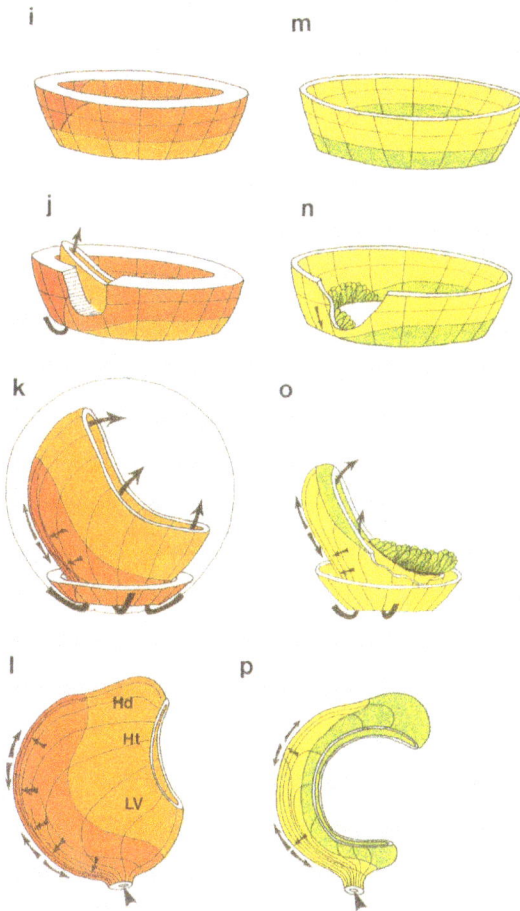

Color Plate 8. (*See Chapter 9, pp. 154–155*) Prospective fates and morphogenetic movements of the gastrula and neurula are shown in several views. Four horizontal rows show development in chronological order, from top to bottom, at the late blastula (*a, e, i,* and *m*), early gastrula (*b, f, j,* and *n*), late gastrula (*c, g, k,* and *o*), and the late neurula (*d, h, l,* and *p*). The vertical columns show different views which are, from left to right, the lateral surface, with dorsal to the left (*a–d*), a midsagittal view (*e–h*), the ring of deep prospective mesodermal cells (*i–l*), and the overlying ring of suprablastoporal endoderm (*m–p*). Special features illustrated include the animal pole (AP), archenteron (A), the blastocoel (BLC), the bottle cells (BC), the blastopore (pointers), the head mesoderm (*Hd*), the heart mesoderm (Ht), the involuting marginal zone (IMZ), composed of deep, nonepithelial mesoderm (*i–l*) and superficial, epithelial endoderm (⌈*m–p*⌉), the lateral and ventral mesoderm (LV), the noninvoluting marginal zone (NIMZ), and the vegetal pole (VP). Prospective tissues shown are epidermis (*light blue*), neural tissue (*darker blue*), dorsal NIMZ (*blue green*), notochord (*red*), somitic mesoderm (*red orange*), migrating mesoderm at the leading edge of the mesodermal mantle (*orange*), suprablastoporal endoderm (*yellow*), a special region of suprablastooral endoderm known as the bottle cells (*green*), and vegetal, subblastoporal endoderm (*yellow*, divided into cells). Movements are indicated by arrows. Dorsal is to the left in all cases. (Reprinted, with permission, from Keller 1991.)

CHAPTER 6

Preparing Embryos for Manipulation

Presented in this chapter are two protocols for preparing embryos for manipulation. Embryos are surrounded by a series of thick, protective jelly membranes. Removal of these membranes is the first step in most micromanipulation procedures, which is achieved by bathing the embryos in 2% (w/v) cysteine as described in Protocol 6.1. The membranes must be completely removed for embryo dissection. For microinjection, membranes can be completely or partially removed. The other important technique in embryo preparation is removal of the vitelline membrane. This is essential for embryo dissection experiments and is described in Protocol 6.2.

PROTOCOL 6.1

Dejellying Embryos

1. Dejelly the embryos by removing buffer and swirling gently in 1x modified Marc's Ringer (**MMR**), or water, with 2% (w/v) **cysteine** at pH 8.0.

   MMR, cysteine (see Appendix for Caution)

2. Gently swirl the eggs for 2–4 minutes until the jelly membranes are visible in the solution and the eggs have started to pack. Dejellying is usually complete in 4 minutes, but is frog-dependent and must be titrated for each animal. It is safest to monitor each dish of embryos during the process and not to depend on timing alone.

   Note: *Prolonged exposure to cysteine solution will damage embryos. In addition, cysteine solutions become less potent during the day and each batch of embryos must monitored during dejellying.*

3. When the eggs begin to pack together closely and fragments of jelly can be seen floating in the buffer, promptly decant the cysteine and rinse the fertilized eggs 10 times in 0.1x modified Barth's saline (**MBS**), over a period of approximately 10 minutes. It is essential to rinse embryos thoroughly in a *clean* beaker to remove all traces of cysteine.

   Note: *The vitelline membrane thickens during the first 30 minutes postfertilization and thus, if possible, it is best not to dejelly until after this stage; the thickened membrane offers some protection to the fragile embryo. Embryos are particularly sensitive to the dejellying process during gastrula and neurula stages and must be closely monitored to avoid excessive treatment.*

   MBS (see Appendix for Caution)

4. After dejellying, place embryos in a clean dish. Keep embryos at low density (100 per 80-mm petri dish) in 0.1x MBS and remove dead embryos promptly.

PROTOCOL 6.2

Removing the Vitelline Membrane

Removal of the vitelline membrane is a technique that becomes easier with practice. The easiest embryos on which to learn this technique are those with a large space between the membrane and the embryo, i.e., late neurula stages.

A membrane that is excessively snug can be loosened by digestion with proteinase K (5 µg/ml) for a few minutes. Note that a longer enzymatic treatment results in the destruction of the vitelline membrane and damage to the embryo. For this reason, manual removal of the vitelline membrane is recommended for micromanipulation procedures.

1. Place the dejellied embryos in a clean petri dish containing 0.1x **MBS**. This procedure may be facilitated by coating the bottom of the dish with 1% agarose (electrophoresis grade) in water.

   MBS (see Appendix for Caution)

2. Use a pair of fairly sharp watchmaker's forceps (with points bent toward each other like pincers) to take hold of the vitelline membrane (see Chapter 4).

3. Use a second pair of forceps (with tips that are slightly blunt to avoid puncturing the membrane) to steady the hold on the embryo.

4. With the sharp pair of forceps, make a tear in the vitelline membrane. If the tear is large enough, the embryo can be popped out. Otherwise, use the blunt pair of forceps to enlarge the tear.

5. Store the naked embryo in 0.1x MBS.

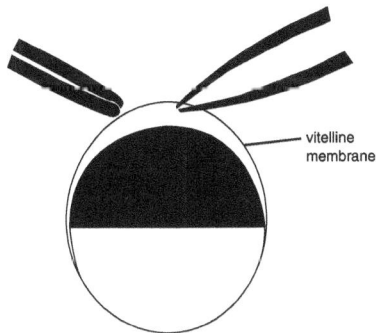

Figure 6.1. Removal of vitelline membrane. When embryo dissection will follow devitellinizing, avoid damage to the area of interest by initiating membrane removal elsewhere. For example, for animal cap dissections, begin removal at the marginal zone. If whole intact embryos are required, begin removal at the animal cap, since damage to this region heals very quickly.

vitelline membrane

Tape 1 of the video series "Manipulating the Early Embryo of Xenopus laevis*" presents a demonstration of vitelline envelope removal.*

CHAPTER 7

Embryo Perturbations

Several physical and chemical treatments used to perturb *Xenopus* axis formation act at various stages during axis formation and give rise to a range of altered (dorsalized or ventralized) phenotypes.

Since classic genetic experiments cannot be performed on *Xenopus laevis*, axis perturbation by nongenetic means is a useful tool for determining the nature and sequence of events that comprise the developmental process. Axis perturbation can also be used to investigate gene function; for example, one can ask whether a particular gene is able to compensate for the effects of a particular treatment. Knowing what the treatment does, or at least where it acts, in the specification of axial patterning provides valuable information on the role of the compensating gene. Dorsalized and ventralized embryos have also been useful as sources of mRNA for use in differential cDNA cloning and as substrates in gene expression screens. mRNAs with dorsalizing (see, e.g., Smith and Harland 1991) and ventralizing (see, e.g., Mead et al. 1996) capacities have been isolated.

The range of ventralized phenotypes was originally classified by Gerhart and colleagues (Scharf and Gerhart 1983) in the *Index of Axial Deficiency* (IAD). On this scale, 0 is normal and 5 is completely ventralized (i.e., completely lacking in dorsal structures). More recently, dorsalized phenotypes (i.e., those lacking ventral structures) have been included in a modified classification system called the Dorsoanterior Index (DAI; Kao and Elinson 1988), illustrated in Figure 7.1. On this scale, 0 is assigned to embryos lacking dorsoanterior structures and 10 is assigned to embryos with the most extremely enhanced dorsoanterior structures. The DAI scale is currently the accepted standard, although the IAD may appear in older publications.

Note that although different treatments often result in embryos with superficially similar phenotypes, when examined at a molecular level, embryos resulting from these treatments may not be equivalent. Nevertheless, these perturbations provide a useful battery of pseudogenetic tools with which to explore *Xenopus* embryogenesis and a somewhat standardized reference scale.

For all perturbations, a number of untreated embryos must be set aside as controls. These embryos must be maintained under the same conditions, most particularly at the same temperature, as the experimental embryos. This will demonstrate that any differences in the development of the embryos are due to the effects of the treatment, rather than to the influence of different temperature regimes.

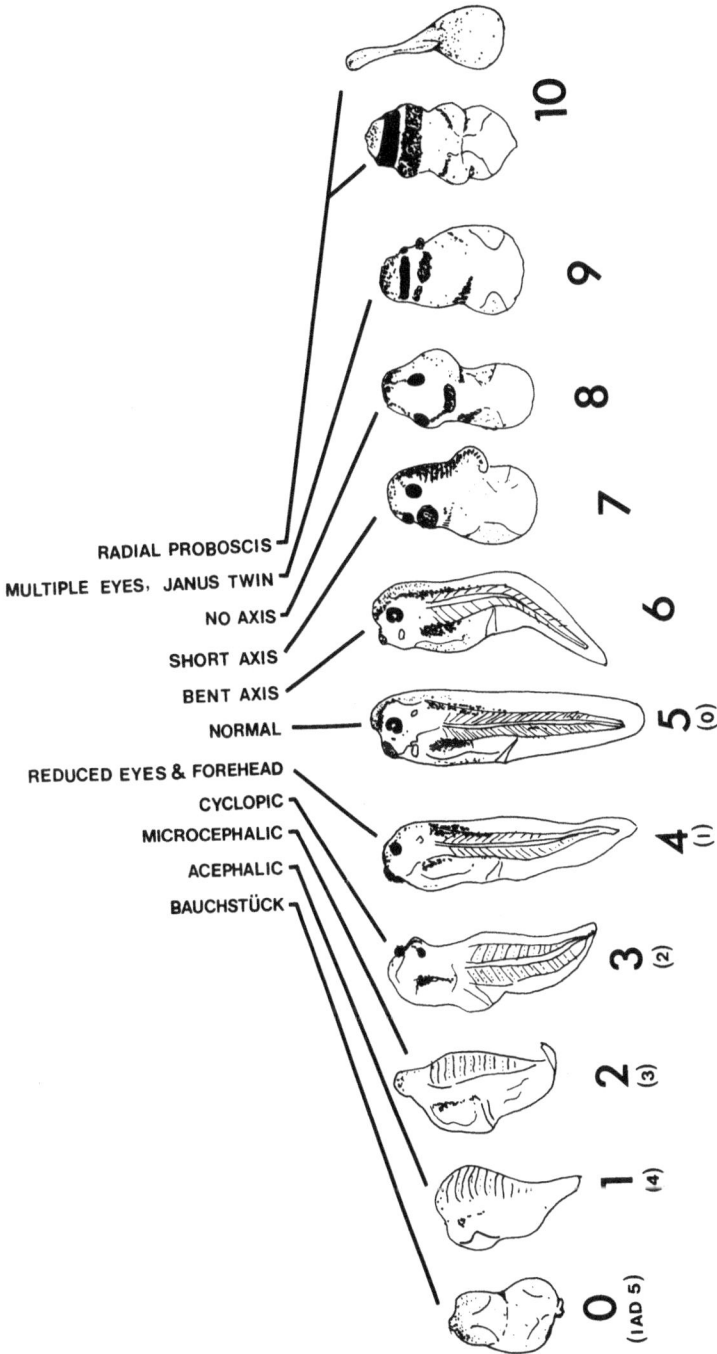

Figure 7.1. The Dorsoanterior Index (DAI). Numbers in parentheses refer to the index of axis deficiency (IAD) used previously for dorsoanterior-deficient embryos. Designations for various DAIs are indicated in Table 7.1. (Reprinted, with permission, from Kao and Elinson 1988.)

PROTOCOL 7.1

Axis Perturbation by UV Treatment

Immediately after fertilization, the *Xenopus* embryo enters the first cell cycle, during which dorsoventral polarity is established. Formation of dorsal tissues is closely correlated with rotation of the egg cortex (cortical cytoplasm) relative to the cytoplasmic core. This corticocytoplasmic rotation (CCR) is mediated by a microtubule-associated motor and takes place on arrays of microtubules located in the vegetal hemisphere (Elinson and Rowning 1988). The rotation is believed to be responsible for the concentration of dorsal determinants, e.g., β-catenin (Heasman 1997; Harland and Gerhart 1997), on the dorsal side of the embryo. Treatment with ultraviolet light (UV) during the first cell cycle (see below) prevents CCR and results in the formation of ventralized embryos (Scharf and Gerhart 1983).

Be certain not to confuse the CCR with the rotation of the whole egg within its membranes, which occurs within the first 30 minutes after fertilization. The latter rotation results from the egg being released from its attachment to the vitelline membrane. This release, which occurs at fertilization, allows the egg to rotate and settle with the heavy yolk of the vegetal pole downward and the animal pole upward. The CCR occurs later during the first cell cycle and involves rotation of only part of the egg (the cortical cytoplasm) relative to the rest of the cytoplasm.

The effects of UV treatment on developing embryos are observed most easily when compared to untreated control embryos. At early gastrula, successfully treated embryos fail to develop a dorsal blastopore lip. Ventralized embryos develop a circumferential blastopore lip about stage 11, when the ventral lip of the blastopore normally forms. UV-treated embryos can be sorted at stage 10.5, when those lacking a blastopore lip are deemed severely ventralized and will later be classified as DAI 0 or DAI 1.

For UV treatment, embryos are dejellied a few minutes after fertilization. The vegetal hemispheres are then exposed to short-wave UV light (254 nm) for 30–90 seconds, as determined empirically. Embryos must be irradiated before the CCR has been completed, i.e., during the first 60% (or 0.6) of the first cell cycle. Since the first cell cycle (until first cleavage begins) takes 90 minutes at 20–25°C, this allows about 55 minutes during which UV irradiation can be performed effectively.

1. Dejelly fertilized eggs as soon as the pigment of the animal hemisphere begins to contract (~1–2 minutes after fertilization).

Note: *For details on obtaining embryos, see Chapter 5.*

2. Rinse the embryos well and transfer healthy individuals to 0.1× modified Barth's saline (**MBS**) in quartz dishes (Figure 7.2A), or to 50-ml orange-cap tubes (Corning) containing approximately 45 ml of 0.1× MBS, sealed with Saran Wrap and a rubber band (Figure 7.2B).

    Notes: *For details on how to prepare quartz dishes or tubes, see Helpful Hints and Figure 7.2.*

    Early dejellying increases the chance of rupturing the vitelline membrane. Discard any embryos with ruptured membranes. Do not overcrowd the embryos. Fill the space available, but make sure that the embryos are not touching one another.

    MBS (see Appendix for Caution)

3. Place each dish or inverted tube on a UVGL-25 lamp (VWR Scientific Products) and irradiate the different batches of embryos for 30, 60, or 90 seconds with short-wave (254 nm) **UV** light.

    UV radiation (see Appendix for Caution)

    Note: *Sensitivity to UV varies from frog to frog so it is necessary to perform a UV titration for each experiment.*

4. Assess embryo mortality in each dish or tube before they reach blastula stage. Choose survivors from a background of 50% mortality.

    Note: *These embryos will have received the highest tolerable dose of UV and should be completely ventralized.*

5. Leave embryos in place until the first cleavage has been completed, then gently transfer to agarose-coated dishes and incubate below 20°C.

    Note: *The effect of UV irradiation on embryos can be nullified by gravity-induced rotation. Thus, after irradiation, leave embryos in place until after completion of first cleavage. Remove any embryos that remain tipped after transfer. This is especially important if irradiation is performed in quartz dishes. Embryos in tubes filled with buffer can be moved immediately after irradiation and little perturbation will occur. Anecdotal evidence suggests that the UV treatment is more effective if the embryos are left to develop at temperatures below 20°C.*

Helpful Hints

- Since the strength of UV lights can vary, and since different batches of embryos respond differently to irradiation, the investigator should empirically determine what treatment is appropriate for the experiment, using the measures given as a guide.

A

embryo well

cement

25-mm quartz square

35-mm Petri dish

B

tube set-up

tube inversion over UV

rubber band

plastic wrap

50-ml tube filled
with 45 ml 0.1 xMBS

embryos

000000

UV source

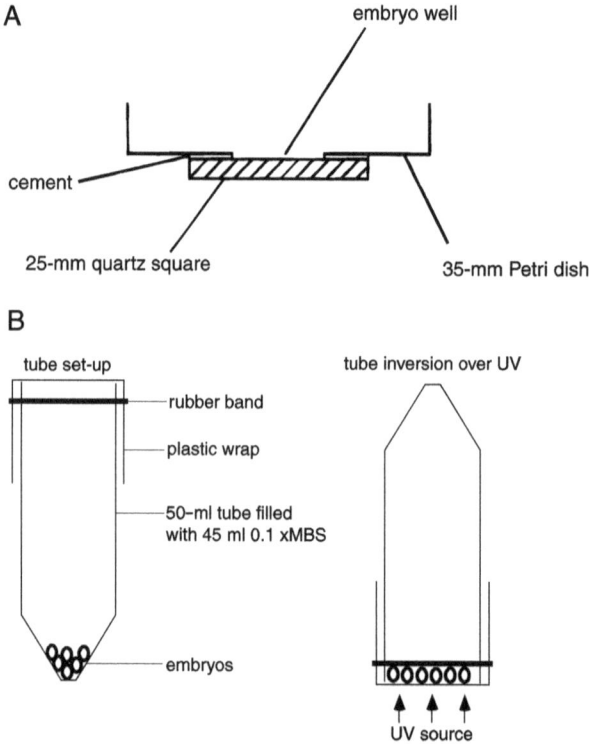

Figure 7.2. (*A*) A quartz dish for use in UV irradiation of embryos. For details of construction, see Protocol 7.1. (*B*) Plastic tubes for use in UV treatment. (*Left*) Tube setup; (*right*) inversion over UV source.

- Quartz dishes are simple to make, convenient, and long-lasting. Good quality quartz is required, but this is not expensive; optical grade distortion-free quartz is not required. Check the quality of cheaper quartz by putting it in the light path of a spectrophotometer. It should show good transmission down past 260 nm. To make a quartz dish, cut a 10 x 25-mm rectangular window in a 35-mm petri dish (e.g., Falcon 1008, Becton Dickinson). Attach a 25-mm square of quartz to the underside of the dish using cement, methylene chloride, or Plexiglas glue. Make sure that the window is completely covered and well-sealed. The corners of the quartz square may need to be filed down to ensure a snug fit. When the adhesive has set, clean the quartz of any cement or glue residue and rinse well. For a representation of the final product, see Figure 7.2.

- An alternative very simple means to irradiate embryos uses 5-ml plastic tubes (e.g., Corning). Fill the tube with 45 ml of 0.1x MBS and add embryos. Cover the tube with plastic wrap (not Parafilm, which attenu-

Table 7.1. The DAI Scale

DAI no.	Designation[a]
0	(IAD 5) no somites present; trace of tail mesenchyme occasionally seen
1	(IAD 4) no otic vesicle(s) present; somites present in trunk or portion thereof
2	(IAD 3) no visible eye pigment; otic vesicles or single vesicle still visible
2.5	cement gland present, but no eye pigment visible
3	(IAD 2) eyes fused or cyclopic, but at least some eye pigment visible; cement gland present.
4	(IAD 1) reduced forehead; eyes smaller than normal and sometimes joined
5	(IAD 0) normal in all externally visible respects
6	mildest effect; embryos show a slight bend in the axis, visible at stage 28
7	severely reduced trunk, ranging from a truncated tail to a small, vestigial axis closely attached to the head; somites visible.
8	complete lack of trunk but relatively normal looking face, although the eyes may be enlarged; no somites visible but muscle tissue can be seen sparsely in histological section; the notochord appears enlarged and irregularly shaped
9	multiple eyes and cement glands
10	radially symmetric embryos which have two forms • embryos either exhibit radial eye (retinal) pigment and cement gland • no eyes/cement gland visible; neural and notochordal tissue develops into a large proboscis extending either inward or outward at the end of gastrulation

The DAI scale is given in whole numbers. Averaging the DAI relative to the number of embryos scored in an experiment will usually give a fractional DAI. This is not meaningful and should be avoided.
[a] Data from Scharf and Gerhart (1983), Kao and Elinson (1988), and Sive et al. (1990).

ates UV) and secure with a rubber band. Invert the tube and allow embryos to settle before irradiation through the plastic wrap.

SCORING UV-INDUCED PERTURBATION OF THE DAI

Axis-deficient embryos can be identified at the onset of gastrulation. Any embryo that forms a dorsal blastopore lip at the same time as untreated controls is not completely ventralized. Good quality ventral embryos develop a blastoporal lip that appears simultaneously around their entire circumference, coincident with the formation of the ventral lip in control embryos. The effectiveness of UV treatment can also be estimated during neurulation. Effective treatment results in the absence of neural folds. For accurate scoring of embryos using the DAI, the otocyst (the white refractile body in the ear of the animal, lying laterally at the hind brain level) is a useful landmark. The otocyst becomes apparent at stages 25–35 but is more obvious at the tadpole stage (about stage 40). A detailed listing of DAI designations is given in Table 7.1.

PROTOCOL 7.2

Axis Perturbation by Lithium Chloride Treatment

Treatment of early blastula embryos (before the 64-cell stage) with lithium chloride results in dorsalization (Kao and Elinson 1988, 1989; Slack 1988). Lithium apparently acts through inhibition of glycogen synthase kinase-3 β (GSK-3 β), which allows activation of the *wnt* pathway that is required for dorsal axis formation (Klein and Melton 1996).

1. Dejelly the required number of early blastula embryos, according to Protocol 6.1, before the 32-cell stage.

2. Rinse the embryos well and transfer healthy individuals to a petri dish containing 0.3 M **lithium chloride** in 0.1x **MBS**. Swirl immediately to equalize the lithium chloride solution over the embryos.

   lithium chloride, MBS (see Appendix for Caution)

3. Incubate embryos at 18–20°C for 10 minutes.

4. Rinse the embryos well and transfer to a clean petri dish containing 0.1x MBS or equivalent.

5. Allow embryos to develop at 18–20°C. Score perturbations at early gastrula.

   Note: *The classical method of postfertilization treatment is carried out as detailed above. Lower concentrations of lithium chloride can be used with extended incubation times, e.g., 0.1 M lithium chloride in 0.1x MBS for 1 hour (recommended for embryos at the 32-cell stage). Different batches of embryos vary in their tolerance to lithium. Some batches can survive treatments lasting more than 1 hour. Since conditions vary between laboratories, the investigator should empirically determine what treatment is appropriate for the experiment, using the measures given as a guide.*

Helpful Hints

- If yolk plug involution is a persistent problem, rescue the embryos by incubation in 5% Ficoll 400 for 1–4 hours during gastrulation (see Appendix 1). This treatment increases pressure on the yolk plug and, in so doing, may encourage involution.

SCORING LITHIUM-CHLORIDE-INDUCED PERTURBATIONS ON THE DAI

The effects of lithium treatment can be observed at early gastrula where, in completely dorsalized embryos, the blastopore lip forms synchronously around the embryo, at the time that the dorsal blastopore lip would normally form. By tailbud or hatching stages, dorsalized embryos have exaggerated heads (DAI >5), which in the most extreme case (DAI 10) are radially symmetrical, with bands of eye (retinal) pigment and cement gland, as well as a cylindrical heart. Another DAI 10 form comprises a long protuberance of notochord and neural tissue with no head. No tail or blood will be present (i.e., the embryos will be dorsoanteriorized and lack both posterior and ventral tissues). DAI 10 embryos are difficult to obtain, but DAI 7 or 8 embryos can be produced routinely. (Note: When applied during gastrula stages, lithium chloride prevents head formation [Fredieu et al. 1997], possibly by activating the *wnt* pathway.)

PROTOCOL 7.3

Axis Perturbation by Retinoic Acid Treatment

Retinoic acids (RA) are derivatives of vitamin A. These compounds bind to nuclear receptors (retinoic acid receptors [RARs] and retinoid X receptors [RXRs]) to alter gene expression. Its effects vary according to time of application. *Xenopus* embryos are most sensitive to RA during late blastula and gastrula stages. Treatment during gastrulation with either low concentrations of RA or short periods of exposure to higher concentrations causes repatterning of the hindbrain (Papalopulu et al. 1991). Exposure to higher concentrations or more prolonged RA treatment causes anterior truncations. The eyes are the most sensitive organs and are lost first, followed by the cement gland and the heart (Durston et al. 1989; Sive et al. 1990). The fore- and midbrain are also affected. Treatment with RA at later stages leads to inhibition of tail formation (Ruiz i Altaba and Jessell 1991). Analysis of gene expression in aberrant individuals has shown that expression of posterior *Hox* genes, particularly of the labial family (*HoxA1, HoxB1,* and *HoxD1*), extends anteriorly (Kolm and Sive 1995), whereas expression of anterior marker genes, such as *otx2*, is inhibited (Pannesse et al. 1995; Gammill and Sive 1997). Embryos are most sensitive to RA during gastrulation, which is the time during which such changes in gene expression occur. Multiple lines of evidence suggest that RA is a normal embryonic inducer required for posterior patterning (for review, see Kolm and Sive 1997).

RA is hydrophobic and readily penetrates the epithelial layer of dejellied embryos. As detailed below, RA should be prepared as a 0.1 M stock in DMSO, and diluted in 0.1x MBS just before use. Exposure of embryos for 5 minutes to a final solution of 1 μM RA will produce mild phenotypes, whereas exposure for 1 or more hours will produce severe phenotypes. Mild (hindbrain) phenotypes can be obtained using a 0.1 μM final solution, with treatments for 10–30 minutes.

1. Make a 0.1 M stock of **RA** in **DMSO** (all trans retinoic acid, Sigma) and store in small aliquots at –20°C. Avoid exposure to light.

   RA, DMSO (see Appendix for Caution)

2. Dejelly embryos as described in Protocol 6.1 and immerse in 1 μM RA solution for several minutes to hours, depending on the desired phenotype.

3. Transfer the embryos to a petri dish containing 0.1x **MBS** and incubate until they reach neurula. Score the embryos.

Note: *The exact time of incubation will depend on the severity of defects desired. By treating embryos at the tailbud stage for 5 minutes in 1 μM RA, and then allowing them to develop to stage 35 in 0.1x MBS, morphological defects (DAI 4) can be observed. More severe treatments can produce more severe phenotypes. The most severe phenotypes resemble DAI 1 embryos.*

MBS (see Appendix for Caution)

Helpful Hints

- RA is hydrophobic and insoluble in water. Upon dilution in 0.1x MBS, it will come out of solution, making it difficult calculate the true concentration of RA in the media. However, by vigorously shaking the dilution, the RA precipitate can be resuspended. A suspension of 1 μM RA in 0.1x MBS has powerful teratogenicity.

SCORING PERTURBATIONS

The effects of RA treatment can be observed at neurula stages, when, after extensive exposure to RA, the anterior neural folds are much smaller than those of controls, and the archenteron is flat and small. The precise phenotype obtained depends on the amount of RA to which embryos are exposed, as well as the timing of the application. Treatment beginning during early gastrula stages will truncate the head, whereas exposure beginning at late gastrula or early neurula stages will truncate the tail and have little effect on the development of the head. DAI values of 1–4 can be obtained, but embryos are never completely ventralized.

PROTOCOL 7.4

Inducing Exogastrulation

The process of gastrulation normally proceeds by involution of mesoderm and endoderm, such that the emerging embryo is eventually entirely surrounded by ectoderm. It is possible to induce abnormal gastrulation such that the mesoderm moves normally but is never covered by ectoderm. In these individuals, the ectoderm tends to evaginate, and a protrusion of mesendoderm forms attached to the ectoderm in a small region. Embryos in which this has occurred are called "exogastrulae." They can be useful for determining whether close contact of ectoderm and mesendoderm is required for a particular process or activation of a particular gene (Holtfreter and Hamberger 1955; for discussion, see Gerhart 1996).

The process of exogastrulation can be induced by treating devitellinated embryos, at the late blastula and gastrula stages, with high-salt solution. It should be noted at the outset, however, that *Xenopus* is reluctant to make good exogastrulae (see Helpful Hints below). In *Xenopus*, gastrulation movements begin internally, before the blastopore lip forms. This process is called "cryptic gastrulation" which makes it very difficult to assess whether any normal gastrulation has occurred prior to the onset of exogastrulation. As a consequence, the results of exogastrulation experiments in *Xenopus* can be difficult to interpret. The difficulties can be overcome by using other amphibian species that exogastrulate more reliably, e.g., *Triturus*.

The following protocol can be used to produce exogastulae in *Xenopus* and, more reliably, in other amphibian species.

1. Dejelly embryos (Protocol 6.1) and remove the vitelline membrane (see Protocol 6.2) from the required number of embryos. Incubate in 0.1x **MBS** until they reach early blastula.

   MBS (see Appendix for Caution)

   Note: *It is essential to remove the vitelline membrane prior to high-salt treatment so that the yolk plug is free to push out without the counteracting pressure from the vitelline membrane.*

2. Transfer the embryos to a fresh petri dish containing a high-salt buffer (1x to 1.3x MBS) and incubate at 18°C until control sibling embryos have completed gastrulation.

3. When the treatment is complete, transfer the embryos to 0.5x MBS. Score the exogastrulae at early neurula.

   Note: *Observe the embryos carefully during gastrulation. In some embryos, part of the mesoderm undergoes transient or partial involution during this stage, leading to a clearly recognizable miniature neural plate. Embryos in which this occurs should be discarded.*

SCORING EXOGASTRULAE

Successful exogastrulation can be scored at early neurula by the appearance of a wrinkled ectoderm (animal hemisphere) and protruding, unpigmented mesendoderm. No cement gland should form in the ectoderm. Animals with obvious neural folds or cement glands cannot be considered complete exogastrulae as some mesoderm involution likely occurred before exogastrulation.

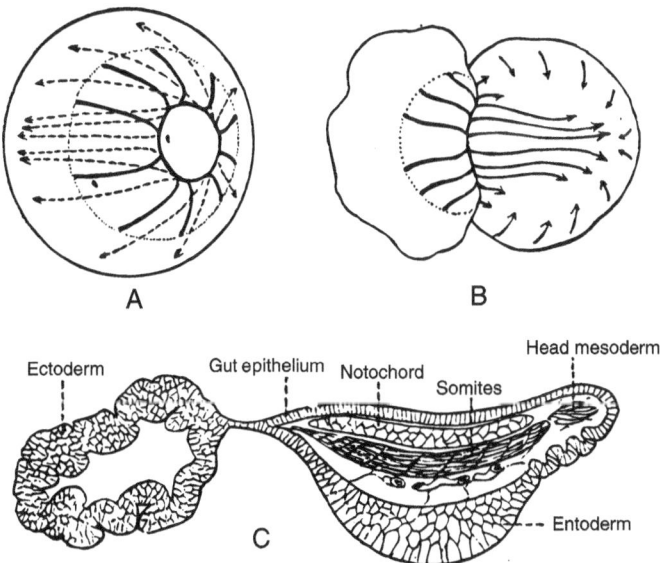

Figure 7.3. Diagrammic illustration of the morphogenetic movements of normal gastrulation (*A*) and of exogastrulation (*B*). Animal hemisphere (ectoderm) is to the left, and mesoderm plus endoderm (entoderm) to the right in both *A* and *B*. (*C*) Main differentiations in an exogastrulated embryo. (Reprinted, with permission, from Holtfreter and Hamberger 1995.)

PROTOCOL 7.5

Tipping and Staining

Tipping and staining is a procedure used to determine the position of the future dorsal midline. It can be extremely useful to know precisely where the dorsal midline of the embryo will form in order to analyze the fate or activity of particular groups of cells before the midline becomes obvious (Stewart and Gerhart 1990).

In pigmented embryos, the dorsal side can be identified at the 1–2-cell stage by a decrease in pigment of the animal hemisphere on the dorsal side, relative to the ventral side of the embryo. However, this approach does not indicate precisely where the midline will form and is not applicable to albino embryos. In pigmented embryos, the future midline can also be located by identifying the sperm entry point (SEP), a darkly pigmented spot in the animal hemisphere. The midline will form opposite the SEP, i.e., at approximately 180° (+/– 30°). This may be sufficiently precise for some applications, but in general, more accuracy is required. Since the SEP is only visible in pigmented embryos, this approach is again not applicable to albino embryos. In both pigmented and albino embryos, the dorsal midline can be determined at early gastrula by noting where the blastopore lip first forms. This approach is accurate, but in many cases, it is desirable to identify the location of the midline at an earlier stage.

The tip and stain method allows the position of the dorsal midline to be determined to within approximately 10°. The method takes advantage of the fact that the natural corticocytoplasmic rotation (CCR) can be overridden by manually rotating or tipping the embryo. The general idea is that after fertilization, but before the natural CCR of the embryo has occurred (<35 minutes postfertilization), embryos are tipped 60–90° from their resting position, so that the animal pole faces sideways instead of upward. In this position, under the influence of gravity, the heavy yolky cytoplasm will move. This movement constitutes an artificial CCR, and the new upper pole, which can be marked with a Nile blue spot, will become the dorsal side (see Figure 7.4).

Success of the tip and stain method can be determined by following sibling embryos until gastrulation, when the dorsal lip of the blastopore should initiate formation right at the dye mark. This method can also be used to rescue the dorsal axis after treatment with UV light (see Protocol 7.1).

1. Dejelly embryos (see Protocol 6.1) 6–10 minutes postfertilization. Rinse well and transfer to an injection dish (see Protocol 8.4) containing tipping solution (0.1x **MBS** containing 4% Ficoll) (Figure 7.4D).

   Note: *The purpose of the Ficoll in tipping solution is to shrink the perivitelline space so that a tipped embryo will not be able to rotate back to its natural resting position (with the animal pale facing up).*

   MBS (see Appendix for Caution)

2. Use a hairloop or blunt forceps to tip each embryo 60–90° from its resting position and set aside at room temperature (Figure 7.4E).

   Note: *Tipping must be carried out within 35 minutes after fertilization.*

3. Meanwhile, prepare a precipitate of Nile blue.

   a. Place a drop (~50 µl) of 1% Nile blue (Sigma) in water on the lid of a small petri dish, next to an equal volume of Na_2CO_3 (1 M).

   b. Immediately before use, mix the drops and collect the resulting orange precipitate on the fire-polished tip of a very finely drawn-out sealed pasteur pipette.

   Note: *The pipette tip should comprise a tiny ball.*

4. Immerse the coated tip in buffer for approximately 1 minute, until the Nile blue precipitate begins to dissolve. The tip is then ready to mark the embryos.

   Note: *A blue haze around the tip of the pipette is apparent when the precipitate begins to dissolve.*

5. Mark each embryo by gently pressing the coated pipette tip against the equator on the uppermost surface (this will become the dorsal side, see Figure 7.4E).

6. After marking, leave the embryos undisturbed in tipping solution until the first cell cycle is approximately 80% complete and then transfer to a new petri dish containing 0.1x MBS or equivalent.

7. Allow the embryos to develop at 18–20°C until they reach stage 10.

8. Test the effectiveness of the technique by checking the position of the Nile blue spot with respect to the initial dorsal blastopore lip (bottle cells) at stage 10–10.25 (Figure 7.4F).

   Note: *The spot should be on the same meridian as the dorsal blastopore lip, or within 10°.*

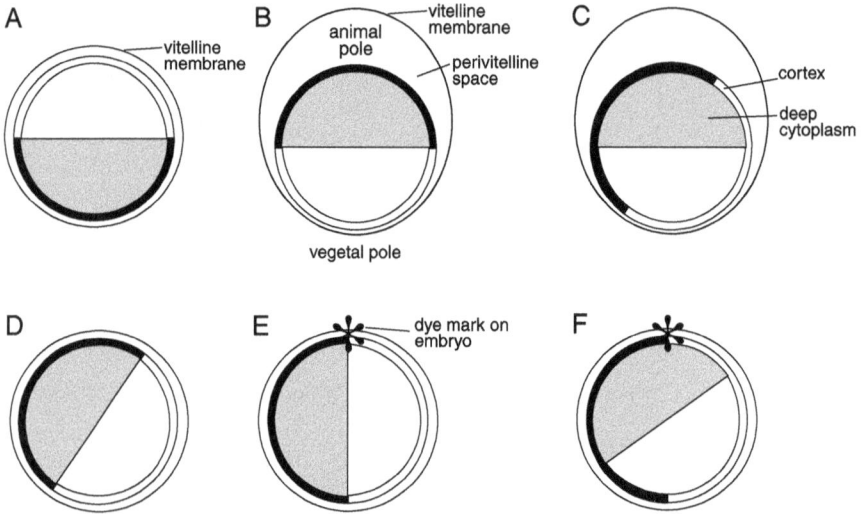

Figure 7.4. Corticocytoplasmic rotation (CCR) in *Xenopus* embryos. (*A–C*) Normal events; (*D–F*) tip and stain events. (*A*) Egg before fertilization lies in random orientation. (*B*) Egg after fertilization. Perivitelleine space swells, animal pole faces upward. (*C*) Fertilized egg soon after *B*. Rotation of cortex relative to deep cytoplasm has occurred (CCR). (*D*) Egg after fertilization (*B*) placed in 4% Ficoll. Perivitelline space shrinks, preventing animal pole upward orientation. (*E*) Manual rotation of egg from *D*, leading to artificially induced CCR. Asterisk indicates dye spot, which marks the future dorsal midline. (*F*) Fertilized egg soon after *E*. An artificial CCR has been induced.

REFERENCES

Durston A.J., Timmermans J.P., Hage W.J., Hendriks H.F., de Vries N.J., Heideveld M., Nieuwkoop P.D. 1989. Retinoic acid causes an anteroposterior transformation in the developing central nervous system. *Nature* **340:** 140–144.

Elinson R.P. and Rowning B. 1998. A transient array of parallel microtubules in frog eggs: Potential tracks for a cytoplasmic rotation that specifies the dorsoventral axis. *Dev. Biol.* **128:** 185–197.

Fredieu J., Cui Y., Maier D., Danilchik M., and Christian J. 1997. Xwnt-8 and lithium can act upon either dorsal mesodermal or neurectodermal cells to cause a loss of forebrain in *Xenopus* embryos. *Dev. Biol.* **186:** 100–114.

Gammill S. and Sive H.L. 1997. Identification of otx2 target genes and restrictions in ectodermal competence during *Xenopus* cement gland function. *Development* **124:** 471–481.

Gerhart J. 1996. Johannes Holtfreter's contributions to ongoing studies of trhe organizer. *Dev. Dyn.* **205:** 245–256.

Harland R. and Gerhart J. 1997. Formation and function of Spemann's organizer. *Annu. Rev. Cell Dev. Biol.* **13:** 611–667.

Heasman J. 1997. Patterning the *Xenopus* blastula. *Development* **124:** 4179–4191.

Holtfreter J. and Hamburger V. 1955. Amphibians. In *Analysis of development* (ed. Willier B.H., Weiss P.A., and Hamburger V.), Chapter 1. Saunders, Philadelphia.

Kao K.R. and Elinson R.P. 1988. The entire mesodermal mantle behaves as Spemann's organizer in dorsoanterior-enhanced *Xenopus laevis* embryos. *Dev. Biol.* **127:** 64–77.

——— 1989. Dorsalization of mesoderm induction by lithium. *Dev. Biol.* **132:** 81–90.

Klein P.S. and Melton D.A. 1996. A molecular mechanism for the effect of lithium on development. *Proc. Natl. Acad. Sci.* **93:** 8455–8459.

Kolm P.J. and Sive H.L. 1995. Regulation of the *Xenopus* labial homeodomain genes, HoxA1 and HoxD1: Activation by retinoids and peptide growth factors. *Dev. Biol.* **167:** 34–49.

——— 1997. Retinoids and posterior neural induction: A reevaluation of Neuwkoop's two-step hypothesis. *Cold Spring Harbor Symp. Quant. Biol.* **62:** 511–521.

Mead P.E., Brivanlou I.H., Kelley C.M., and Zon L.I. 1996. BMP-4-responsive regulation of dorsal-ventral pattening by the homeobox protein Mix.1. *Nature* **382:** 357–360.

Pannesse M., Polo C., Andreazzoli M., Vignali R., Kablar B., Barsacchi G., and Boncinelli E. 1995. The *Xenopus* homolog of *Oxt2* is a maternal homeobox gene that demarcates and specifies anterior body regions. *Development* **121:** 81–90.

Papolopulu N., Clarke J.D.W., Bradley L., Wilkinson D., Krumlauf R., and Holder N. 1991. Retinoic acid causes abnormal development and segmental patterning of the anterior hindbrain in *Xenopus* embryos. *Development* **113:** 1145–1158.

Ruiz i Altaba A. and Jessell T. 1991. Retinoic acid modifies mesodermal patterning in early *Xenopus* embryos. *Genes Dev.* **5:** 175–187.

Scharf S.R. and Gerhart J.C. 1983. Axis determination in eggs of *Xenopus laevis*: A critical period before first cleavage, identified by the common effects of cold, pressure and ultraviolet irradiation. *Dev. Biol.* **99:** 75–87.

Sive H., Draper B., Harland R., and Weintraub H. 1990. Identification of a retinoic acid-sensitive period during primary axis formation in *Xenopus laevis*. *Genes Dev.* **4:** 932–942.

Slack J.M.W., Isaacs H.V., and Darlington B.G. 1988 Inductive effects of fibroblast growth factor and lithium ion on *Xenopus* blastula ectoderm. *Development* **103:** 581–590.

Smith W.C. and Harland R.M. 1991. Injected Xwnt-8 RNA acts early in *Xenopus* embryos to promote formation of a vegetal dorsalizing center. *Cell* **67:** 753–765.

Stewart R. and Gerhart J. 1990. The anterior extent of dorsal development of the *Xenopus* embryonic axis depends on the quantity of organizer in the late blastula. *Development* **109:** 363–372.

Tape 1 of the video series "Manipulating the Early Embryo of Xenopus laevis" presents illustrations of axial defects in embryos treated with UV light and LiCl, and demonstration of the "tipping and staining" method for determining the dorsal/ventral axis of 1-cell embryos.

CHAPTER 8

Microinjection

Many experiments use microinjection (see, e.g., Chapter 3), and the procedure can be performed on oocytes or embryos. Although the procedures for microinjection are broadly similar, differences exist in the way in which oocytes and embryos are prepared and in their tolerance of injected materials. Therefore, these aspects are addressed separately. Forward planning is particularly important for embryo injection, since the frogs lay eggs only for a limited time, and once the eggs are fertilized, there is a very short period during which the embryos are suitable for injection. In contrast, oocytes remain in good condition for a relatively long time, and injections can be continued, if necessary, over a period of days. Oocyte preparation is described in Protocol 8.1, and embryo preparation in Chapters 5 and 6.

INJECTION OF OOCYTES

Oocytes can be injected with mRNA or DNA constructs into the cytoplasm or nucleus, respectively. The cytoplasm can withstand the introduction of up to 50 nl of injected material, and the nucleus can tolerate up to 20 nl.

It is possible to microinject oocytes that are still contained within their ovarian follicles (see Protocol 8.7), but most researchers find it more convenient to work with defolliculated oocytes. Defolliculation can be carried out enzymatically (see Protocol 8.2) by treatment with collagenase, or it can be performed manually (see Protocol 8.3). Enzymatic treatment is recommended for preparation of large numbers of oocytes (>1000); however, this treatment causes complications because it often damages the quality of the oocytes. The manual procedure, which requires some practice, is recommended for experiments requiring only a few hundred oocytes.

INJECTION OF EMBRYOS

Embryos can be microinjected with mRNA or DNA constructs at the single-cell stage or later. RNA injection has the advantage of allowing uniform expression over a large region of the embryo (up to about one fifth, with a single injection). However, translation of exogenous mRNAs generally begins as soon as the material is injected. This may not coincide with the timing of normal gene expression and the results can be misleading. Promoter-driven DNA constructs are transcribed only after the mid blastula transition, which may be preferable. However, expression from these constructs is mosaic, with many cells not producing any exogenous protein (for further discussion, see Chapter 3).

Injection of embryos with more than 5 ng of mRNA and 100 pg of DNA leads to abnormal development and death during gastrulation. Therefore, it is important to keep quantities of injected material to a minimum. Embryos are also more sensitive than oocytes to the volume of nucleic acid injected. Volumes for the one-cell to four-cell stage should be kept below 10 nl, or even less for injection of later-cleavage-stage blastomeres.

After fertilization, the dorsal side of the embryo can be distinguished reasonably accurately by the lighter pigmentation characteristic of the animal hemisphere (a consequence of the corticocytoplasmic rotation). This allows specific targeting of injection. The future dorsal side can be predicted very accurately as described by Gerhart et al. (1981; see Protocol 7.5), but this accuracy is generally not necessary for the purposes of microinjection. By the 32-cell stage, some embryos (up to 70% in a good batch) display the "perfect" cleavage pattern that has allowed fate maps to be drawn (Dale and Slack 1987; Moody 1987). At this point, it is quite possible to inject specific blastomeres; however, the volume injected should be kept below 10 nl. (For each blastomere, there is an equivalent blastomere across the axis of bilateral symmetry, and unless asymmetric expression is desired, both may be injected.)

Embryos are sensitive to perturbation and must be treated gently during the injection process. However, the jelly coat is extremely tough and must be loosened or removed entirely prior to injection (see Protocol 6.1). Completely dejellied embryos should be transferred to injection dishes containing a solution of Ficoll 400 (2–5% in 1/3 MMR). This collapses the vitelline space, reducing the pressure on the embryo, and therefore preventing leakage. As with oocytes (see Protocol 8.6), excess buffer is drawn off until surface tension holds the embryos in place. Following injection, the dish can be topped up with buffer without Ficoll.

An alternative to complete removal is to soften the embryo's jelly coat. Although the ideal degree of softening requires some practice to achieve, this approach has the advantage of leaving some jelly to provide support for the embryo and to help to prevent leakage. The softening process is equivalent to a shortened dejellying treatment (see Protocol 6.1) lasting between 30 seconds and 1 minute. The treatment must be followed by extensive washing. After this, the jelly will be intact, but expanded, and the embryo will be ready for transfer to an injection dish.

INJECTION CHECKLIST

All of the reagents and equipment for injection and manipulation must be assembled before beginning any microinjection procedure. The following checklist gives the materials required for embryo injection to ensure that there is no last minute rush. The hardware is the same for oocyte injections, but only a single medium is needed for oocyte incubation, usually MBS-H (modifed Barth's saline, buffered to pH 7.8 with HEPES) or OR-2 (Oocyte Ringer's).

2% **cysteine** (w/v) in 1/3x **MMR** (modified Marc's Ringer) or water; adjust pH to 8.0 (for dejellying embryos). Prepare fresh on day of injection.

70% **ethanol** (for sterilization of instruments)

4% Ficoll (w/v) in 1/3x MMR (or equivalent)

Hairloop or blunt eyebrow knife (for pushing embryos around)

Microinjector (see Chapter 4)

1/3x MMR (or equivalent)

1x MMR (or equivalent buffer)

mRNA or DNA (for injection). Prepare fresh dilutions on day of injection.

Injection needles (drawn-out micropipettes)

Nile blue in phosphate buffer or MMR (for albino embryos)

Injection dishes (see Protocol 8.4)

Watchmakers forceps, sharp and blunt (for removing the vitelline membrane)

cysteine, MMR, ethanol (see Appendix for Caution)

PROTOCOL 8.1

Isolation of Oocytes

Oocytes are obtained from sexually mature females by surgically removing parts of the ovary. The operation is not fatal and can be performed on an anaesthetized frog several times during its lifetime. However, a recovery period of 2 weeks is recommended between operations. As with induction of ovulation (see Protocol 5.2), keep a careful record of all operations performed and include details of oocyte quality in this record. A frog that produces one good batch of oocytes, e.g., those that translate injected mRNAs efficiently, should be noted and used again. Oocyte quality is generally frog-dependent.

1. Anesthetize a mature female frog by submerging it in 0.02% benzocaine (see Appendix 1) until unconscious (~10 minutes).

   Note: *An unconscious frog will give no response when turned on its back.*

2. Place the frog bellyside up on a piece of plastic-coated benchtop paper (plastic side up).

3. Make a small incision (~1 cm long) through the skin and muscle layers on one side of the abdomen (see Figure 8.1).

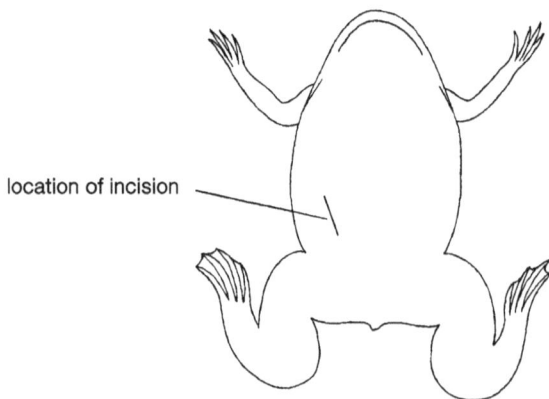

location of incision

Figure 8.1. Location of incision. (Courtesy of Mark Curtis.)

Note: *Use scissors rather than a scalpel because the latter is less likely to cause peripheral damage. Be careful to avoid the major blood vessel which lies along the mid-line. Although the skin is difficult to pick up, even with forceps, it can be picked up easily with a paper tissue. The initial small cut can be made through paper tissue and skin, and then lengthened to 1 cm.*

4. Remove a small quantity of ovarian tissue using a pair of forceps and a pair of scissors.

   Note: *The lobes of the ovaries should be readily visible once the incision is made. If the ovary is not obvious, the frog is either juvenile, moribund, or male.*

5. Examine the tissue under a dissecting microscope to ensure that the ovaries are healthy.

   Note: *In healthy ovaries, the oocytes should show a distinct pigmentation difference between the animal and vegetal hemispheres, the pigment should not be mottled, and only a low percentage of oocytes should be undergoing resorption. The latter are easily recognized by their heavy covering of blood vessels.*

6. Assuming that the ovaries are healthy, remove enough oocytes for the experiment in hand and place them in oocyte culture medium (see Appendix 1).

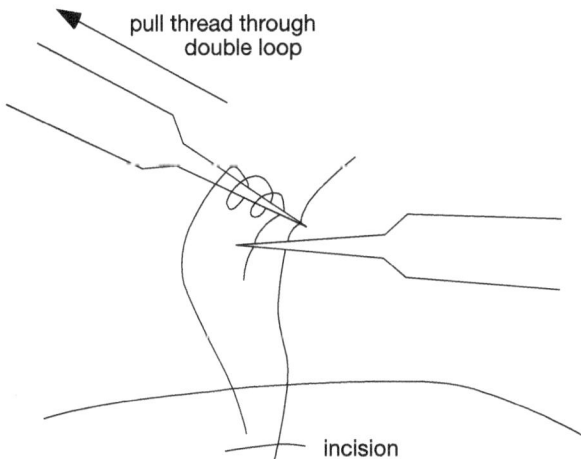

Figure 8.2. Suture the incision site.

7. Suture the incision site immediately (one suture in the muscle wall and one in the skin) using Ethicon 3-0 Silk thread and a surgeon's knot (see Figure 8.2).

    Note: *To tie a surgeon's knot, use two pairs of forceps. Use the right pair to wrap one end of the suture twice around the left pair to make a double loop, and use that same (left) pair to pick up the other end of the suture, pulling it through the double loop. The procedure is repeated to complete the knot (see Figure 8.2).*

8. Allow the frog to recover in a bucket in a small volume of 20 mM NaCl. Support the frog so that her nose is above the surface of the water. She should regain consciousness in a matter of minutes.

    Note: *The frog should bleed very little during this procedure, if at all. Only superficial blood vessels should be damaged. Profuse bleeding indicates irreparable damage to a major blood vessel. If this occurs, the frog will not recover and should be humanely sacrificed by reimmersion in benzocaine for 30 minutes, or in ice water for 30 minutes before freezing for a minimum of 24 hours.*

PROTOCOL 8.2

Enzymatic Defolliculation of Oocytes

1. Use a pair of sharp forceps to tear a sample of excised ovarian tissue (see Protocol 8.1) into small pieces and transfer the pieces to a 50-ml polypropylene tube (Corning) containing 40 ml of oocyte culture medium (see Appendix 1).

2. Add powdered collagenase to a final concentration of 0.2% (w/v) and incubate, with gentle rocking, until the oocytes are well separated from each other (1–1.5 hours).

   Note: *Prolonged exposure to collagenase can damage the vitelline membranes of the oocytes and should be avoided.*

3. Wash oocytes in fresh oocyte culture medium until collagenase is completely removed (solution should be clear with no brown discoloration).

4. Transfer oocytes to a glass petri dish and set aside overnight at 15–20°C.

5. The next day, select healthy survivors and transfer them to a fresh petri dish containing oocyte culture medium.

   Note: *If they are not crowded, and if the medium is changed daily, the oocytes can remain healthy for several days. Healthy oocytes retain a clear border between animal and vegetal hemispheres. Do not use mottled or unevenly pigmented oocytes. Oocytes in which the nucleus has floated to the surface, displacing the pigment into a circle, should also be discarded. If large numbers of oocytes appear to be unhealthy, discard the entire batch.*

PROTOCOL 8.3

Manual Defolliculation of Oocytes

Manual isolation of oocytes offers several advantages over collagenase treatment: (1) The resulting oocytes have a higher rate of protein synthesis (Wallace et al. 1973), (2) they mature more reliably (Smith et al. 1991), (3) they can be fertilized after defolliculation (Heasman et al. 1991), and (4) they can be injected immediately after defolliculation. There is no need for the lengthy enzymatic treatment (1–3 hours) and recovery times (up to overnight) required after enzymatic defolliculation.

The main disadvantages of manual defolliculation are that fewer oocytes can be isolated and that the procedure requires some practice. However, since most experiments do not require more than 250 oocytes, it may be preferable to learn and practice the technique of isolating oocytes manually.

1. Wash the excised ovarian tissue thoroughly with **MBS** buffered to pH 7.8 with HEPES (see Appendix 1) and transfer a piece of ovary containing several hundred large oocytes to a 90-mm petri dish.

   MBS (see Appendix for Caution)

2. Use watchmaker's forceps (Dumont #5) to grab the stalk that connects the oocyte to the rest of the ovarian tissue.

3. With the other hand, use a second pair of forceps to grab the stalk of the same follicle, nearer to the oocyte, and pull up and around the oocyte.

   Notes: *If the defolliculation is successful, a constriction will be visible in the oocyte. The constriction enlarges as the pairs of forceps are pulled further apart.*

   Occasionally, the oocyte will roll completely free from the follicle covering, but in most cases, the follicle covering will be only partially removed. In this case, the oocyte can be teased out of the covering as long as at least one third of the covering is removed (evidenced by the constriction of the oocyte extending at least one-third around the oocyte).

Helpful Hints

- The most common mistake made when manually defolliculating oocytes is severing the oocyte from the ovarian tissue without defolliculating it. If no constriction was observed when the oocyte was isolated (see Note in step 3 above), or vascular tissue remains associated with the oocyte, this usually means that the oocyte was not defolliculated. If this is suspected, there is no point in trying to rescue the individual since defolliculating the isolated oocyte is virtually impossible. If oocytes are continually severed rather than defolliculated, too much of the stalk is being held and a more delicate grip should be used.

- Another common mistake when defolliculating is rupturing the oocyte, which can occur when the stalk is grabbed too close to the oocyte. In this situation, when the forceps are pulled apart, the oocyte is ruptured. The chance of rupturing the oocyte can be reduced by using blunt forceps.

- Manual defolliculation of oocytes requires practice. In the interest of speed, it is not necessary for every attempt to be successful. If an oocyte is severed from the ovary or ruptured, it should be discarded and the next follicle tackled. With practice, 100–300 oocytes can be isolated routinely in 1 hour.

PROTOCOL 8.4

Preparation of Injection Dishes

Injection dishes are modified petri dishes to which a nylon mesh has been fixed to hold oocytes or embryos in place. The mesh also helps to keep track of which oocyte has been injected since it holds the oocytes in regular rows. Ensure that the grid fits snugly to the edge of the dish and lies flat on the surface, so that the oocytes do not become trapped underneath.

1. Cut a sheet of 800-μm mesh Nitex screen (Tetko) into circles that fit snugly into petri dishes.

2. Fix the mesh in place by melting the bottom of the plastic petri dish with five drops of **methylene chloride** (or **chloroform**).

   Note: *Use a silicone rubber cork of about the same diameter as the dish to hold the grid flat while the plastic sets.*

   methylene chloride, chloroform (see Appendix for Caution)

3. After complete evaporation of the solvent, rinse the dishes thoroughly with 95% **ethanol** and then distilled water before use.

   ethanol (see Appendix for Caution)

PROTOCOL 8.5

Calibration of the Injection Volume for Oocytes and Embryos

The injection volume must be calibrated carefully. The following protocol describes the calibration procedure for a pressure injector.

1. Backfill the needle using an automatic pipettor fitted with a long narrow tip (of the kind used for loading sequencing gels) and mount on the injector (see Chapter 4).

2. Break the needle tip to produce an orifice of approximately 10 μm.

3. Place a small drop of paraffin oil on a microscope slide and mount the slide on the stage of a dissecting microscope.

4. Calibrate the eyepiece micrometer for the appropriate magnification and perform a trial injection into the drop of oil. The injected liquid forms a sphere within the oil droplet.

5. Measure the diameter of the sphere using the eyepiece micrometer and calculate the injected volume ($v = 4/3\pi r^3$, where v is the volume and r is the radius of the sphere).

Helpful Hints

- The different methods for calibrating the injection volume include deposition of the drop directly onto the stage micrometer (note that the micrometer must be siliconized for the drop to be near spherical).

- The injected volume can be calculated as described in step 5 above.

- Alternatively, allow the drop to hang at the end of the needle, where its diameter can be measured using an eyepiece micrometer. The drop must be measured quickly because evaporation causes it to shrink rapidly.

- Alternatively, inject the drop into a dish of paraffin oil. Approximate the volume of the drop by using the equation $v = 4/3 \pi r^3$ (where v is the volume and r is the radius of the drop) or using the figures in Table 8.1. Once the appropriate pressure has been established, the duration of the pressure burst can be used to control the volume precisely.

Table 8.1. Calculating Microinjection Volumes

Diameter of drop (μm)	Radius of drop (μm)	Volume $4/3\pi r^3$ (nl)
125	62.5	1.03
140	70	1.44
150	75	1.77
160	80	2.15
170	85	2.58
180	90	3.06
200	100	4.20
225	112.5	5.90
250	125	8.20
275	137.5	11.04
300	150	14.18
325	162.5	18.02
350	175	22.51
375	187.5	27.69
400	200	33.60
450	225	47.84
500	250	65.63

PROTOCOL 8.6

Injection of Defolliculated Oocytes

1. Fill an injection dish with O-R2 (see Appendix 1) and transfer the required number of oocytes to the mesh (see Protocol 8.4) using a short pasteur pipette attached to a 10-ml pipette pump (Bel-Art Products).

2. Ensure that the oocytes are oriented with their animal (pigmented) poles uppermost.

   Note: *To aid orientation, swirl the dish gently. The oocytes tend to settle in the correct orientation; rebels can be manipulated with a hair loop.*

3. Load the injection needle, as described in Chapter 4, and draw off the O-R2 buffer to a level at which the oocytes begin to flatten. At this point, surface tension holds them firmly in place and increases their turgor so that they are easily penetrated with the injection needle.

   Note: *Do not allow the oocytes to remain in this position for more than a few minutes, since their exposed surface will dry out and they will die.*

   a. For nuclear injections (e.g., when injecting DNA constructs), move the dish into the path of the needle so that it penetrates the animal pole.

   Note: *The nucleus is very large and therefore difficult to miss; however, for the first attempt, it may be advisable to practice by injecting a dye such as trypan blue. This, followed by coarse dissection of the oocyte, reveals the location of the injected material, nuclear or otherwise. The nucleus is only about 5% of the oocyte volume, but it can tolerate a surprisingly large injection volume, although it should not be necessary to inject more than 20 nl.*

   b. For cytoplasmic injections (e.g., when injecting mRNA), inject the oocyte near its equator, aiming down into the center of the oocyte.

   Note: *This avoids the nucleus located just below the animal pole. Volumes up to 50 nl are readily tolerated in the cytoplasm.*

4. After injection, gently withdraw the needle and move on to the next oocyte.

5. Refill the dish with O-R2 before the oocytes dry out.

Helpful Hints

- The injection solution should be under slight positive pressure, so that the oocyte contents do not enter and clog the needle. The pressure can be adjusted so that the flow stops immediately when the needle is withdrawn from the buffer.

- The oocytes should be sufficiently firm, and the needle sufficiently fine, so that the oocyte is penetrated with minimum deformation. There should be no leakage at the site of injection after the needle has been withdrawn, and the injection wound should be scarcely visible.

PROTOCOL 8.7

Injection of Folliculated Oocytes

As an alternative to defolliculated oocytes, intact oocytes within their follicles can be injected. However, very fine injection needles cannot be used because of the toughness of the follicle. The volumes injected into oocytes tend to be quite large (20–50 nl), and thus a coarse needle that has been manually broken to a sharp angle is quite suitable.

1. Transfer a piece of ovarian tissue, containing small bunches of oocytes, to a glass microscope slide and select an oocyte for injection.

    Note: *Blot excess medium from the bunch of oocytes with the corner of a tissue, but work quickly to prevent drying.*

2. Gently orient the oocyte with a pair of blunt forceps, squeezing slightly to increase the turgor and ease the penetration of the injection needle through the thecal layer.

3. After injection, gently withdraw the needle and move on to the next oocyte.

4. Transfer the injected oocytes back to O-R2 before they dry out.

PROTOCOL 8.8

Preparation of Secreted Proteins from Oocytes

1. Isolate and prepare defolliculated oocytes as described in Protocol 8.3.

   Note: *Enzymatic defolliculation is not recommended for this procedure since collagenase tends to inhibit protein synthesis. Although the effect of collagenase is batch-dependent, translation is often severely affected.*

2. Rinse oocytes in 0.5x Leibovitz's L15 medium (Life Technologies), supplemented with 0.5 mg/ml bovine serum albumin (BSA), 100 units/ml penicillin, and 100 μg/ml streptomycin sulfate (Life Technologies). Culture oocytes overnight at 19°C.

3. The next day, transfer the required number of healthy oocytes to injection dishes containing O-R2 (see Appendix 1).

4. Inject each oocyte with approximately 50 ng of RNA as described in Protocol 8.6.

5. Allow the oocytes to recover in injection dishes for a minimum of 2 hours. Alternatively, incubate overnight to allow the injected RNA to take over the translation machinery more completely and produce larger quantities of exogenous protein synthesis, thus effectively reducing the background of endogenous protein.

6. Again select healthy-looking oocytes and transfer them to a sterile tube containing O-R2 with 0.5 mg/ml BSA, 100 units/ml penicillin, 100 μg/ml streptomycin, and 0.5 mCi/ml [^{35}S]methionine. Allow 10 μl of solution per oocyte.

   radioactive substances (see Appendix for Caution)

7. Culture oocytes in these tubes overnight or for 2–4 days at 19°C. Check the culture at least once a day and remove any ailing oocytes.

   Note: *Overnight incubation is sufficient to produce ^{35}S-labeled protein, but longer incubations will produce a larger mass of secreted protein.*

8. At the end of the culture period, collect media and assay for the presence of exogenous proteins.

REFERENCES

Dale L. and Slack J.M. 1987. Fate map for the 32-cell stage of *Xenopus laevis*. *Development* **99:** 527–551.

Gerhart J., Ubbels G., Black S., Hara K., and Kirschner M. 1981. A reinvestigation of the role of the grey crescent in axis formation in *Xenopus laevis*. *Nature* **292:** 511–516.

Gurdon J.B. 1974. *The control of gene expression in animal development.* Oxford and Harvard University Press, Cambridge, Massachusetts.

———— 1977. Methods for nuclear transplantation in amphibia. *Methods Cell Biol.* **16:** 125–139.

Heasman J., Holwill S., and Wylie C.C. 1991. Fertilization of cultured *Xenopus* oocytes and use in studies of maternally inherited molecules. *Methods Cell Biol.* **36:** 213–230.

Moody S.A. 1987. Fates of the blastomeres of the 16-cell stage *Xenopus* embryo. *Dev. Biol.* **119:** 560–578.

Smith L.D., Xu W.L., and Varnold R.L. 1991. Oogenesis and oocyte isolation. *Methods Cell Biol.* **36:** 45–58.

Wallace R.A., Jared D.W., Dumont J.N., and Sega M.W. 1973. Protein incorporation by isolated amphibian oocytes III. Optimum incubation conditions. *J. Exp. Zool.* **184:** 321–334.

The video series "Manipulating the Early Embryo of Xenopus laevis*" demonstrates how to dissect pieces of ovary from adult* Xenopus *and how to subsequently isolate oocytes by manual defolliculation. Pigment differences indicating dorsal and ventral sides of the 4-cell embryo and the bastula are also illustrated, and the technique for injection into 4-cell embryos is described.*

CHAPTER 9

Fate Mapping and Lineage Labeling

Embryologists since the early part of the century have followed the ontogeny of particular regions of amphibian embryos by fate mapping, marking parts of the embryo and monitoring the subsequent development of those areas, or using lineage labels, tracers that mark the cell or cells of interest and all descendants. Mapping the locations and phenotypes of the descendants of labeled cells or tissues through developmental stages establishes the normal fate of cells and tissues in their natural embryonic environment. Such observations have yielded important insights into the generation of intricate structures in the embryo and the nature of tissue movements during complex events such as gastrulation. Early studies involved the application of vital dyes to particular regions of the embryo, which could then be observed at later stages. For descriptions of these methods, see Rugh (1948) and Hamburger (1960).

Newer techniques, described in detail below, have permitted a more refined analysis of fate. Because the molecules used are far more stable, they demarcate labeled cells more intensely and, in the case of lineage labels, are more definitively cell-autonomous. The data from fate maps are now often used to target misexpression of RNAs, and other molecules, to cells in the young embryo that will give rise to particular structures later in development. Lineage labeling also provides the means by which the fate determination and commitment of a cell can be tested. This is accomplished by marking a cell and then testing for fate changes after overexpression or deletion of a gene product, or after changing its developmental environment (either in vivo or in vitro).

Finally, lineage labeling allows all of the cells of an embryo to be marked, by injecting tracer at the one- or two-cell stages, and using the tissues of these marked embryos in transplantation and in vitro recombination experiments with unlabeled host embryos. The lineage tracer can be used to track the migration and differentiation of the transplanted population.

FATE MAPS AT CLEAVAGE, GASTRULA, AND NEURULA STAGES

Although fate-mapping techniques have been used to examine the generation of almost every tissue in the embryo, there are several stages at which
they have been particularly valuable during early *Xenopus* development.
Maps made at early cleavage, gastrula, and neural plate stages are described
here because each illustrates different technologies and addresses an important subject for investigators studying *Xenopus* embryogenesis.

Marking each of the cells of *Xenopus* cleavage stages has provided a number of detailed fate maps for all of the organ systems (Hirose and Jacobson
1979; Jacobson and Hirose 1981; Jacobson 1983; Masho and Kubota, 1986;
Dale and Slack 1987; Moody 1987a,b; Takasaki, 1987; Masho 1988; Moody
and Kline 1990). The maps presented here illustrate the tissues derived from
each blastomere, in the 16-cell embryo (Figure 9.1) and 32-cell embryo
(Figures 9.2 and 9.3). The dramatic movements occurring during early development in the descendent cells derived from blastomeres labeled at the 16-cell
stage are shown in Figure 9.4. Nomenclature schemes for identifying particular blastomeres are illustrated in Figures 9.5 and 9.6. The identity of early blastomeres can be established by following dorsoventral pigment differences in
early embryos (see Lineage Labeling, below). These normal fate maps were
produced using the horseradish peroxidase (HRP) lineage labeling method
described below. Such maps are particularly useful for targeting injection of
molecules into particular regions of embryos. They have also been an extremely useful resource in blastomere transplant studies to assess the degree of commitment of these cells to their fates (see, e.g., Gallagher et al. 1991).

Gastrulation so transforms the arrangement of cells within the early
embryo that it would be exceedingly difficult to follow these morphogenetic changes without fate maps. The maps compiled by Keller and colleagues
(see, e.g., Keller 1975, 1976) using vital dye mapping illustrate the extent of
narrowing and elongation that occurs in both dorsal mesoderm and presumptive neural tissue during this period (Figure 9.7). The extent of cell
movement during gastrulation is especially evident in that part of the Keller
fate map illustrating the movement of presumptive neural tissue (see Chapter
2, Figure 2.5) and in the fate map of Bauer et al. (1994), using the HRP-
labeling technique to illustrate the fate of particular blastomeres during gastrulation (see Figure 9.4).

Between the time the neural plate forms on the dorsal side of the embryo
and neural tube closure occurs, there is still further elongation of neural tissue.
The extensive morphogenesis within regions of presumptive neural tissues
during gastrulation and neurulation is shown in Figure 2.13 in Chapter 2. The

complex array of neural tissues in the brain and spinal cord (also introduced in Chapter 2) becomes morphologically evident during tadpole stages. The fates of regions of the neural plate have been mapped in *Xenopus* by Eagleson and Harris (1990), who transplanted pieces of labeled tissue from one embryo to the same site in an unlabeled embryo (as described below). The neural plate fate map is illustrated in Figure 9.8.

PROCEDURES FOR FATE MAPPING OF EMBRYONIC TISSUES

A number of methods have been used by developmental biologists for labeling and following the fate of particular tissues in the embryo. The methods described here are useful for following later embryonic stages when single cells are small and for establishing the fate of populations of cells. Lineage-labeling techniques for following the fates of single cells in early embryos are described in the next section. Several methods have been devised for fate mapping tissues in the embryo. The classic technique utilizes vital dyes, such as Nile blue sulfate, to label patches of tissue on the embryo. A simple method for labeling embryos by this technique is described in Protocol 7.5.

The fate map of the *Xenopus* neural plate (Eagleson and Harris 1990) uses two approaches. In one, a particular tissue is removed from an unlabeled embryo, soaked for 7–8 minutes in a 200 μM solution of the fluorescent Hoechst DNA-binding dye (33258), washed for 5 minutes, and then returned to the unlabeled host. After fixation, at a later stage, the labeled implant is viewed by epifluorescence.

The other approach labels particular regions of the embryo by applying small crystals of DiI or DiO, which can also be monitored at later stages by virtue of their fluorescence. It can be difficult to localize the application of crystalline DiI and DiO, and an alternative procedure has been developed (G. Eagleson, unpubl.), whereby the DiI is first dissolved in DMSO at a concentration of 20 mg/ml. This stock solution is then diluted 1:30 to 1:50 in 250 mM sucrose and loaded into a glass micropipette with a 25–50-μm tip, attached to a microinjection apparatus (see Chapter 8). Dye application may be less injurious to the embryo if the tip of the micropipette is fire-polished. A very slight positive pressure from a microinjection apparatus will minimize the chance of the needle becoming clogged with the DiI solution. The rate of injection that is appropriate for the size of the DiI spot to be applied must be determined empirically. Embryos to be injected can be in any one

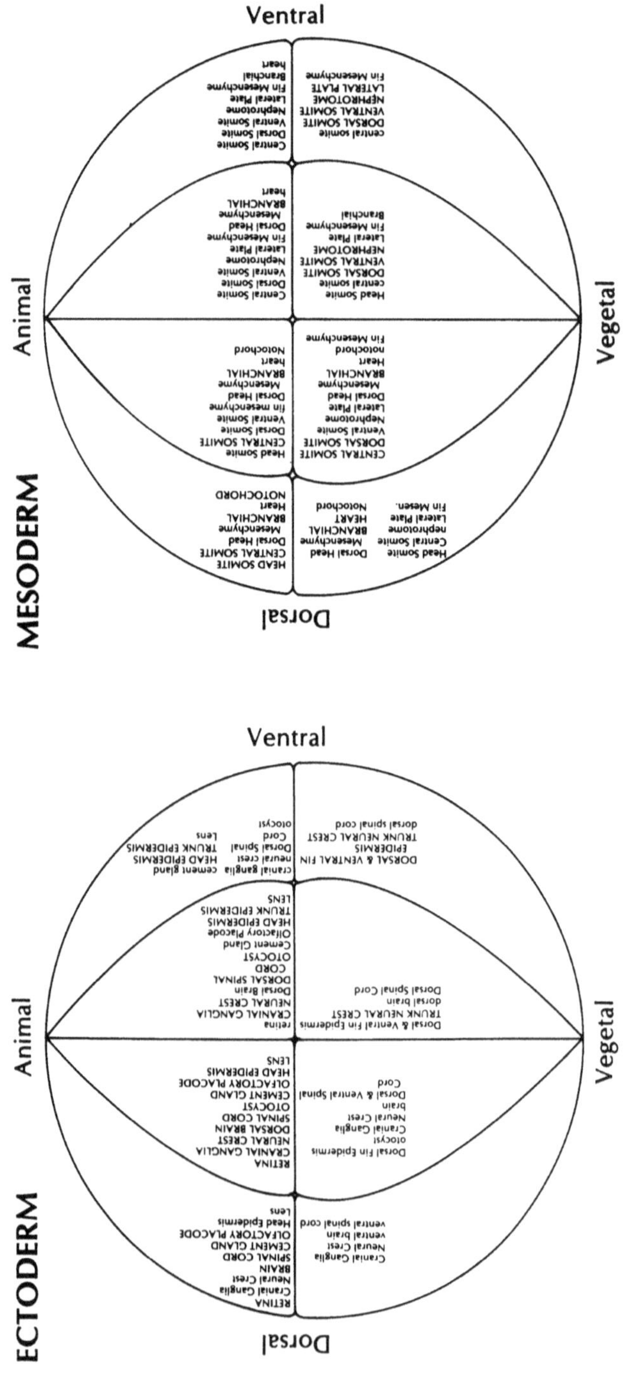
MESODERM
Ventral
Dorsal
Animal
Vegetal
ECTODERM
Ventral
Dorsal
Animal
Vegetal

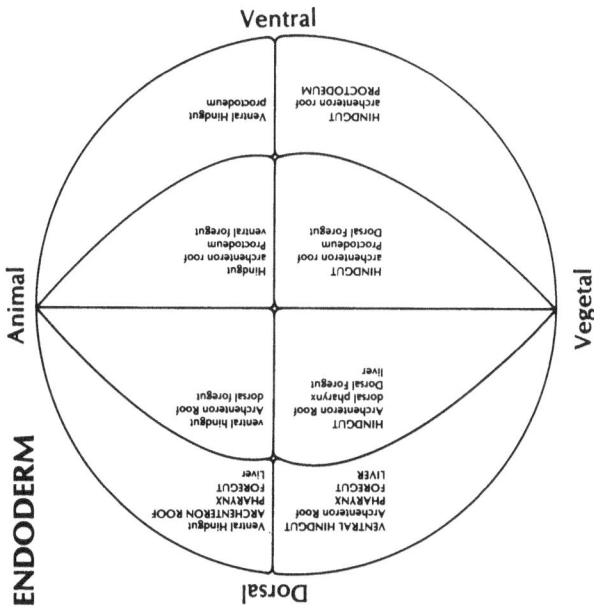

Figure 9.1. Summary of the fates of the blastomeres of the 16-cell-stage embryo. Structures written *all in capital letters* receive major contributions from the cell in which that structure is written. Structures *starting with a capital letter* receive small contributions from the cell in which that structure is written. Structures written *all in lowercase letters* receive contributions from the cell in which that structure is written in only 50% of the embryos studied. Structures that contained labeled cells in fewer than 50% of the embryos studied were not included in these summary diagrams. (Reprinted, with permission, from Moody 1987a.)

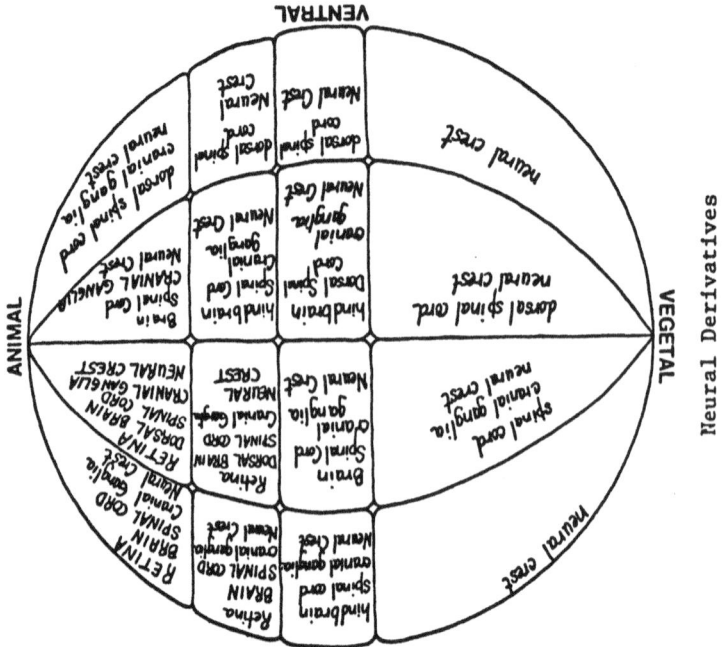

Neural Derivatives

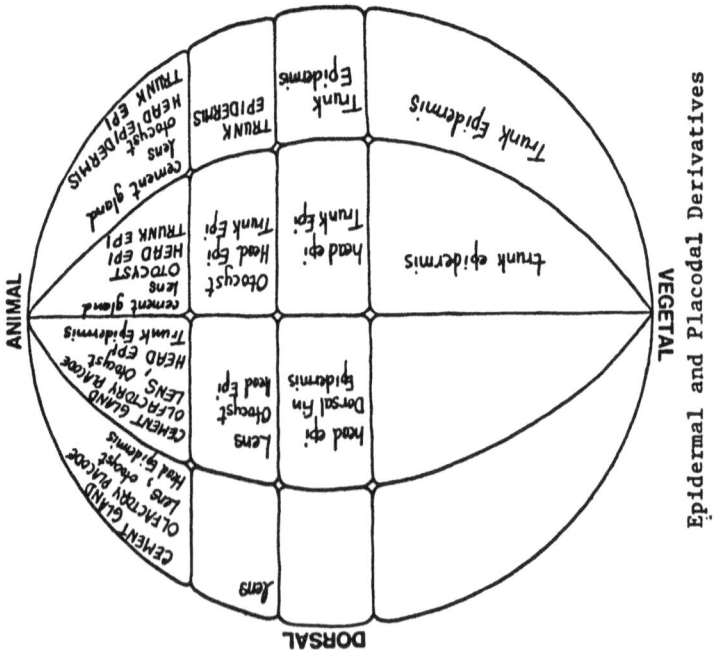

Epidermal and Placodal Derivatives

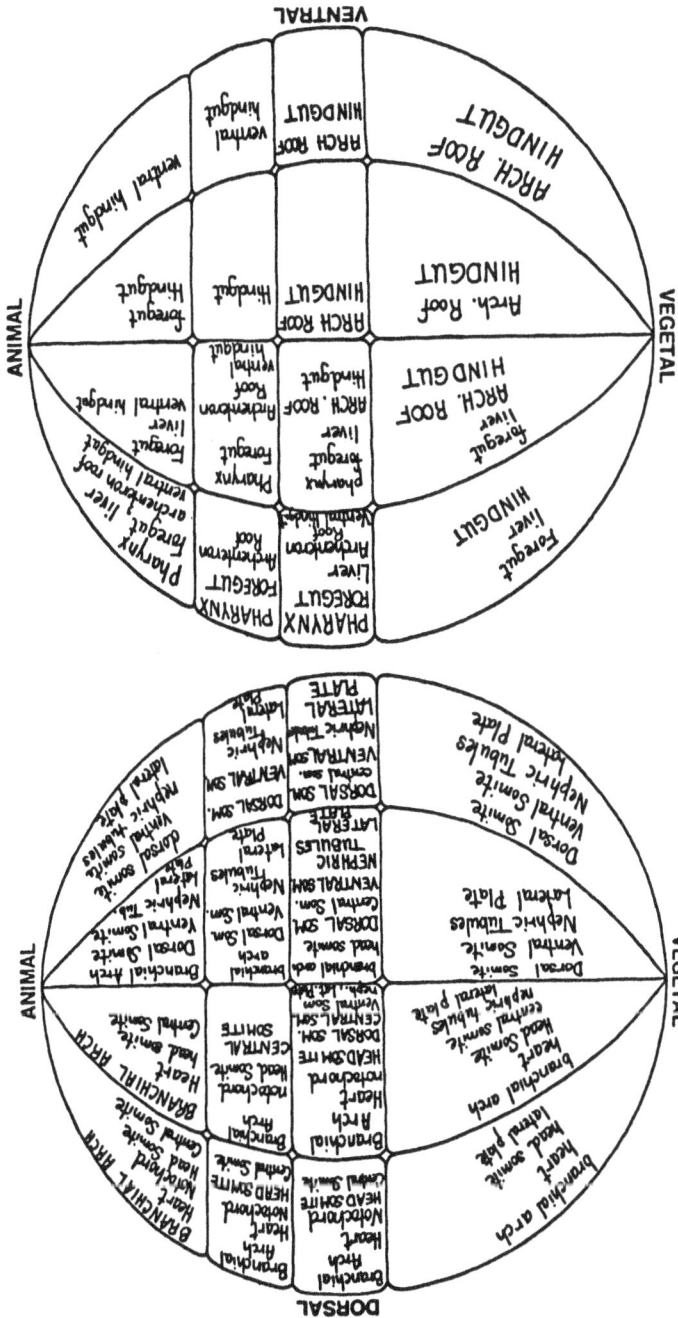

FIGURE 9.2. Summary of the fates of the blastomeres of the 32-cell-stage embryo. For explanation of frequency of formation of different tissues by given blastomeres, see Figure 9.1 legend. (Reprinted, with permission, from Moody 1987b.)

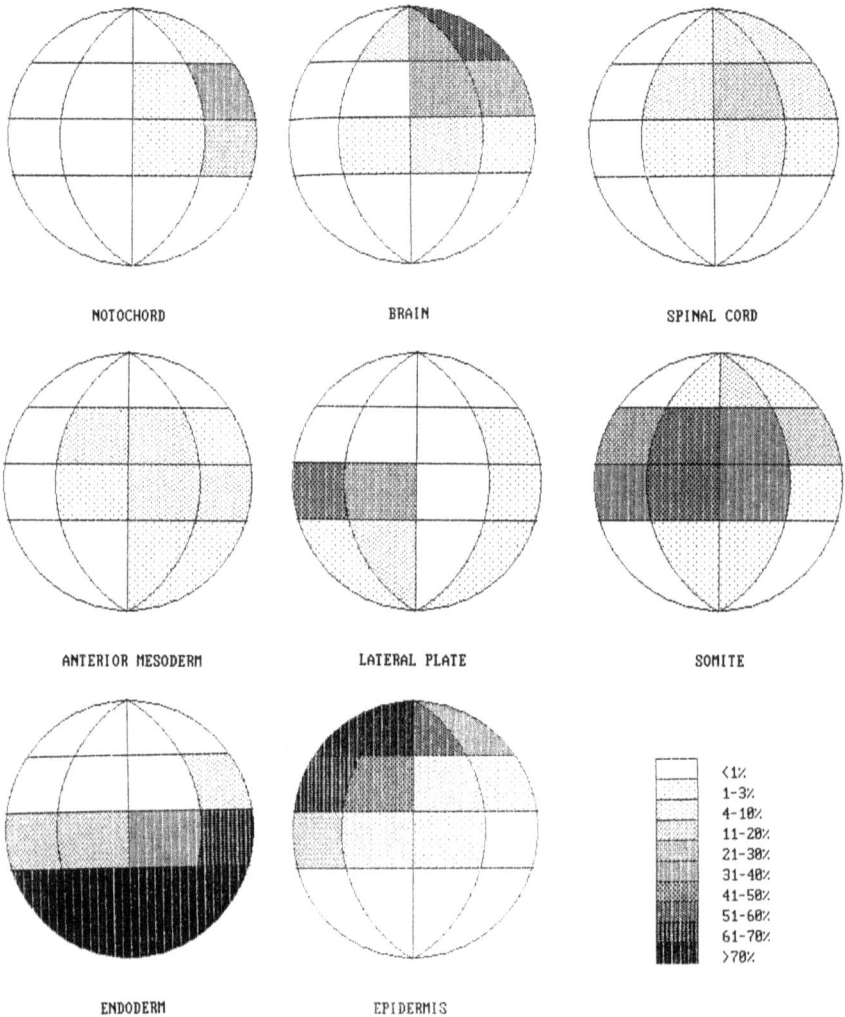

Figure 9.3. Prospective regions for each tissue at the 32-cell stage. In each diagram, the animal pole is up and the dorsal side on the right. Percentages indicate the proportion of each blastomere that contributes to named tissues. (Reprinted, with permission from Dale and Slack 1987 [© Company of Biologists Ltd.].)

of the standard amphibian saline solutions, e.g., 1x modified Marc's Ringer (MMR), but addition of 4% Ficoll is advisable to prevent the DiI solution from leaking out of the needle too quickly.

A very useful modification of the DiI technique allows observation of marked regions of the embryo after in situ hybridization (based on the method of Izpisua-Belmonte et al. 1993), thus permitting the correlation of

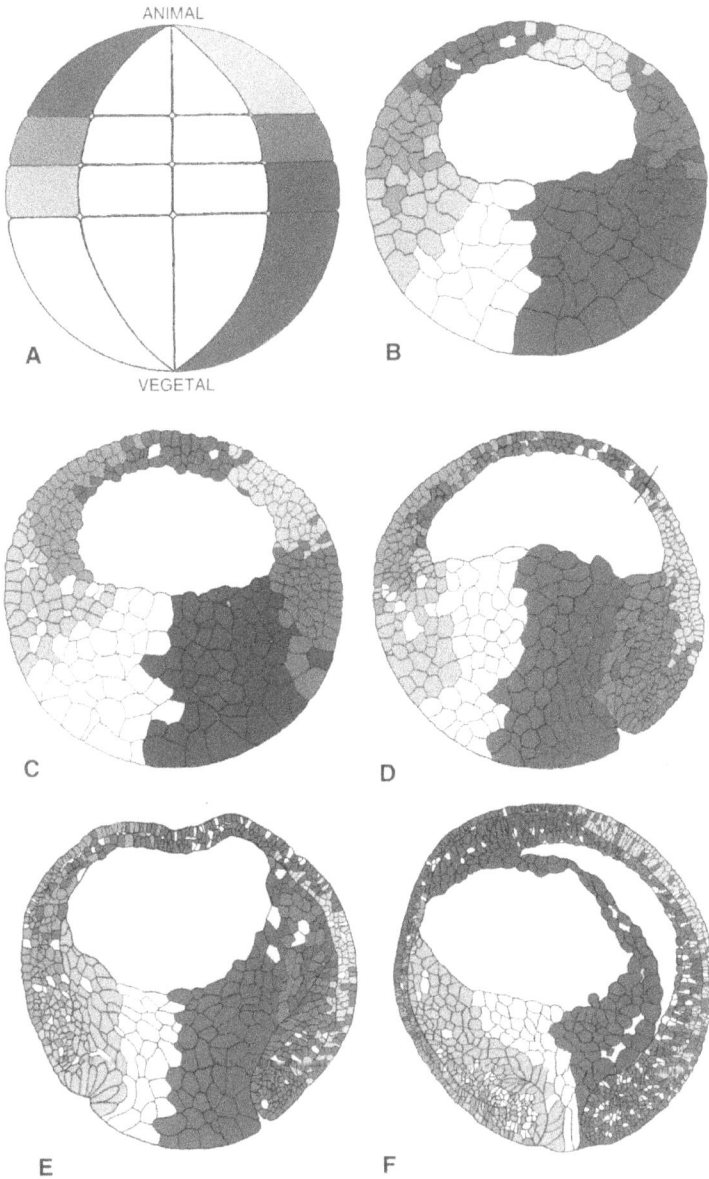

Figure 9.4. (*See Color Plate 7*) Summary diagrams illustrating the locations of the clones derived from the midline blastomeres of the 32-cell embryo (*A*) at stage 8 (*B*), stage 9 (*C*), stage 10 (*D*), stage 11 (*E*), and stage 12.5–13 (*F*). Data were derived from tissue sections in which two adjacent blastomere clones were labeled with different lineage dyes. Diagrams are oriented as in Figure 9.6, top. Clones are represented by the following colors: (*lilac*) C4; (*orange*) B4; (*green*) A4; (*yellow*) A1; (*red*) B1; (*blue*) C1; (*purple*) D1. (*A*, modified, with permission, from Bauer et al. 1994 [© company of biologists Ltd.]; *B–F*, modified, with permission, from Hausen and Riebesell 1991.)

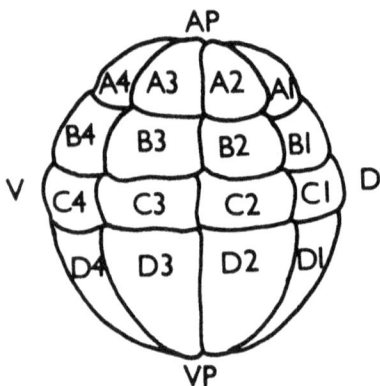

Figure 9.5. Nomenclature of blastomeres at the 32-cell stage. (D), dorsal; (V), ventral; (AP), animal pole; (VP), vegetal pole. (Reprinted, with permission, from Dale and Slack 1987 [© Company of Biologists Ltd.].)

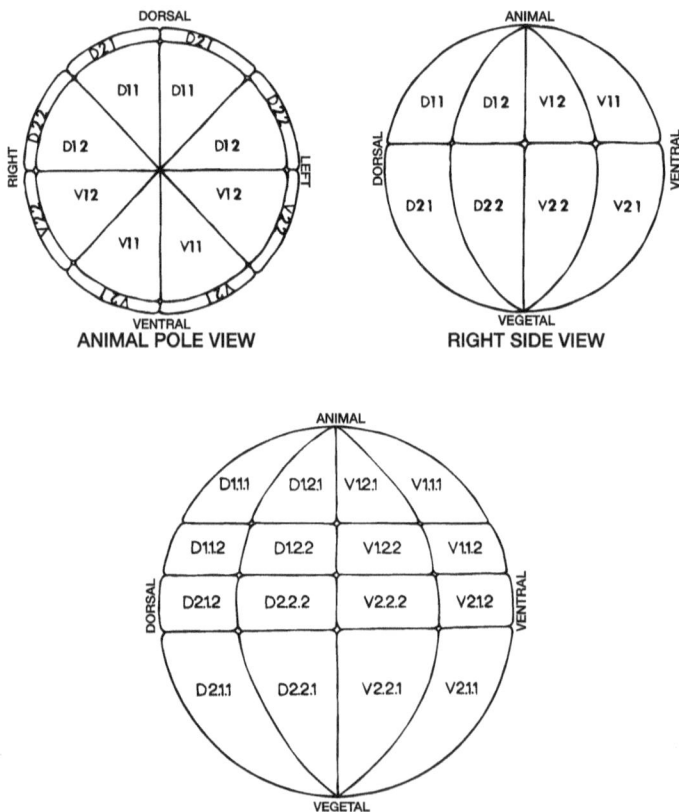

Figure 9.6. Lineage-based nomenclature of blastomeres from early stage embryos. (*Top*) 16-cell stage embryo, as assigned by Hirose and Jacobson (1979). (*Bottom*) 32-cell stage embryo, as assigned by Jacobson and Hirose (1981).

a particular morphological region with a particular domain of gene expression. Either before or after fixation, the region of the embryo to be marked is labeled by brief application of DiI. At some point before beginning the in situ hybridization procedure, a photoconversion process is used to create an insoluble product at the site of application of DiI (DiI itself is lipid-soluble and would be lost during the hybridization procedure). The DiI-treated embryo is incubated in the dark with diaminobenzidine (1 mg/ml) in 0.1 M Tris-HCl (pH 7.4) for 30 minutes to 1 hour. Embryos are then illuminated at 547 nm (the wavelength used to monitor rhodamine fluorescence) until the red fluorescence begins to fade. The embryos are then washed two to three times for 30 minutes in 0.1 M Tris-HCl.

LINEAGE LABELING

Tracer Molecules

A successful lineage tracer molecule must meet several requirements: (1) it must be nontoxic and nonreactive so that it does not change the developmental fate of the labeled cell; (2) it must be small enough to diffuse quickly through the injected cell, before the cell divides, so that all of the descendants are evenly and completely labeled, but large enough not to pass through the numerous gap junctions between adjacent blastomeres (Guthrie et al. 1988); (3) it must remain detectable throughout development and not be diluted by cell mitosis or intracellular degradation; and (4) it should be easily detectable by simple histological procedures. These requirements have been fulfilled by two classes of molecules: horseradish peroxidase (HRP) and fluorescent dextrans (Weisblat et al. 1978; Jacobson 1985; Gimlich and Braun 1985; Stent and Weisblat 1985). HRP is a plant enzyme that has no natural substrate in animal cells, and it is not recognized by the embryonic lysosomal compartment until the late tadpole stages. Dextrans are hydrophilic polysaccharides that are biologically inert and resistant to a cell's endogenous glycosidases. At low concentrations, both tracers appear to be nontoxic, but at high concentrations, they may cause labeled cells to die or to stop dividing. HRP has a molecular weight of about 44,000 and diffuses rapidly. Approximately 5 minutes after injection, cells are completely and evenly filled with the enzyme. The fluorescent dextrans are available in a variety of molecular weights, but the 10,000-molecular-weight and 40,000-molecular-weight compounds are most useful because they diffuse rapidly throughout the injected cell; the 10,000-molecular-weight dextrans may label the nucleus as well as the cytoplasm of the cell. Both markers are

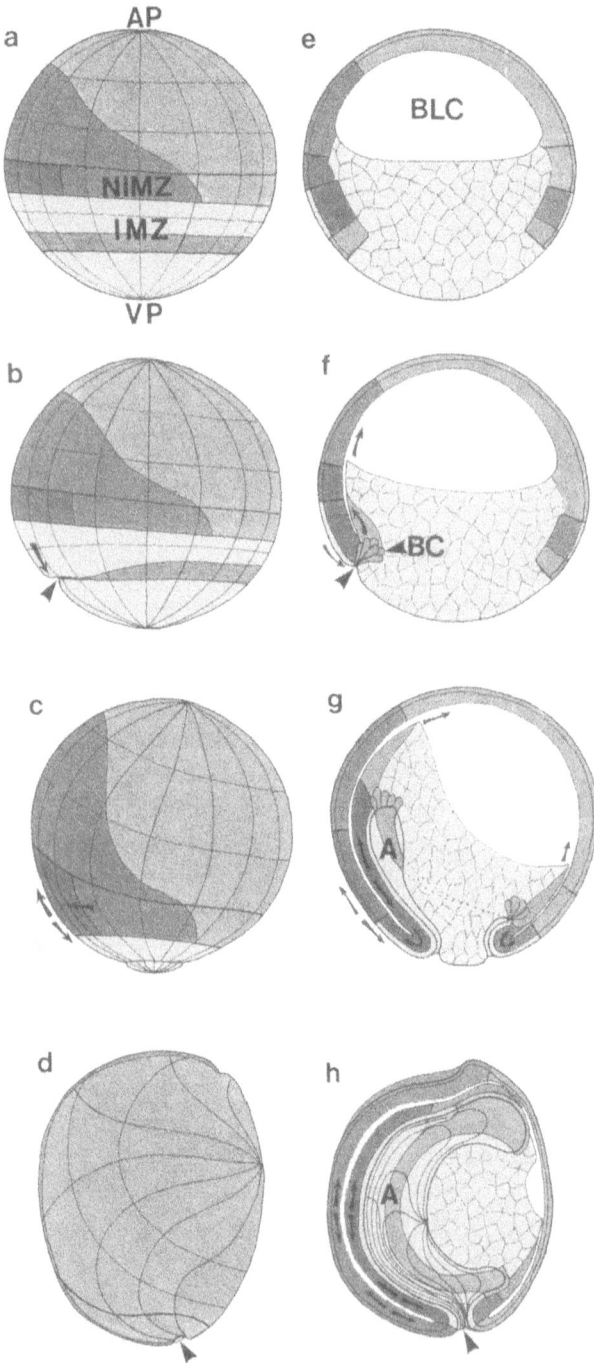

Figure 9.7. (*See facing page for legend.*)

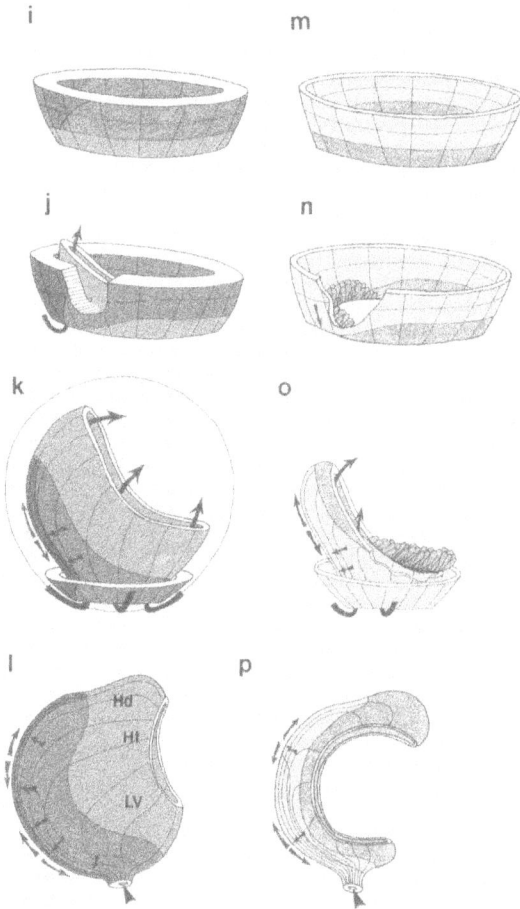

Figure 9.7. (*See Color Plate 8*) Prospective fates and morphogenetic movements of the gastrula and neurula are shown in several views. Four horizontal rows show development in chronological order, from top to bottom, at late blastula (*a, e, i,* and *m*), early gastrula (*b, f, j,* and *n*), late gastrula (*c, g, k,* and *o*), and late neurula (*d, h, l,* and *p*). The vertical columns show different views which are, from left to right, the lateral surface, with dorsal to the left (*a–d*), a midsagittal view (*e–h*), the ring of deep prospective mesodermal cells (*i–l*), and the overlying ring of suprablastoporal endoderm (*m–p*). Special features illustrated include the animal pole (AP), archenteron (A), the blastocoel (BLC), the bottle cells (BC), the blastopore (pointers), the head mesoderm (Hd), the heart mesoderm (Ht), the involuting marginal zone (IMZ), composed of deep, nonepithelial mesoderm (*i–l*) and superficial, epithelial endoderm (*m–p*), the lateral and ventral mesoderm (LV), the noninvoluting marginal zone (NIMZ), and the vegetal pole (VP). Prospective tissues shown are epidermis (*light blue*), neural tissue (*darker blue*), dorsal NIMZ (*blue green*), notochord (*red*), somitic mesoderm (*red orange*), migrating mesoderm at the leading edge of the mesodermal mantle (*orange*), suprablastoporal endoderm (*yellow*), a special region of suprablastooral endoderm known as the bottle cells (*green*), and vegetal, subblastoporal endoderm (*yellow*, divided into cells). Movements are indicated by arrows. Dorsal is to the left in all cases. (Reprinted, with permission, from Keller 1991.)

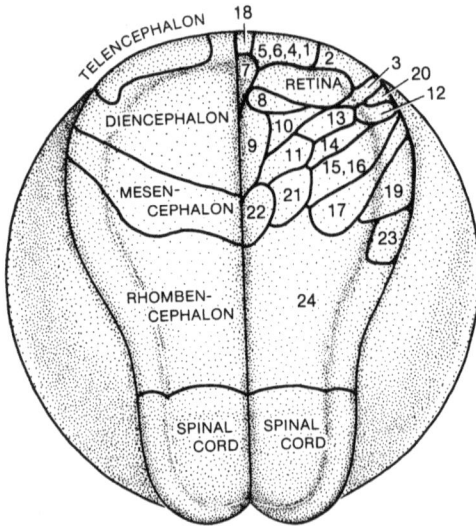

Figure 9.8. Fate map of the presumptive brain areas of the *Xenopus* neural plate (stage 15). The numbers to the right correspond to the key in Table 9.1. Shown on the left side are the main subdivisions of the brain. (Reprinted, with permission, from Eagleson and Harris 1990.)

Table 9.1. Key to Brain Areas

1. Area olfactoria primitiva	13. Anterior thalamic nucleus
2. Primordium piriforme	14. Central thalamic nucleus
3. Primordium hippocampi	15. Posterior thalamic nucleus
4. Lamina terminalis/nervus terminalis	16. Praectectum
5. Anterior preoptic area	17. Optic tectum
6. Magnocellular preoptic nucleus	18. hypophysis (anterior pituitary)
7. Supra chiasmatic nucleus	19. Cerebellum
8. Chiasmatic ridge	20. Epiphysis
9. Ventral hypothalamic nucleus/infundibulum	21. Tegmentum dorsale
10. Ventromedial thalamic nucleus	22. Tegmentum ventrale
11. Posterior tuberculum	23. Choroid plexus
12. Habenular comissure	24. Medulla oblongata
a. Dorsal habenular nucleus	
b. Ventral habenular nucleus	

Reprinted, with permission, from Eagleson and Harris (1990).

too large to pass through gap junctions. However, sometimes a heavily labeled clone and a very lightly labeled clone are observed in an embryo, as though the tracer has passed from the injected blastomere. It seems that the heavily labeled clone is derived from the injected cell and the lightly labeled clone is derived from an adjacent sister cell that received some label via a cytoplasmic bridge remaining from the previous cell division. This labeling pattern can be avoided by injecting blastomeres late in the cell cycle when these bridges are closed by cytokinesis (Moody 1987a).

Two mRNAs that encode tracer molecules, β-galactosidase (β-gal) and green fluorescent protein (GFP; Chalfie et al. 1994), are also useful as cellular markers. Both proteins are derived from nonvertebrates (β-gal from bacteria and GFP from jellyfish), can be distinguished from endogenous vertebrate proteins, and have no known deleterious effects on developing vertebrate cells. Like HRP and dextrans, the mRNAs can be toxic to the injected cell if delivered at too high a concentration. mRNAs are too large to diffuse through gap junctions, but they do not always mark the entire lineage of the injected cell. Clones derived from a blastomere injected with an mRNA tracer are often smaller than those derived from the same blastomere labeled with HRP or dextran. Clones labeled with mRNA may also lack some of the tissues present in the HRP- or dextran-labeled clone. For example, GFP-labeled cells do not extend throughout the full axial domain of tissues normally derived from 32-cell animal tier blastomeres, suggesting that the mRNA does not diffuse evenly throughout the blastomere prior to the next cell division (S.A. Moody, unpubl.). The mRNAs are either too large and highly charged to diffuse completely through the cell (see also Chapter 3), or they are loaded onto polysomes as they diffuse from the injection site, and become anchored before reaching the boundaries of the injected cell. Although this makes tracer mRNAs less useful as straightforward lineage markers, they are ideal as markers for the location of exogenous mRNAs. Mixing β-gal or GFP mRNA with a test mRNA will identify those regions of the embryo that express the protein product of the exogenous mRNA when the tissue is analyzed for the tracer protein at a later point in development (assuming that the tracer and the test mRNAs diffuse to the same extent). However, the tracer mRNA and the test mRNA may not be distributed and/or expressed equally in all descendants of the injected blastomere. For example, immunohistochemistry has shown that not every GFP-expressing cell expresses the fibroblast growth factor protein derived from a coinjected mRNA, as detected by immunocytochemistry (S.A. Moody, unpubl.). Thus, if it is important to know exactly which cells express the exogenous protein, it is best to include an internal epitope tag in the test mRNA (e.g., *myc* or hemagglutinin; see Vize et al. 1991), although there are also caveats to using such tags (see Ferreiro et al. 1998).

All four tracer molecules (HRP, dextran, GFP, and β-gal) can be detected in the descendants at least through tadpole stages 45–48. Because *Xenopus* cells decrease in size by cell division up through blastula stages (during which there are basically no G phases in the cell cycle), the originally inject-ed concentration of tracer remains stable. The growth of the embryo after neurulation does not appear to dilute the tracers significantly. However, once tadpoles begin to feed, HRP and dextran labeling can become granular and uneven, suggesting that the molecules are being packaged into lysosomes. Injected tracer mRNAs are probably degraded by the end of gastrulation, but both β-gal and GFP are very stable and can be detected at least through stage 45, well after their RNAs are gone.

The four tracer molecules are easy to detect by simple histological proce-dures. Enzymes, such as HRP and β-gal, are fixed in place with aldehydes and then are detected by applying a substrate that undergoes a colorimetric reaction, producing an insoluble, colored precipitate. The location of the pre-cipitate indicates the cellular location of the enzyme, which was either injected directly or encoded by the injected mRNA. An advantage of this technique is that the preparations, which can be made either as whole mounts or on tissue sections, are essentially permanent and can be used as reference material for years. In addition, the preparations can be cleared in organic solvents for improved three-dimensional visualization of the stain-ing pattern. The advantages of fluorescent tracers are that they can be viewed in the live embryo (although great care must be taken not to damage the flu-orescent cells by photoablation [Shankland 1984]), and they can be com-bined with other fluorescent methods, such as UV-excitable nuclear markers, fluorescent streptavidin, to detect biotin-labeled compounds, and the immunological detection of cell-type-specific proteins. To visualize fluores-cently labeled cells, either a compound microscope, equipped with epifluo-rescence, or a laser confocal microscope must be used. Clones at or near the surface can be viewed as whole mounts. However, because of the opacity of the early embryos, imparted by the intracellular yolk platelets, fluorescence from the deeper tissues is quenched in the whole-mount preparations, neces-sitating sectioning the tissue.

Microinjection

To perform lineage tracing, a very small volume of the tracer molecule must be delivered into the blastomere without damaging that cell. To perform this delicate procedure, a microinjection apparatus is required. Several different types of apparatuses are available, but the most important feature to consider is the typical injection volumes that will be used. Injection volumes into oocytes, fertilized eggs, and two-cell embryos can be as large as 10 nl (or sometimes larger), but for any older stages, it is wise to keep injection volumes to approximately 1 nl per blastomere. Although some laboratories report injecting 4–10 nl of mRNA into each blastomere of a two-cell embryo, these volumes are cytotoxic if using HRP or dextran tracers at the concentrations recommended here (see below and Chapter 3). In addition, at stages later than two cells, large-volume injections of either mRNAs or tracers are not well tolerated by the cells and can result in artifactual fate changes. Injecting more than 10 nl of tracer into some 16-cell blastomeres can have significant effects on cell fate (e.g., drive epidermal lineages into brain) and should be avoided (Hainski and Moody 1992). For more information on microinjection of cells, see Chapters 8 and 11.

The concentration of the tracer to be injected is also an important consideration. The label must be bright enough to be detected easily, but not so concentrated that it inhibits normal development. The following concentrations are recommended for the labels listed: 1% fluorescein-dextran, 0.5% Texas Red-dextran, 1% rhodamine-dextran, 5% HRP, and 300–500 pg per cell of β-gal mRNA or of GFP mRNA (S.A. Moody, unpubl.).

Great care must be taken not to damage the injected cell. Many laboratories put embryos in a 3–5% solution of Ficoll during microinjection. This prevents leakage of the tracer through the puncture wound and enhances healing. Ficoll must not be used if the embryos are to be devitellinated to perform a transplantation or dissection. In the presence of Ficoll, the vitelline membrane will collapse so tightly against the blastomeres that its manual removal will be virtually impossible. If a large hole is punctured into the cell, cytoplasm (and label) may leak out, or a bleb may form at the injection site. If the volume or concentration of tracer is too large, the injected cell or some of its descendants stop dividing. At the extreme, shortly after injection, there will be one or two large cells in a field of smaller ones. However, damaged cells may not be noticed until later in development. Larger than normal labeled cells may be incorporated into the organs of the embryo, or there may be labeled cells that are the correct size, but spherical, rather than differentiated, in shape. This can happen to the entire clone or to a subset of the clone. These cells will often move to the correct regions of the embryo, but never differentiate. Another sign of damage is the accumulation of labeled cells in the spaces within the embryo, i.e., the central canal and ventricles of the nervous system and the lumen of the gut, liver, and heart. These damaged cells probably dissociated from the rest of the embryo during gastrulation movements, and accumulated wherever space appeared.

If any of these signs of damage occur, discard the embryos and inject a smaller volume or concentration of tracer in the next experiment.

Identifying Specific Blastomeres

If targeting of the marker is not important, the embryos can be injected earlier (one, two, or four cells) without worrying about specific blastomeres. However, if a study requires localization of the marker to specific tissues or regions, it is essential to know where the dorsal side of the embryo will be. In-vitro-fertilized eggs should be tipped and/or the sperm entry point (SEP) marked as described in Protocol 7.5 (Tipping and Staining). For naturally fertilized eggs, and for eggs on which the SEP has not been marked, the future dorsal side of the embryo can be predicted very accurately (>90%) by noting the orientation of the first cleavage furrow. At fertilization, the animal hemisphere pigmentation begins to contract toward the SEP on the ventral side, causing the dorsal equatorial region to become less pigmented. If the first cleavage furrow bisects this lighter area (what would be the gray crescent in *Rana* species) equally between the two daughter cells, then that lighter area can be used as the indicator of the dorsal side, and the first cleavage furrow, the midsagittal plane. In some embryos, however, the first cleavage furrow will separate a darkly pigmented half from a lightly pigmented half. In these cases, the furrow still coincides with the midsagittal plane, but the pigmentation bears no relationship to the dorsoventral axis (Klein 1987; Masho 1990). By selecting embryos at the two-cell stage with the first furrow bisecting the lightly pigmented region of the animal hemisphere, there is a high degree of consistency in the fate maps, but only about a 70% consistency if eggs are chosen simply by separation of pigmentation. Embryos can also be selected at the early part of the four-cell cleavage, when the first and the second furrows can be distinguished at the vegetal pole; the first furrow should be complete and the second furrow not yet complete. If, however, embryos are not selected until the end of the four-cell stage, when it is no longer possible to discriminate between the first and second cleavage furrows, the lightly pigmented cells may be dorsal in only about 70% of embryos. It is important to note that the cleavage plane is not a determinant of the dorsoventral axis, and may be an unreliable marker under some circumstances (Danilchik and Black 1988). Vigorous dejellying during cortical rotation might uncouple the cleavage plane from axil formation completely.

Embryos must also be selected for regular cleavage furrows. These are found in a smaller and smaller percentage of embryos as cell divisions proceed. Again, if precise localization of the marker is not essential for the experimental design, then cleavage patterns are not a serious concern.

However, if localization is important, e.g., for quantitative analyses, embryos should be selected at each cleavage stage for their adherence to the patterns used for the published fate maps. However, not every blastomere in the embryo has to be "perfect." If a specific cell is being targeted, only that cell need cleave according to the ideal pattern, not every cell in the embryo. For example, often either the ventral or the dorsal animal cells have the stereotypic pattern, but not both; the same is true for animal versus vegetal cells.

A note on blastomere nomenclature: Nakamura and Kishiyama (1971) presented a very simple nomenclature for the 32-cell embryo (also utilized by Dale and Slack 1987; see Figure 9.5). When Hirose and Jacobson (1979; Jacobson and Hirose 1981) began to map the nervous system lineages at all of the different cleavage stages, they devised a plan, similar to those used in sea urchins and ascidians, that would relate the cells to their mothers, grandmothers, and descendants. Although these numbers and letters are difficult to remember if the system is not used on a daily basis, there is a logic to the system that communicates lineal relationships. All cells starting with D are on the dorsal side of the embryo, and all cells starting with V are on the ventral side. A number is then added at each cleavage stage that denotes the position of the blastomere with regard to poles and the midlines (see Figure 9.6).

Raising and Processing Embryos

Embryos injected with enzyme tracers can be raised under normal laboratory conditions, whereas embryos injected with fluorescent tracers should be raised in the dark, as a precaution against photoablation of the labeled clone or photobleaching of the label. The embryos can be raised in an 18–20°C incubator or on the bench beneath a darkened plastic cover, such as the top of a slide box. Embryos are reported to grow faster under red safe lights (Wetts and Fraser 1989). Enzyme tracers can be visualized after fixation with a simple histochemical reaction (HRP, Moody 1987a; β-gal, Vize et al. 1991). In addition, anti-HRP and anti-β-gal antibodies are commercially available if double-labeling with another protein marker is desired. Fluorescent dextrans can be viewed immediately with epifluorescence or laser-confocal microscopy in fixed or living embryos. However, excitation of the fluorophore releases free radicals that can damage living cells (Shankland 1984). Therefore, it is better to view living embryos for very short periods, or under low-light conditions. Detectable levels of GFP can be observed approximately 2 hours after injection of the appropriate mRNA. Again, precautions must be taken to avoid photoablation of living cells. Enzymes and GFP can be fixed in place by aldehydes. To fix the dextran label in place, it is necessary to use a label that is conjugated to lysine; the amino acid side chains allow the dextran to be cross-linked to intracellular proteins by formaldehyde. Different laboratories have favorite aldehyde fixative recipes, and a few of these are listed in the following protocols. One caution is that because formaldehyde is unstable in solution, and its breakdown products greatly enhance autofluorescence, freshly prepared paraformaldehyde (or freshly thawed from a frozen stock) must be used for fluorescent dextran- or GFP-labeled specimens. As noted above, enzyme histochemical reactions provide long-lasting specimens. Fluorescent lineage tracers can also be extremely hardy if the specimens are stored in the dark and refrigerated or frozen. For example, dextran- and GFP-labeled tissue sections can be stored at −80°C for 2 years with no detectable diminution of the signal (S.A. Moody, unpubl.). Samples must be mounted in a buffered, aqueous medium, of which there are several commercially available. Fluorescein absorption is especially sensitive to acidic pH, and most organic solvents used to store or clear tissues/embryos prepared with a histochemical reaction (HRP, β-gal, alkaline phosphatase) will destroy fluorescence.

PROTOCOL 9.1

Preparation of Tracers

FLUORESCENT DEXTRANS

1. Make 100 ml of 0.2 N **KCl**. The pH will be approximately 5.8. Adjust the pH to 6.8 by titrating with 0.05 M **KOH**.

   Note: *The pH will rise transiently and then gradually fall; it can take about 3 hours to reach a stable pH of 6.8. Be careful not to overshoot; the final pH is important for the intracellular health of the cell being injected. This solution is stable, and it can be filter-sterilized and stored at room temperature for several months.*

   KCl, KOH (see Appendix for Caution)

2. In a microcentrifuge tube, make up a 0.5–1.0% solution of dextran in the 0.2 N KCl.

   Note: *Dextran-amines conjugated to fluorochromes can be obtained from Molecular Probes. It is convenient to make up 100 µl at a time, which should last for about 100 experiments. Vortex vigorously to dissolve.*

3. Transfer the dye solution to a microfiltration tube, e.g., Costar Spin-X devices (cellulose acetate) for aqueous solutions (Costar 8160; Fisher 07-200-385). Spin at 14,000 rpm for 15–20 minutes at room temperature. The flow-through is now sterile, particle-free, and ready for use. Remove the filter unit from the tube and aliquot the dye into small volumes.

   Note: *A convenient way to store the dye is to prepare 5-µl aliquots in sterile capillary tubes, sealed at the ends with hematocrit tube-sealing compound (to prevent evaporation during storage). Aliquots can be stored at –20°C for up to 6 months. Longer storage is possible, but the dye can become toxic, probably because it becomes too concentrated.*

HRP

1. Prepare a 5% solution of HRP using sterile distilled water.

   Note: *Many isoforms of HRP are commercially available, but some preparations contain toxic contaminants. Boehringer Mannheim 814 393 (EIA grade, 90% isoenzyme C) is recommended (S.A. Moody, pers. comm.).*

2. Dissolve in small quantities, microfilter, aliquot, and store as described above in the note to step 3.

β-GAL AND GFP

These vectors can be purchased from a number of companies (e.g., Clontech 6047-1 and 6089-1).

1. Subclone the coding region into a *Xenopus*-appropriate expression vector so that mRNA can be transcribed with upstream and downstream sequences, allowing efficient translation after injection (see Chapter 3).

2. Alternatively, *Xenopus* vectors containing the β-gal-coding sequence or GFP-coding sequence can be obtained from a number of laboratories (see http://vize222.zo.utexas.edu). Prepare injection RNA according to the method of Vize et al. (1991), dissolve in sterile, RNase-free TE, and inject approximately 300–500 pg per blastomere. To coinject with a test RNA, simply mix the two RNAs at the appropriate concentrations.

PROTOCOL 9.2

Fixatives

FLUORESCENT-DEXTRAN AND GFP-LABELED SPECIMENS

A good fixative is 4% paraformaldehyde in PBS (0.1 M sodium phosphate at pH 7.4, 0.9% NaCl).

1. Add 4 g of **paraformaldehyde** in 40 ml of distilled water. Stirring constantly, heat to 60°C in a beaker, covered with foil, in a *chemical fume hood*. Do not allow the temperature to rise above 65°C.

   paraformaldehyde (see Appendix for Caution)

2. Add 1 N **NaOH** dropwise until the solution clears. Cool solution on ice to room temperature. Add 0.9 g of NaCl only if the tissues are to be stained by immunological techniques.

   NaOH (see Appendix for Caution)

3. Add 40 ml of 0.2 M **Na$_2$HPO$_4$** (monobasic), 10 ml of 0.2 M **NaH$_2$PO$_4$** (dibasic). Adjust the volume to 100 ml with distilled water. Store at –20°C in small aliquots for weeks.

   Na$_2$HPO$_4$, NaH$_2$PO$_4$ (see Appendix for Caution)

HRP-LABELED SPECIMENS

A good fixative is 4% paraformaldehyde, 1% glutaraldehyde, and 0.5% DMSO in 0.1 M phosphate buffer (pH 7.4).

1. Add 4 g of paraformaldehyde to 40 ml of distilled water. Stirring constantly, heat to 60°C in a beaker, covered with foil, in a *chemical fume hood*. Do not allow the temperature to rise above 65°C.

2. Add 1 N NaOH dropwise until the solution clears. Cool solution on ice to room temperature.

3. Add 8 ml of 50% EM-grade **glutaraldehyde**, 0.5 ml of **DMSO**, 40 ml of 0.2 M Na$_2$HPO$_4$ (monobasic), and 10 ml of 0.2 M NaH$_2$PO$_4$ (dibasic). Adjust the volume to 100 ml with distilled water. Mix thoroughly and store in the refrigerator for several months.

   glutaraldehyde, DMSO (see Appendix for Caution)

β-GAL FIXATION

1. Collect β-gal-expressing embryos, at appropriate stages, into glass vials and, if necessary, remove membranes.

2. Rinse the embryos several times with 1x PBS and fix for 1 hour at room temperature in **MEMFA** (Appendix 1), or for 1 hour on ice in freshly made 1x PBS containing 2% formaldehyde, 0.2% gluteraldehyde, 0.02% NP-40, and 0.01% **sodium deoxycholate**. After fixation, rinse the embryos three times in 1x PBS.

MEMFA, sodium deoxycholate (see Appendix for Caution)

PROTOCOL 9.3

Histochemical Reactions

β-GAL

1. Stain embryos at 30°C in 1x PBS containing 5 mM **K₃Fe(Cn)₆**, 5 mM K₄Fe(CN)₆, 1 mg/ml **X-gal**, 2 mM MgCl₂. The X-gal stock is made in dimethylformamide (DMF).

    Note: *The length of the staining reaction depends on the amount of β-gal activity in the embryos. If the stain is localized near the surface, blue cells should develop within 1–2 hours. However, if the β-gal activity is localized deeper in the embryo, staining will probably not be apparent until the embryo has been cleared. Start out by staining the embryos for 12 hours, but remember that longer staining periods lead to higher background staining.*

    K₃Fe(Cn)₆, X-gal (see Appendix for Caution)

2. When staining is complete, refix the embryos as described above (β-gal fixation) and then replace the fixing buffer with 100% **methanol**. Store embryos in 100% methanol.

    methanol (see Appendix for Caution)

HRP

Because diaminobenzidine (DAB) is a carcinogen, it is safest to purchase it in premeasured Isopacs and make up the stock solutions in these bottles (e.g., Sigma D 9015, 3,3 -diaminobenzidine tetrahydrochloride). Make up a 1% stock using sterile distilled water. For easy handling, freeze the stock in aliquots and store for several months. Always wear gloves, work on bench paper with a beaker of bleach nearby to deactivate any spilled DAB, and rinse all containers with bleach after the DAB reaction is completed.

1. For tissue sections, dilute the 1% **DAB** stock to 0.0125% (2.5 ml of stock into 200 ml of 0.1 M phosphate buffer at pH 7.4), and add 6 ml of 30% hydrogen peroxide.

    DAB (see Appendix for Caution)

2. For whole mounts, dilute the 1% DAB stock to 0.1%.

3. For slides:

   a. Hydrate to aqueous solutions and rinse in phosphate buffer. Submerge slides in 0.0125% DAB solution containing hydrogen peroxide (6 µl of 30% hydrogen peroxide to 200 ml of DAB solution) and allow the reaction to proceed at room temperature (usually 7–10 minutes). Check the reaction under a microscope after about 5 minutes.

   b. Stop the reaction by rinsing in two changes of 0.1 M phosphate buffer for 2 minutes each. Dehydrate through a graded series of ethanols and clear in two washes of toluene. Coverslip with Permount.

4. For whole mounts:

   a. Rinse embryos with phosphate buffer in a dram vial with screw cap. Add 1.0 ml of 0.1% DAB solution and agitate for 30 minutes at room temperature. Dilute 30% peroxide to 0.3%, and add 3.3 ml to the vial. Agitate for another 15–30 minutes, checking the reaction frequently under the microscope.

   b. Stop the reaction by washing five times (5 minutes each) with phosphate buffer. Dehydrate through a graded series of **ethanols**. If desired, clear as described for immunochemistry (Protocol 12.4, step 8) or whole-mount in situ hybirdization (Protocol 13.6, step 12).

   Note: *Liquid DAB waste should be deactivated with fresh bleach and stored in a designated waste container. Any solid touched by the DAB solution should be rinsed with bleach before disposal. The DAB solutions should turn dark brown when oxidized by the bleach, and will eventually clear when the DAB has been completely broken down.*

   ethanol (see Appendix for Caution)

REFERENCES

Bauer D.V., Huang S., and Moody S.A. 1994. The cleavage stage origin of Spemann's Organizer: Analysis of the movements of blastomere clones before and during gastrulation in *Xenopus. Development* **120:** 1179–1189.

Chalfie M., Tu Y., Euskirchen G., Ward W.W., and Prasher D.C. 1994. Green fluorescent protein as a marker for gene expression. *Science* **263:** 802–805.

Dale L. and Slack J.M.W. 1987. Fate map of the 32-cell stage of *Xenopus laevis. Development.* **99:** 527–551.

Danilchik M.V. and Black S.D. 1988. The first cleavage plane and the embryonic axis are determined by separate mechanisms in *Xenopus laevis. Dev. Biol.* **128:** 58–64.

Eagleson G.W. and Harris W.A. 1990. Mapping of the presumptive brain regions in the neural plate of *Xenopus laevis. J. Neurobiol.* **21:** 427–440.

Ferreiro B., Artinger M., Cho K., and Niehrs C. 1998. Antimorphic goosecoids. *Development* **125:** 1347–1359.

Gallagher B.C., Hainski A.M., and Moody S.A. 1991. Autonomous differentiation of dorsal axial structures from an animal cap cleavage stage blastomere in *Xenopus. Development* **112:** 1103–1114.

Gimlich R.L. and Braun J. 1985. Improved fluorescent compounds for tracing cell lineage. *Dev. Biol.* **109:** 509–514.

Guthrie S., Turin L., and Warner A.E. 1988. Patterns of junctional communication during development of the early amphibian embryo. *Development* **103:** 769–783.

Hainski A.M. and Moody S.A. 1992. *Xenopus* maternal RNAs from á dorsal animal blastomere induce a secondary axis in host embryos. *Development* **116:** 347–355.

Hamburger V. 1960. *A manual of experimental embryology.* University of Chicago Press, Chicago.

Hausen P. and Riebesell M. 1991. *The early development of* Xenopus laevis. Springer-Verlag, Berlin.

Hirose G. and Jacobson M. 1979. Clonal organization of the central nervous system of the frog. I. Clones stemming from individual blastomeres of the 16-cell and earlier stages. *Dev. Biol.* **71:** 191–202.

Izpisua-Belmonte J.C., DeRobertis E.M., Storey K.G., and Stern C.D. 1993. The homeobox gene *goosecoid* and the origin of organizer cells in the early chick Blastoderm. *Cell* **74:** 645 659.

Jacobson M. 1983. Clonal organization of the central nervous system of the frog. III. Clones stemming from individual blastomeres of the 128-, 256-, and 512-cell stages. *Dev. Biol.* **71:** 191–202.

———1985. Clonal analysis and cell lineages of the vertebrate nervous system. *Annu. Rev. Neurosci.* **8:** 71–102.

Jacobson M. and Hirose G. 1981. Clonal organization of the central nervous system of the frog. II. Clones stemming from individual blastomeres of the 32- and 64-cell stages. *J. Neurosci.* **1:** 271–284.

Keller R. 1975. Vital dye mapping of the gastrula and neurula of *Xenopus laevis* I. *Dev. Biol.* **42:** 222–241.

────1976. Vital dye mapping of the gastrula and neurula of *Xenopus laevis* II. *Dev. Biol.* **51:** 118–137.

────1991. Early embryonic development of *Xenopus laevis*. *Methods Cell Biol.* **36:** 61–113.

Klein S.L. 1987 The first cleavage furrow demarcates the dorsal-ventral axis in *Xenopus* embryos. *Dev. Biol.* **120:** 299–304.

Masho R. 1988. Fates of animal-dorsal blastomeres of eight-cell stage *Xenopus* embryos vary according to the specific patterns of the third cleavage plane. *Dev. Growth Diff.* **30:** 347–359.

────1990. Close correlation between the first cleavage plane and the body axis in early *Xenopus* embryos. *Dev. Growth Diff.* **32:** 57–64.

Masho R. and Kubota H.Y. 1986. Developmental fates of blastomeres of eight-cell stage *Xenopus laevis* embryos. *Dev. Growth Differ.* **28:** 113–123.

Moody S.A. 1987a. Fates of the blastomeres of the 16-cell stage *Xenopus* embryo. *Dev. Biol.* **119:** 560–578.

────1987b. Fates of the blastomeres of the 32-cell stage *Xenopus* embryo. *Dev. Biol.* **122:** 300–319.

Moody S.A. and Kline M.J. 1990. Segregation of fate during cleavage of frog *(Xenopus laevis)* blastomeres. *Anat. Embryol.* **182:** 347–362.

Nakamura O. and Kishiyama K. 1971. Prospective fates of blastomeres at the 32-cell stage of *Xenopus laevis* embryos. *Proc. Jpn. Acad.* **47:** 407–412.

Rugh R. 1948. *Experimental embryology: A manual of techniques and procedures.* Burgess Publishing Co., Minneapolis.

Shankland M. 1984. Positional determination of supernumerary blast cell death in the leech embryo. *Nature* **307:** 541–543.

Stent G.S. and Weisblat D.A. 1985. Cell lineage in the development of invertebrate nervous systems. *Annu. Rev. Neurosci.* **8:** 45–70.

Takasaki H. 1987. Fates and roles of the presumptive organizer region in the 32-cell embryo in normal development of *Xenopus laevis*. *Dev. Growth Differ.* **29:** 141–152.

Vize P.D., Melton D.A., Hemmati-Brivanlou A., and Harland R.M. 1991. Assays for gene function in developing *Xenopus* embryos. *Methods Cell Biol.* **36:** 367–387.

Weisblat D.A., Sawyer R.T., and Stent G.S. 1978. Cell lineage analysis by intracellular injection of a tracer enzyme. *Science* **202:** 1295–1298.

Wetts R. and Fraser S.E. 1989. Cell lineage analysis reveals multipotent precursors in the ciliary margin of the frog retina. *Dev. Biol.* **136:** 254–263.

Tape 1 of the video series "Manipulating the Early Embryo of Xenopus laevis*" presents a discussion of pigment differences indicating dorsal and ventral sides of the 4-cell embryo.*

CHAPTER 10

Microdissection

When analyzing any developmental process, one set of important questions are those relating to formation of different cell types. What makes a cell decide to become a particular cell type? What cell interactions are involved in this decision? The "commitment" of cells to a particular lineage is also called "specification" or "determination." Analysis of the cell interactions, or inductions, required for tissue-type determination is a critical step in the identification of the genes that control the process. Lineage commitment and inductive interactions can be assessed by explant or transplant assays. Such assays led to the initial discovery of induction (Spemann 1938), and they continue to provide extremely powerful tools for investigating when and how cells acquire their fate.

This chapter presents several assays, including methods to investigate lineage commitment and induction. One of the most useful aspects of *Xenopus* embryos is the ease with which the various tissues can be dissected and used in these assays. Assays are simplified by the large size of the embryos, the ability of cells to survive without added nutrients (due to the yolk contained within embryonic cells), the ability to distinguish and physically separate different germ layers, and the excellent fate maps that exist for *Xenopus* (see, e.g., Dale and Slack 1987; Moody 1987a,b; Keller 1975, 1976).

Both explant and transplant assays are used to assess lineage commitment. Explant assays involve the removal of groups of cells or single cells from the embryo and their culture in a medium believed to be free from inductive factors. Transplant assays involve moving groups of cells or single cells (donor cells) from a specific region of one embryo to a different region of another embryo (the host embryo), which may be similar or different in age. Tissue is transplanted to regions that are believed to exert little or no influence on the fate of the transplanted cells. In both explant and transplant assays, the fate of the donor cells is assayed morphologically, histologically, and at the molecular level, using cDNA probes or antibodies. Transplanted cells must be distinguishable in some way from the host embryo, and this is achieved by lineage labeling either the host or donor tissue. Lineage commitment, as defined by explant assays, is sometimes called specification, to distinguish it from lineage commitment as assessed by transplantation assays, which is called determination.

Explant and transplant analyses are also used to investigate inductive interactions. In both cases, the fate of the test cells is measured after their juxtaposition to a heterologous group of cells that are thought to have inductive ability. In explants, the assay is performed by placing the two groups of cells in close apposition in a "conjugation" assay. Transplant assays for inductive interactions are similar to those for lineage commitment, although here the donor cells are moved to a part of the host embryo that is expected to alter their fate.

EXPLANTS

The first step in all microdissections is dejellying of embryos and removal of the vitelline membrane (see Protocols 6.1 and 6.2). Use sharp tools to carry out all procedures; dishes should be coated with 1% agarose in sterile water (the coating prevents the cells from sticking to the dish and can be very thin, or several millimeters thick, depending on the worker's preference). To prevent dissociation of the tissue after the initial puncturing of the outer cell layer, all dissections should be performed in isotonic solution (0.5–1x MBS or equivalent; see Appendix 1). It is advisable to transfer the tissue to a clean dish prior to culture to cut down the risk of bacterial contamination. For long-term culture (more than a few hours), 0.5x MBS should be used. To prevent explanted tissue from sticking to dissecting tools or transfer pipettes, perform dissections in 100 µg/ml bovine serum albumin (Sigma) added to the isotonic dissection and culture medium. Culture medium should be supplemented with appropriate antibiotics, usually penicillin, gentamycin, or streptomycin, and great care must be taken to avoid contamination because fungi and non-susceptible bacteria quickly compromise the experiment. Most investigators find it convenient to keep a squirt bottle of 70% ethanol close at hand to sterilize dissecting instruments by wiping frequently.

Explanted tissue should be maintained in clean buffer, since culture with embryo debris may lead to contamination. Explants can be moved using a pipettor and sterile tip or a wide-bore plastic pipette. Do not allow the explants to come into contact with the air—they will explode immediately. Always keep them covered in buffer. Include several intact embryos cultured in parallel with explants to allow accurate staging.

TRANSPLANTS

As for explants, embryos to be used in transplant assays must be dejellied and the vitelline membrane removed before beginning (Protocols 6.1 and 6.2). Transplantion of a tissue from a donor embryo to a host embryo is most readily accomplished by placing both embryos in a petri dish, lined with a thin layer of modeling clay, into which the embryos are embedded and held in place. Clay available in retail stores is adequate if it is does not dissolve in water (e.g., Permoplast, American Art Clay). A blue color clay provides a good background for dissections. Donor embryos should be labeled (e.g., by injection at the one-cell stage with horseradish peroxidase or fluoresceinated dextran [fldx]) to enable unambiguous identification of the transplanted tissue. fldx is a particularly useful marker since it persists through in situ hybridization procedures.

Glass bridges made from small pieces of glass are used to weigh down transplanted tissue (or explants), keeping it in place until healing is complete. The bridges are made from thin strips of glass, cut from microscope coverslips after scoring with a diamond pencil. These may be used as is or heated gently over a small-bore Bunsen burner to produce a slight curve. The curve is designed to match the curvature of the embryo. The bridges are larger than the explants or embryos and thus will rest on the explants or embryos while their ends are embedded in the clay lining the dissection dish.

PROTOCOL 10.1

Animal Cap Isolation

The animal cap refers to tissue situated around the animal (pigmented) pole of a blastula or very early gastrula stage embryo. This tissue is fated to become cement gland/neurectoderm on the dorsal side of the embryo and epidermis on the ventral side. Animal caps are extremely useful because they are composed of pluripotent cells that can be induced to form endodermal, mesodermal, or ectodermal cell types (see, e.g., Lamb et al. 1993; Henry et al. 1996). Animal caps can therefore serve as a useful substrate to assess the activity of various inducing factors. Caps from embryos that have been injected with expression constructs can also be removed and analyzed to assess the activity of various genes. In addition, caps can be used in conjugation (induction) assays (see Protocols 10.4 and 10.5).

Removal of the animal cap is a relatively easy dissection, but care must be taken to remove pure ectodermal cell populations, without contaminating mesodermal (equatorial) cells. In general, smaller caps (comprising <50% of the animal hemisphere) contain a more homogeneous population of cells than larger caps, and, for this reason, small caps are preferable. The competence of animal cap cells changes with time, such that mid-blastula caps are highly competent to form mesoderm, whereas early gastrula caps can be refractory to mesoderm induction, but competent for neural induction. Therefore, for any one experiment, it is advisable to use caps from embryos of the same age (within ~0.5 stage of each other; Nieuwkoop and Faber 1994) to ensure that the tissue, and hence its response to the assay, is as homogeneous as possible.

1. Place the embryos in 1x modified Barth's saline (**MBS**), dejelly them, and remove the vitelline membrane just below the equator as described in Protocol 6.2.

   Note: *If several embryos are prepared before cap removal, take care to avoid puncturing the blastocoel since this will allow contact between the cap and the vegetal yolk, resulting in possible induction of the cap.*

   MBS (see Appendix for Caution)

2. Use a sharp pair of forceps, eyebrow knife, glass or tungsten needle (Chapter 4) to cut out the cap (see Figure 10.1).

Note: *It is useful to leave the remainder of the embryo intact, so that the area from which the cap has been removed can be seen.*

3. Check that only animal cap cells have been removed. If this is the case, the tissue will be of uniform thickness. Uneven thickness suggests that marginal zone cells have been included in the dissection and the cap should be discarded or trimmed.

4. Remove any adhering vegetal, yolky cells immediately and culture the explant in 0.5x MBS or equivalent.

Note: *Caps round up within minutes and completely heal over within a few hours. Low calcium/low magnesium buffer (LCMR; Appendix 1) slows healing but may partially neuralize the tissue.*

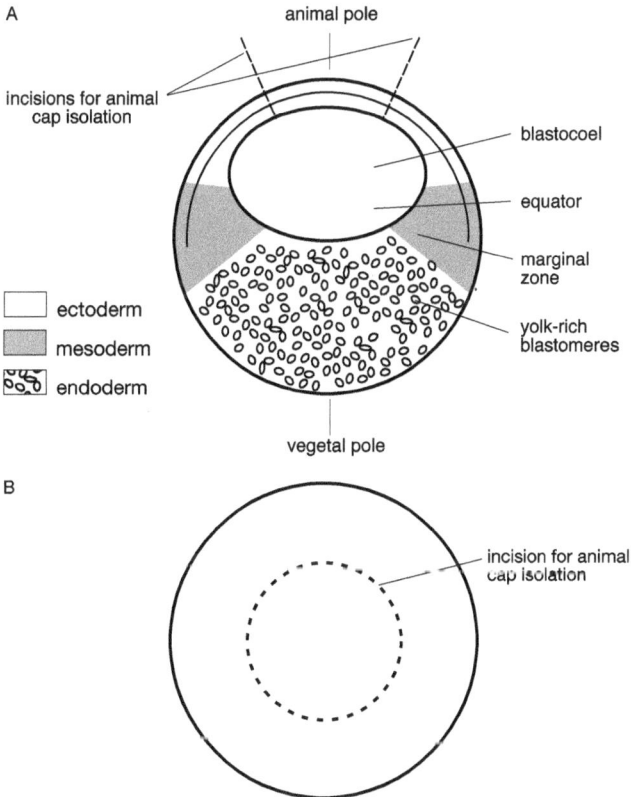

Figure 10.1. Diagrams of a blastula-stage embryo showing regions and incisions (*dashed lines*) for animal cap isolation. (*A*) Cross section; (*B*) animal pole view. (*Unshaded*) Ectoderm; (*gray*) mesoderm; (*circles*) endoderm.

PROTOCOL 10.2

Dissociation and Reaggregation of Animal Caps

Preparations of single cells derived from animal caps have proven to be useful substrates for assaying the activity of inducing molecules such as activin (see, e.g., Green et al. 1992). The idea behind the assay is that batches of dissociated cells can be exposed to more uniform concentrations of inducing factors than the multilayered intact cap. Single cells derived from caps have also been used to analyze the role of cell-to-cell contact in neural determination (Grunz et al. 1975; Wilson and Hemmati-Brivanlou 1995) and competence (Grainger and Gurdon 1989). Animal caps require divalent cations for their integrity and thus can be dissociated by exposure to medium lacking Ca^{++} and Mg^{++} ions. With this treatment, the cells comprising the inner layers of the cap dissociate within about 20 minutes. Cells can be reaggregated after this treatment and should survive well. The outer layer of cells is more recalcitrant to dissociation and requires a more severe treatment.

1. Remove animal caps (see Protocol 10.1) and transfer to 0.5x **MBS**.

   MBS (see Appendix for Caution)

2. Place the caps in agarose-coated dishes containing Ca^{++}/Mg^{++}-free saline. A 12-well tissue culture dish works well. Approximately five caps should be placed into a single well or dish.

   Notes: *The agarose should be thick enough (~2 mm) to make a concave surface into which the cells will fall during the reaggregation process.*
   If a significant volume of 0.5x MBS is carried over, remove half of the buffer in the well/dish and replace with fresh Ca^{++}/Mg^{++}-free saline. This ensures that any divalent cations carried over from the dissecting dish are removed.

3. Incubate at room temperature for approximately 20 minutes. During this time, the inner cap cells separate from the outer layer.

   Note: *Cells may remain in a clump, but these can be eased apart by gently pipetting the buffer over them using a pipettor and disposable tip.*

4. Discard the sheet of outer epithelial cells at this stage. These cells are much more adhesive than those of the inner layer and will not dissociate with this treatment.

Note: *If necessary, the outer cells can be dissociated by treatment with PhoNaK buffer (see Appendix 1).*

5. Maintain the dissociated embryonic cells for up to 5 hours in Ca^{++}/Mg^{++}-free saline at 15–20°C or treat according to the chosen assay.

6. If reaggregation of dissociated cells is required, replace the Ca^{++}/Mg^{++}-free saline with normal saline and incubate the cells at room temperature. Reaggregation should be complete in less than 2 hours.

   Note: *Reaggregation can be accelerated by gently swirling the dish to concentrate the cells or by brushing the cells into the center of the dish using an eyebrow knife.*

Helpful Hints

- When reaggregating cells, do not attempt to make more than one aggregate per culture well, since aggregates stick to one another very efficiently.

PROTOCOL 10.3

Ectodermal (Animal Cap) Layer Separations

The blastula animal cap comprises two morphologically distinct cell layers, an outer monolayer termed the epithelial layer, which consists of tightly adherent pigmented cells, and an inner layer several cells thick termed the sensorial layer, which consists of loosely adherent cells. These layers persist throughout the larval stages and have different fates and potentials (see, e.g., Drysdale and Elinson 1993; Bradley et al. 1996). With the use of isolated layers, it is feasible to test their developmental potential and response to induction.

Separation of cell layers is most easily achieved by tackling the intact embryo, rather than an isolated animal cap.

1. Dejelly and devitellinize the blastula/early gastrula embryos (Protocols 6.1 and 6.2). Insert a very fine eyebrow knife just below the surface of the embryo (at late blastula/early gastrula), between the inner and outer ectodermal layers, close to the marginal zone (see Figure 10.2).

    Note: *It is important to insert the knife horizontally, parallel to the surface of the embryo.*

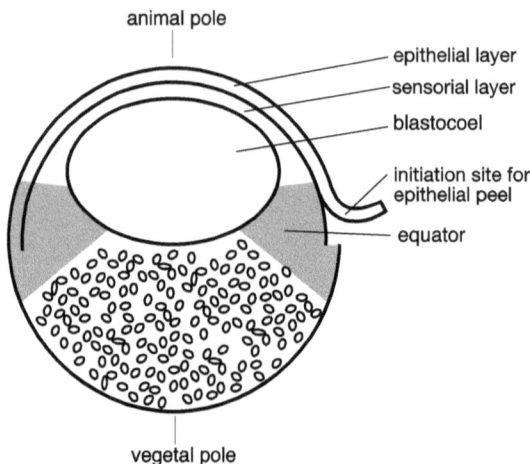

Figure 10.2. Diagram of a blastula-stage embryo in cross section. The epithelial (*outer*) and sensorial (*inner*) layers of the embryo are indicated, with incisions for epithelial layer isolation. The sensorial layer can be cut as for an animal cap (Protocol 10.1) subsequent to epithelial layer removal. (*Unshaded*) Ectoderm; (*gray*) mesoderm; (*circles*) endoderm.

2. Gently lift the knife up and bring with it the thin outer layer of pigmented cells.

   Note: *If the outer layer only has been isolated, it should curl up immediately. If more than the outer layer of cells has been removed, the onset of curling will be delayed.*

3. Use the eyebrow knife in a gentle scraping motion to peel back the outer layer of cells over the animal cap.

   Note: *Once the outer layer cells over the blastocoel have been removed, it is easy to check that the correct layer has been taken because a layer of (unpigmented) cells will still cover the blastocoel.*

4. Trim off the outer layer of cells to the correct size, removing the marginal zone cells.

5. Remove the inner layer of cells as described for animal cap isolation (Protocol 10.1).

PROTOCOL 10.4

Animal Cap/Vegetal Endoderm Conjugates

Mesoderm can be induced in animal cap cells by contact with vegetal cells (presumptive endoderm). This juxtaposition of tissues was the first indication that the process of mesoderm induction could be separated from neural induction (Nieuwkoop 1952a,b). Nieuwkoop also showed that dorsal vegetal cells can induce dorsal mesoderm and that ventral vegetal cells can induce ventral mesoderm in animal caps (Nieuwkoop 1952a,b). With the use of vegetal tissue taken from embryos in which various gene products have been deleted or perturbed, animal cap conjugates can be used to test the requirement for certain components in mesoderm determination (e.g., Heasman et al., 1994). The conjugates can also be used to test whether a particular gene or promoter construct is responsive to signals that arise from the vegetal cells.

Isolated vegetal cap tissue can be cultured independently and used in a manner similar to that of animal caps. Vegetal caps are used to test the activity of genes on a presumptive endodermal substrate (e.g., Cornell and Kimelman 1995).

This protocol describes how to set up a conjugate using animal cap and vegetal tissue. In theory, a single embryo can provide both animal and vegetal cells; however, in practice, this is difficult to achieve.

1. Dejelly and devitellinize blastula-stage embryos (Protocols 6.1 and 6.2). Isolate vegetal cells, either from the vegetal "cap" region (centered around the vegetal pole and extending approximately half-way toward the equator of the blastula-stage embryo) or from the dorsal or ventral sides of the embryo (see Figure 10.3).

   Note: *Care must be taken not to damage the large vegetal cells. They are full of yolk and are very fragile.*

2. Remove an animal cap (see Protocol 10.1).

3. Assemble the conjugate by gently pushing the inner (sensorial) surface of the animal cap toward the inner surface of the vegetal tissue.

4. Let the conjugate stand for 30 minutes or so, before moving it to a clean dish. After 30 minutes, tissues will adhere and the conjugate can be moved using a wide-bore plastic pipette.

Helpful Hints

- It is generally best to isolate all of the required samples of vegetal tissue and set them aside before isolating animal caps, since the vegetal tissue will not round up as animal caps do. Move the vegetal caps to a clean dish before making the conjugates. This will cut down the amount of debris present and reduce the chance of contamination. When assembling the conjugate, the animal cap should initially be balanced on top of the vegetal cells (or *vice versa*), but tissues should adhere very quickly (within 15–30 seconds).

- If animal caps round up before conjugates can be made, cut them open with an eyebrow knife before apposing to vegetal tissue.

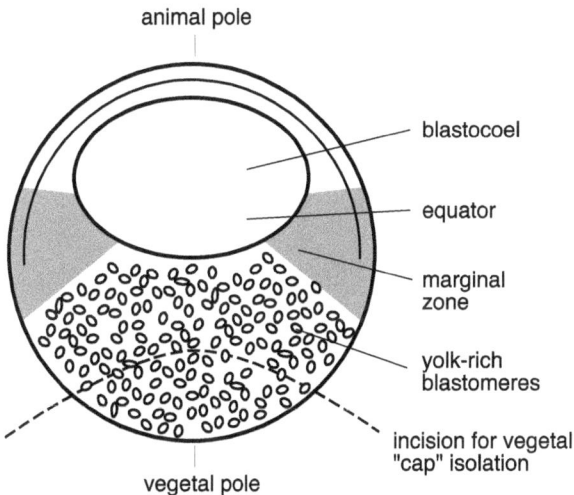

animal pole

blastocoel

equator

marginal zone

yolk-rich blastomeres

incision for vegetal "cap" isolation

vegetal pole

Figure 10.3. Diagram of a blastula-stage embryo in cross section. Incisions (*dashed line*) for vegetal tissue (endoderm) isolation ("vegetal cap") are shown. Animal caps are isolated as detailed in Protocol 10.1. (*Unshaded*) Ectoderm; (*gray*) mesoderm; (*circles*) endoderm.

PROTOCOL 10.5

Animal Cap/Dorsal Mesoderm Conjugates

Dorsal mesoderm is also called "organizer" tissue or Spemann's organizer. It is fated to form the prechordal plate and notochord, as well as some somitic tissue. The organizer tissue is able to induce a secondary axis when transplanted into the ventral side of a host embryo, or into the blastocoel (Protocol 10.7). It is also able to induce neural tissue in animal caps (see, e.g., Sharpe et al. 1987; Sive et al. 1989). More lateral mesoderm also has inducing capacity and can be tested in this assay (see, e.g., Bang et al. 1997). The animal cap/dorsal mesoderm conjugate can be useful in assessing whether test genes or promoters are responsive to dorsal mesodermal signals and, if so, by what region of the mesoderm they are induced. The example presented here is an animal cap/dorsal mesoderm conjugate.

1. Select mid-gastrula embryos, dejelly, and remove the vitelline membrane (see Protocols 6.1 and 6.2).

   Note: *Embryos at this stage contain dorsal mesoderm that will strongly induce neural tissue in animal caps. Early gastrula embryos can also be used (see, e.g., Zoltewicz and Gerhart 1997).*

2. Make two superficial slits, one on either side of the dorsal blastopore, to mark the dorsal region (~60° apart) (Figure 10.4, slits 1 and 2).

3. Extend these slits anteriorly to the blastocoel and make a third slit in the blastocoel (Figure 10.4, slit 3). Peel the dorsal ectoderm as far back as the blastopore, to expose the mesoderm, being careful not to leave any sensorial (inner layer) ectoderm adhering to the mesoderm. Cut off the ectodermal flap (slit 4).

   Note: *This allows the position of the involuted mesoderm and advancing archenteron to be located. Mesoderm with the strongest inducing potential lies above the archenteron (see Figure 10.4B), which is easily identified as a darkly pigmented slit, visible through the overlying mesoderm (in Nile-blue-stained albinos, the archenteron is visible in the intact embryo).*

4. From the blastopore, make two additional cuts into the mesoderm, one on either side of the dorsal midline, 60° apart (Figure 10.4B, slits 5 and 6).

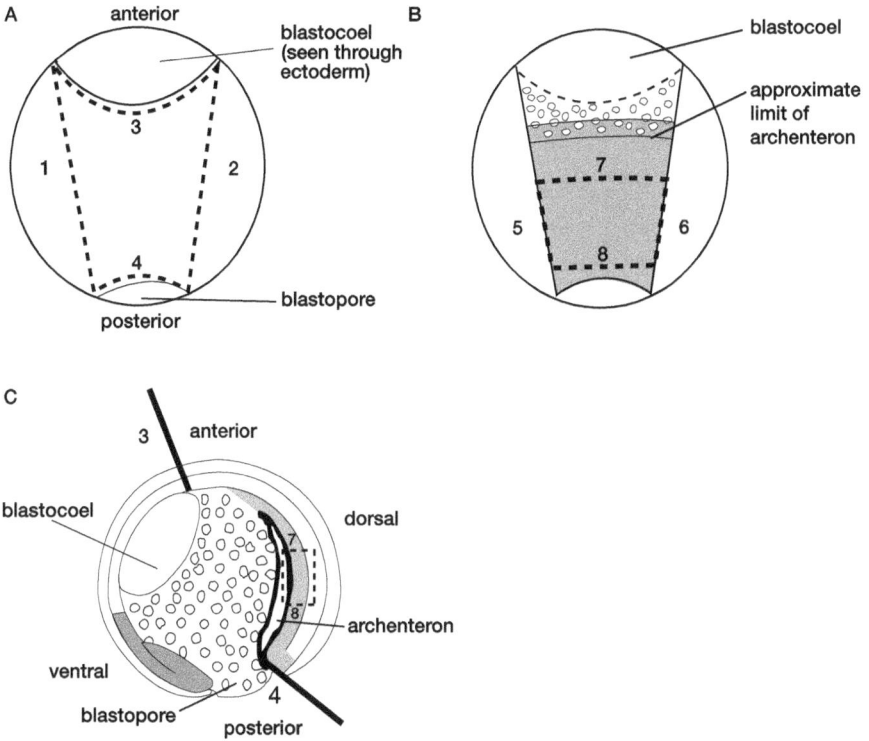

Figure 10.4. Isolation of posterior dorsal mesoderm from a mid-gastrula stage embryo. (*A*) Dorsal view of intact embryo. Incisions to remove surface ectoderm are numbered 1–4, on dashed lines. (*B*) Dorsal view of embryo with mesendoderm exposed. Subsequent incisions to remove posterior mesoderm are numbered 5–8, on dashed lines. (*C*) Sagittal section through intact mid-gastrula showing position of selected incisions from *A* and *B*. Position of posterior dorsal mesoderm removed is boxed. (*Unshaded*) Ectoderm; (*gray*) mesoderm; (*circles*) endoderm.

5. Make a traverse slit at the level of the blastopore (Figure 10.4C, continuation of Slit 4 into mesoderm). Peel the dorsal mesendoderm as far back as the anterior extent of the archenteron. Tissue below the archenteron is endoderm and should be discarded. Cut off a piece of tissue overlying the archenteron (Figure 10.4C, slits 7 and 8) and keep this piece of tissue in 0.5× **MBS.**

   MBS (see Appendix for Caution)

6. Perform animal cap/mesoderm conjugation as for animal/vegetal conjugates (see Protocol 10.3).

Helpful Hints

- The separation of mesoderm from ectoderm is easy to achieve until mid gastrula, after which mild trypsin or collagenase treatment is necessary (see Protocol 10.9).

- If dorsal mesoderm is isolated an hour or so before use, it will elongate and curl. The concave side corresponds to the mesoderm and the convex side corresponds to endoderm that originally comprised the marginal zone outer layer (Chapter 2, Figure 2.3).

PROTOCOL 10.6

Keller Explants

The purpose of Keller explants was initially to allow observation of gastrulation movements, particularly convergent extension, in culture. This is difficult to do when explants curl up, but in Keller sandwiches, the explants are kept flat, and instead of involuting beneath the ectoderm, mesoderm elongates in a plane with adjacent ectoderm (Keller and Danilchik 1988). Explants are made at the onset of gastrulation before significant vertical juxtaposition of ectoderm and mesoderm has occurred. The Keller explant has also been used to look at the role of vertical versus planar signaling in patterning the neural plate (Doniach et al. 1992). Since it can be very difficult to isolate explants before any vertical contact between ectoderm and mesendoderm has taken place, conclusions drawn from such experiments must be tempered by this caveat.

The basic Keller explant is a rectangle of dorsal mesendoderm and ectoderm that is approximately 60–90° wide. It extends from the bottle cells to the animal pole. Involuted head mesoderm is removed and the explant is then cultured flat, either as a single sheet (open-face sandwich) or more frequently as two sheets sandwiched together with their inner surfaces apposed (closed sandwich). Explants are cultured beneath a coverslip fragment or a glass bridge (see introduction to transplants) resting on silicone vacuum grease until the desired stage, usually during or after neurulation.

1. Dejelly embryos and remove the vitelline envelope from early gastrula (stage-10–10+) embryos (as described in Protocols 6.1 and 6.2).

   Note: *To avoid damaging the dorsal side or animal cap, remove the membrane from the ventral side of the embryo.*

2. Place the devitellinated embryo animal pole down, with the dorsal side (bottle cells) toward the hand in which the eyebrow knife is held (the "cutting" hand).

3. Using a hair loop or forceps to hold the embryo in place, poke the tip of the eyebrow hair downward at one end of the line of bottle cells and make a radial cut toward the animal pole (Figure 10.6, slit 1).

   Note: *This cut is easily achieved by cutting against the base of the petri dish, or by flicking the knife upward.*

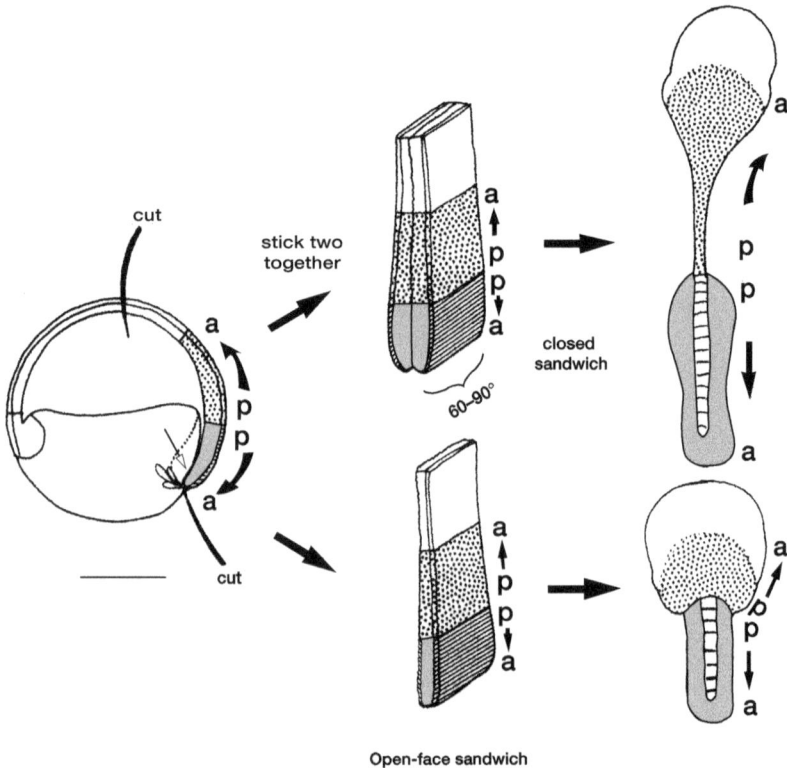

Open-face sandwich

Figure 10.5. Regions of early gastrula used to make Keller explants. Stage 10+ embryo in sagittal cross section, dorsal to the right, animal pole up. The shading approximates tissue types predicted by the fate map; the animal (*upper*) hemisphere consists of ectoderm that will give rise to epidermis (*white*) and future neurectoderm (*light stippling*). (*Dark gray*) Dorsal mesoderm; (*stripes*) archenteron roof endoderm. Explants are made by cutting out a rectangle of tissue reaching from approximately the animal pole to blastopore, as indicated by the upper and lower "cuts," respectively, and about 60–90° wide around the equator. The head mesoderm, bounded by the dotted line (open arrow) and the dorsal mesoderm, is removed from explants. The predicted A-P polarity of the mesoderm and ectoderm is indicated. Sandwich explants are made by putting two of these rectangles together with their inner surfaces apposed; these undergo convergent extension (narrowing and elongating) in both the posterior neurectoderm and the mesoendoderm. Open-face explants undergo convergent extension only in the mesoendoderm. The layer of endoderm is not shown in the explants depicted on the right. In these, the white column down the center of the mesoderm represents the notochord. Bar, 500 μm. (Reprinted, with permission, from Donaich 1992 [©AAAS].)

4. Make a similar cut at the opposite end of the bottle cell line and turn the embryo over, i.e., animal pole facing up (Figure 10.6, slit 2).

   Note: *If the cuts have not reached the animal pole, insert the tip of the eyebrow knife into one of the cuts, under the region that needs to be cut. Then press the hair loop against the tip of the eyebrow hair. This will complete the*

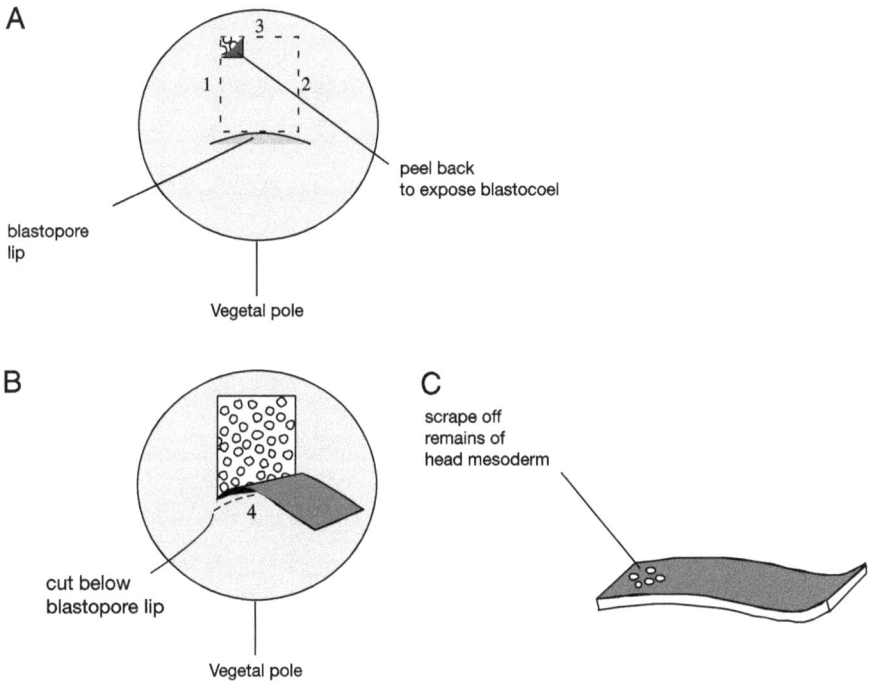

Figure 10.6. Preparation of a Keller explant, for use in sandwich or open-face explant assays. (*A*) Initial cuts, 1–3; (*B*) peel back of mesendoderm and secondary cut 4 (*C*) trimming explant.

cut by pinching the tissue between the two hairs. Alternatively, hold the region adjacent to the cutting edge with the hairloop and gently flick upward with the eyebrow hair. This trick requires some practice, but, once mastered, it is a clean and fast way to cut through thin tissues.

5. Make a horizontal cut at the animal end of the dissection to complete a rectangle with the previous two incisions. (Figure 10.6, slit 3)

   Note: *Use one of the cutting methods discussed in the note to step 4 above.*

6. Hold the embryo down with a hair loop and using an eyebrow knife, peel back the dorsal tissues to the dorsal blastopore lip. Use the edge of the eyebrow hair to deepen the cleft formed by the involuting head mesoderm to allow peeling back to the lip (see Figure 10.5, open arrow).

   Note: *In very early gastrulae, the cleft has yet to form, so it must be cut with the knife. This removes some of the head mesoderm from the noninvoluted chordamesoderm.*

7. Flip the embryo over (vegetal pole up) and, with the edge of the eyebrow hair, cut downward along or just below the bottle cells, all the way through to the blastocoel. This is the bottom edge of the rectangle.

   Note: *Do not cut off the bottle cells.*

8. Remove any pieces of involuting head mesoderm that are left on the inner side of the outer mesoderm cells. Place the rectangle inner surface down, and trim off any material that protrudes beyond the bottle cells. Then, turn the explant over and use the tip of the eyebrow knife to gently pick off any loose head mesoderm cells.

9. Trim the explant into a regular rectangle of the desired dimensions.

10. For an open face explant:

    a. Transfer the explant to a new dish of Sater's Modified Blastocoel Buffer (see Appendix 1) and lay the tissue flat, with the side of interest uppermost.

    b. Put a dab of vacuum grease at each end of a sterile rectangle of coverslip (~5 x 10 mm) or glass bridge.

    Note: *The grease serves to hold the coverslip in place on top of the explant and should be used sparingly. It is convenient to dispense the grease using a 5-cc syringe.*

    c. Gently place the coverslip on top of the explant. Exert slight pressure to flatten the explant but do not damage the cells. Avoid contaminating the explant with vacuum grease.

11. To make a sandwich:

    a. Dissect two rectangular explants of the same size, as described above, and immediately press the inner sides together, making sure that the bottle cells are aligned.

    Note: *If explants are left in medium for more than a few minutes, they will not adhere well.*

    b. Trim the edges of the sandwich so that the explants are exactly the same size and set aside in a dissecting dish.

    Note: *The dissecting dish should be kept relatively free from debris, and thus, it is advisable to prepare sandwiches one at a time.*

    c. Transfer sandwiches to a clean dish and gently press each one under a coverslip (as described for open-faced sandwiches). Do not squash the sandwich!

Note: *Make sure that the layers of the sandwich are well aligned and that the explant is flat. If explants are out of register, the layers will not elongate efficiently.*

Helpful Hints

- Include three to four control embryos with intact vitelline membranes in the same dish as the explants to monitor the developmental stage of the explants. Culture the embryos and explants at 15–23°C, depending on the desired rate of development.

- In the late-teen stages, the mesoderm of the explants begins to dissociate. If this is a problem, remove half the salt concentration of the medium and replace the lost volume with sterile water, supplemented with antibiotics (gentamycin at 0.05 mg/ml). Alternatively, replace the entire volume with 0.5x MBS with the same antibiotic.

- If the incubation time exceeds 24 hours, replace the medium every day.

PROTOCOL 10.7

Einstecks

"Einstecks" refers to a procedure for placing a piece of tissue into the blastocoel of an early gastrula, in order to assess the inductive potential of the introduced tissue. The foreign tissue adheres to surrounding tissue and becomes incorporated into the host embryo. This is a simple transplant procedure that has been used to assess which types of axial tissue can be induced by different regions of the mesendoderm or ectoderm, and how forced expression of genes or treatment with various factors can alter this potential (see, e.g., Ruiz i Altaba and Melton 1989; Sive and Cheng 1991).

1. Dejelly and then carefully remove the vitelline membrane (see Protocols 6.1 and 6.2) of an embryo at late blastula/early gastrula.

2. Make a slit in the animal hemisphere, just large enough to accommodate insertion of the chosen foreign tissue.

   Note: *If using gastrula, make the slit ventrally to avoid perturbing gastrulation.*

3. Insert the foreign tissue and bring the edges of the slit as close together as possible (so that they are at least partially touching).

4. Place the embryo in 1x **MBS** at 15–20°C to heal, for 1 hour or so. During this time, the slit should close completely.

   MBS (see Appendix for Caution)

5. Transfer the healed embryo to 0.1x MBS. If an open wound remains, culture the embryo in 0.5x MBS.

   Note: *Culturing in 0.5x MBS increases the chance of exogastrulation slightly, but unless the embryo heals, the experiment will fail completely.*

PROTOCOL 10.8

Transplantation of Lens Ectoderm

Transplanting a piece of tissue from one site to another (other than into the blastocoel) is a useful way of examining the sequence of inductions required to produce a particular organ. For example, the young primordium of a particular organ can be transplanted from a young individual into an older embryo to determine whether the older embryo still has the capacity to induce the tissue to form the organ. Although this capacity can also be assayed using explants, the complex surroundings of the host embryo cannot be reproduced in an explant.

The example presented here uses the case of transplanting a piece of gastrula stage animal cap ectoderm into the presumptive lens of a neural-plate-stage host embryo, with the aim of assessing the ability of the ectoderm to induce lens formation (Figure 10.6) (Severtnick and Grainger 1991). Note that the technique can be adapted for use in other regions of the embryo. Care must be taken to inflict as little damage on the embryo as possible during this procedure. Well-treated embryos recover rapidly, whereas embryos with extensive injuries do not.

1. Dejelly and devitellinate gastrula-stage donor (stage 10–12) and neurula-stage host (stage 14) embryos (Protocols 6.1 and 6.2).

2. Orient the host embryo with the target site (in this case, the presumptive lens area) uppermost.

3. Use a fine needle, or a needle and an eyebrow knife, to peel back a rectangle of host ectoderm that corresponds to the presumptive lens area.

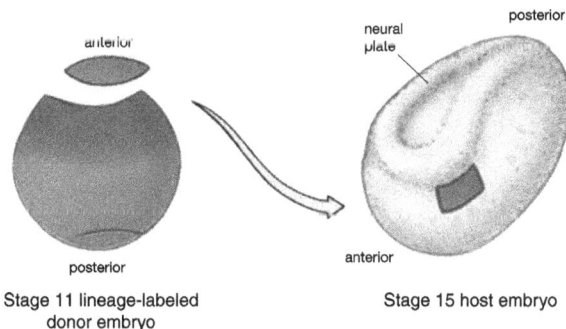

Stage 11 lineage-labeled
donor embryo

Stage 15 host embryo

Figure 10.7. Illustration of transplantation of animal cap ectoderm into the presumptive lens site of a neural plate stage host embryo. (Reprinted, with permission, from Zygar et al. 1998.)

Note: *The embryo should be embedded in a clay-covered dish for ease of manipulation.*

4. Before the host site heals (a matter of minutes), remove a piece of animal cap tissue from the donor embryo by cutting out ectoderm with a fine needle or eyebrow knife.

   Note: *Make sure that the donor tissue is similar in size to the tissue removed from the host.*

5. Insert the donor ectoderm into the target site and hold it in place with a curved glass bridge.

   Note: *The ends of the bridge should rest in the clay (see introduction).*

6. When the transplant site has healed completely (15–30 minutes), remove the glass bridge and transfer the embryos to 0.5x **MBS**, or equivalent, for long-term culture.

   MBS (see Appendix for Caution)

PROTOCOL 10.9

Dissection of Tightly Adhering Tissues by Trypsin Treatment

In older embryos (late gastrula and beyond), tissues begin to stick to one another and cannot be peeled apart. For assays requiring isolated tissues, it is necessary to separate such tissues enzymatically, using trypsin. Enzymatically dissected tissue can be used in quantitative gene expression assays or in specification or induction assays.

Enzymatic dissection can be readily accomplished by mild trypsin treatment. Embryos are treated singly, or in small numbers, to avoid possible toxic effects of excessive trypsin digestion. In the example presented here, neural tissue is separated from a neural-plate-stage embryo, but the technique can be adapted to many different tissue types.

All tools used for these manipulations should be kept separate from tools used to handle subsequent tissues: Tissues deteriorate rapidly if exposed to even small amounts of trypsin for long periods of time.

1. Place a dejellied, devitellinized embryo (Protocols 6.1 and 6.2) in a dissection dish containing 1x **MBS**, freshly supplemented with 0.01% trypsin (Sigma). Use a neurula stage 14–16 embryo before the neural tube has closed.

   Note: *The trypsin concentration can be reduced to as little as 0.001% (w/v) if embryos are particularly sensitive.*

   MBS (see Appendix for Caution)

2. Use a fine needle, or eyebrow knife, to cut around the neural plate, just deeply enough to allow the trypsin to penetrate the ectodermal layer.

3. After a few minutes, peel back the neural plate, which should separate easily from the underlying mesoderm, and transfer the tissue to fresh 1x MBS containing 0.02% soybean trypsin inhibitor (Sigma).

4. Incubate for 5 minutes at 15–20°C and then transfer to a clean culture dish containing 1x MBS.

   Note: *The tissue can be maintained in 1x MBS for a few hours. After this, it should be transferred to 0.5x MBS.*

5. When the neural plate has been removed, the notochord will be exposed and, if required, can also be removed.

   Note: *Once removed, transfer the notochord immediately to medium containing trypsin inhibitor, as described in step 4 above.*

PROTOCOL 10.10

Cortical Isolation from Oocytes and Eggs

The cortex is the layer of gelatinous cytoplasm that lies just below the plasma membrane of the egg. Rotation of the cortex relative to the deeper cytoplasm soon after fertilization is intimately linked to normal dorsal axis specification. Vegetally, dorsal determinants and many interesting RNAs are specifically localized in the vegetal cortex (see, e.g., Rebagliati et al. 1985; Elinson and Rowning 1988; Zhang and King 1996). The cortex can be dissected from the egg to analyze its composition and activity or to clone associated RNAs.

Two types of cortices have been isolated from *Xenopus* eggs: the oocyte cortex and the cortex of the fertilized egg. The vegetal cortex of the full-grown oocyte in the ovary comprises a unique cytoskeletal domain, containing a storehouse of localized mRNAs. Isolated cortices can be used to examine these mRNAs, as described by Elinson et al. (1993). The fertilized egg cortex has been isolated to examine the mechanism of the cortical rotation (Houliston and Elinson 1991). Both isolations are carried out in P10EM, a medium designed to retain microtubules. The procedure for isolating the vegetal cortex of the fertilized egg is presented below.

1. Dejelly embryos with **cysteine**, as described in Protocol 6.1, and place them in a dish containing P10EM (100 mM PIPES, 10 mM EGTA, 1 mM **MgSO$_4$**, pH 6.9)

   cysteine, MgSO$_4$ (see Appendix for Caution)

2. Remove the vitelline membranes manually, as described in Protocol 6.2 and orient the embryos vegetal pole up.

3. Use a fine watchmaker's forceps to clip the cortex (and overlying plasma membrane) repeatedly around the egg at a latitude of 60–80° from vegetal pole (see Figure 10.8).

4. Once a complete circle has been cut, grasp the edge of the cortex with both pairs of forceps and gently peel the cortex from the deeper cytoplasm.

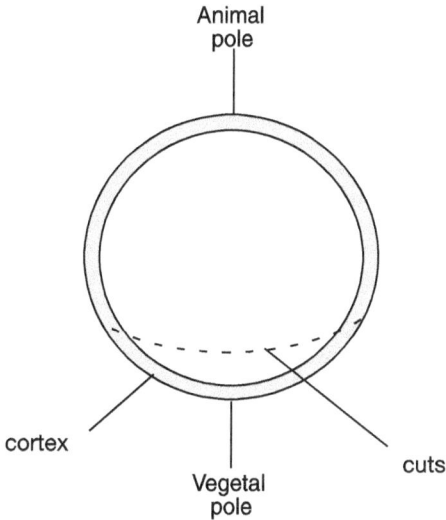

Figure 10.8. Diagram showing incision sites for vegetal cortex isolation from a fertilized egg.

Helpful Hints

• The ease of peeling and the degree of cytoplasmic contamination associated with the cortex depend on the stage of the cell cycle. Cortices that are free from cytoplasmic contamination are almost transparent and are most easily obtained at the time of cortical rotation, i.e., 0.4–0.8 normalized time of the first cell cycle (36–72 minutes postfertilization). The cortices are robust and can be transferred by pipette to other solutions, e.g., 0.5× **MBS**, providing the cortex does not touch the air-water interface. Cortices isolated in this way retain the array of parallel microtubules involved in the cortical rotation.

MBS (see Appendix for Caution)

REFERENCES

Bang A., Papalopulu N., Kintner C., and Goulding M. 1997. Expression of *Pax-3* is initiated in the early neural plate by posteriorizing signals produced by the organizer and by posterior non-axial mesoderm. *Development* **124:** 2075–2085.

Bradley L., Wainstock D., and Sive H. 1996. Positive and negative signals modulate formation of the *Xenopus* cement gland. *Development* **122:** 2739–2750.

Cornell R., Musci T., and Kimelman D. 1995. FGF is a prospective competence factor for early activin-type signals in *Xenopus* mesoderm induction. *Development* **121:** 2429–2437.

Dale L. and Slack J.M.W. 1987. Fate map for the 32-cell stage of *Xenopus laevis*. *Development* **99:** 527–551.

Doniach T., Phillips C.R., and Gerhart J.C. 1992. Planar induction of anteroposterior pattern in the developing central nervous system of *Xenopus laevis*. *Science* **257:** 542–545.

Drysdale T.A. and Elinson R.P. 1993. Inductive events in the pattering of the *Xenopus laevis* hatching and cement glands, two cell types which delimit head boundaries. *Dev. Biol.* **158:** 245–253.

Elinson R.P. and Rowning B. 1988. A transient array of parallel microtubules in frog eggs: Potential tracks for a cytoplasmic rotation that specifies the dorso-ventral axis. *Dev. Biol.* **128:**185–197.

Elinson R.P., King M.L., and Forristall C. 1993. Isolated vegetal cortex from *Xenopus* oocytes selectively retains localized mRNAs. *Dev. Biol.* **160:** 554–562.

Grainger R.M. and Gurdon J.B. 1989. Loss of competence in amphibian induction can take place in single nondividing cells. *Proc. Natl. Acad. Sci.* **86:** 1900–1904.

Green J.B.A., New H.V., and Smith J.C. 1992. Reponses of embryonic *Xenopus* cells to activin and FGF are separated by multiple dose thresholds and correspond to distinct axes of the mesoderm. *Cell* **71:** 731–739.

Grunz H., Multier-Lajous A.-M., Herbst R., and Arkenberg G. 1975. The differentiation of isolated amphibian ectoderm with or without treatment with an inductor. *Wilhelm Roux's Arch.* **178:** 277–284.

Heasman J., Ginsberg D., Geiger B., Goldstone K., Pratt T., Yoshida-Noro C., and Wylie C. 1994. A functional test for maternally inherited cadherin in *Xenopus* shows its importance in cell adhesion at the blastula stage. *Development* **120:** 49–57.

Henry G., Brivanlou I., Kessler D., Hemmati-Brivanlou A., and Melton D. 1996. TGF-β signals and a prepattern in *Xenopus laevis* endodermal development. *Development* **122:**1007-1015.

Houliston E. and Elinson R.P. 1991. Patterns of microtubule polymerization relating to cortical rotation in *Xenopus laevis* eggs. *Development* **112:** 107–117.

Keller R. 1975. Vital dye mapping of the gastrula and neurula of *Xenopus laevis* I. Prospective areas and morphogenetic movements of the superficial layer. *Dev. Biol.* **42:** 222–241.

———1976. Vital dye mapping of the gastrula and neurula of *Xenopus laevis* II. Prospective areas and morphogenetic movements of the deep layer. *Dev. Biol.* **51:** 118–137.

Keller R. and Danilchik M. 1988. Regional expression, pattern and timing of convergence and extension during gastrulation of *Xenopus laevis*. *Development* **103:** 193–209.

Lamb T.M., Knecht A.K., Smith W.C., Stachel S.E., Economides A.N., Stahl N., Yancopolous G.D., and Harland R.M. 1993. Neural induction by the secreted polypeptide noggin. *Science* **262:** 713–718.

Moody S.A. 1987. Fates of the blastomeres of the 16-cell stage *Xenopus* embryo. *Dev. Biol.* **119:** 560–578.

———1987. Fates of the blastomeres of the 32-cell *Xenopus* embryos. *Dev. Biol.* **122:** 300–319.

Nieuwkoop P.D. 1952a. Activation and organization of the central nervous system in amphibians. I. Induction and activation. *J. Exp. Zool.* **120:** 1–32.

———1952b. Activation and organization of the central nervous system in amphibians. II. Differentiation and organization. *J. Exp. Zool.* **120:** 33–81.

Nieuwkoop P.D. and Faber J. 1994. Normal tables of *Xenopus laevis* (Daudin), 3rd Edition. Garland Publishing, Inc., New York.

Rebagliati M., Weeks D., Harvey R., and Melton D. 1985. Identification and cloning of localized maternal RNAs from *Xenopus eggs*. *Cell* **42:** 769–777.

Ruiz i Altaba A. and Melton D.A. 1989. Interaction between peptide growth factors and homeobox genes in the establishment of antero-posterior polarity in frog embryos. *Nature* **341:** 33–38.

Servetnick M. and Grainger R.M. 1991. Changes in neural and lens competence in *Xenopus* ectoder: Evidence for an autonomous developmental timer. *Development* **112:** 177–188.

Sharpe C., Fritz A., DeRobertis E., and Gurdon J. 1987. A homeobox-containing marker of posterior neural differentiation shows the importance of predetermination in neural induction. *Cell* **50:** 749–758.

Sive H., Hattori K., and Weintraub H. 1989. Progressive determination during formation of the anteroposterior axis in *Xenopus laevis*. *Cell* **58:** 171–180.

Sive H.L. and Cheng P.F. 1991. Retinoic acid perturbs the expression of *Xhox.lab* genes and alters mesodermal determination in *Xenopus laevis*. *Genes Dev.* **5:** 1321–1332.

Spellmann H. 1938. Embryonic development and induction. Yale University Press, New Haven.

Wilson P.A. and Hemmati-Brivanlou A. 1995. Induction of epidermis and inhibition of neural fate by Bmp-4. *Nature* **376:** 331–333.

Zhang J. and King M. 1996. *Xenopus Veg1* RNA is localized to the vegetal cortex during oogenesis and encodes a novel T-box transcription factor involved in mesodermal patterning. *Development* **122:** 411.

Zoltewicz J. and Gerhart J. 1997. The Spemann organizer of *Xenopus* is patterned along its anteroposterior axis at the earliest gastrula stage. *Dev. Biol.* **192:** 482–491.

The video series "Manipulating the Early Embryo of Xenopus laevis*" presents demonstrations of all of the dissections described in this chapter. The tapes illustrate animal cap isolation, ectodermal layer separations, dissociation and reaggregation of animal caps, animal cap/vegetal conjugates, animal cap/dorsal mesoderm conjugates, transplants, explants, dissection of tightly adhering tissues by trypsin treatment, and cortical isolation from eggs.*

CHAPTER 11

Transgenesis in *Xenopus* Embryos*

Genetic manipulation of *Xenopus* embryos can be accomplished in a number of ways, as discussed in Chapter 3. However, it is only recently that an effective way of introducing transgenes into *Xenopus* has been developed (Kroll and Amaya 1996). This technology has revolutionized gene expression studies in *Xenopus* and can be used in many applications, for example.

- to misexpress genes during development with much better spatial and temporal control than was previously possible

- to label specific structures in vivo, using the green fluorescent protein as a marker

- to study the regulation of gene promoters from many organisms (e.g., other amphibia, zebrafish, pufferfish, and mice)

- to study the molecular basis of later developmental events such as organogenesis

- to generate mutations in genes using gene trap approaches

Briefly, the protocol involves (1) incubating sperm nuclei with linearized plasmid DNA, (2) incubating the nuclei/DNA reaction with a high-speed *Xenopus* interphase egg extract and a small amount of restriction enzyme (the extract partially decondenses the sperm chromatin but does not promote replication, and the restriction enzyme stimulates recombination and integration by creating double-stranded breaks in the sperm chromatin), and (3) diluting the reaction mix and injecting it into unfertilized eggs to produce transgenic embryos.

This technique permits large-scale transgenesis in *Xenopus* embryos. Unlike embryos injected with plasmids, the transgenic embryos show correct spatial and temporal regulation of integrated promoter constructs. One of the great advantages of this system over transgenesis in mice or zebrafish is that the transgene can be integrated into the male genome prior to fertil-

* Contributions to this chapter were made by Drs. Kristen Kroll and Enrique Amaya, the developers of the *Xenopus* transgenic technology described herein.

ization; therefore, the resulting embryos are not chimeric and breeding of animals is not required. In other words, transgenic embryos can be generated one day and analyzed the next.

EFFICIENCY AND UTILITY OF THE TRANSGENESIS PROCEDURE

The transgenesis procedure described here can be used to generate up to hundreds of normal transgenic embryos per day. After transplantation with swelled sperm nuclei, 20–30% of the eggs cleave and develop normally. In a typical experiment, a single worker can transplant approximately 500 sperm nuclei per hour to produce several hundred to a thousand normally cleaving embryos. As with embryos produced by in vitro fertilization, the frequency of normal advanced development varies somewhat depending on the overall quality of the eggs; typically, 5–40% of the cleaving eggs develop normally beyond feeding tadpole stages. The fraction of embryos carrying a transgene depends on the amount of restriction enzyme used in the restriction-enzyme-mediated integration (REMI) procedure and the concentration of DNA added to sperm nuclei (see below), but commonly, 60–80% of normal tadpoles will be transgenic.

Transgenes introduced into sperm nuclei appear to be integrated into the genome, and some embryos do seem to escape significant chromosomal damage. Reports in both *Xenopus laevis* (B. Knox, unpubl.) and *Xenopus tropicalis* (M. Offield, E. Amaya, and R. Grainger, unpubl.) show that transgenic animals can survive to sexual maturity and transmit transgenes through the germ line and that offspring express the transgene appropriately. The Mendelian segregation pattern of transgenes seen during germ-line transmission (M. Offield and R. Grainger, unpubl.) strongly suggests that there is integration of foreign DNA in the genome. This conclusion is supported by Southern blot analysis of DNA from transgenic animals, which indicates that transgenes are linked to host DNA sequences (Kroll and Amaya 1996) and from identification of host DNA sequences adjacent to transgenic DNA (O. Bronchain and E. Amaya, unpubl.). Transgenic DNA is integrated into the genome as a single copy in some instances and as short concatamers (two to six copies) in other cases (Kroll and Amaya 1996).

Embryos derived from sperm nuclear transplantation express plasmids nonmosaically at high frequency from gastrula through tadpole stages, and promoters introduced into the embryo in this way appear to direct expression consistent with endogenous promoters. Promoters from ubiquitously expressed genes activate reporter gene expression in every cell of transgenic embryos, starting at the late blastula or early gastrula stages (Kroll and Amaya 1996; Pownall et al. 1998; Kroll et al. 1998; Huang et al. 1999). A number of laboratories have produced transgenic embryos that express genes from spatially restricted promoters (Kroll and Amaya 1996; Latinkic et al. 1997; Knox et al. 1998; and many others whose results are as yet unpublished). For example, this method has been used to produce transgenic embryos with plasmids containing a muscle-specific actin promoter (Mohun et al. 1986) linked to chloramphenicol acetyltransferase (pRLCAR) or green fluorescent protein (pCARGFP) (Kroll and Amaya 1996). Approximately 40–60% of tadpoles derived from sperm nuclear transplantations with these plasmids show stable, nonmosaic expression. Expression is restricted to the somites and heart tissue, as expected for this regionally restricted promoter (Kroll and Amaya 1996). Since the transgenes are expressed in all expected cells of the transgenic embryos, it is likely that integration occurs before the first cleavage division, thus ensuring that all cells of the embryo inherit copies of the plasmid.

PREPARATIONS FOR THE TRANSGENESIS PROCEDURE

Before starting the nuclear transplantation procedure, it is necessary to set up an apparatus for the injections, prepare a *Xenopus* egg extract, make sperm nuclei, and prepare the DNA to be used for transgenesis. Preparing an egg extract is the most complex step in the process, but it need only be done occasionally, since very small amounts of extract are used in each experiment and it can be stored frozen for long periods of time. It takes just a few hours to prepare sperm nuclei (a procedure based on the method of Murray 1991), and they can be stored at 4°C and used for up to 3 days. After 3 days, the nuclei appear to deteriorate and the fraction of transgenic animals generated from them is diminished. Sperm nuclei can be frozen and stored for longer periods, although their efficacy is somewhat less than that of fresh nuclei.

PREPARATION OF NUCLEAR TRANSPLANTATION NEEDLES AND INJECTION APPARATUS

Nuclear transplantation needles are unlike standard needles used for DNA and RNA injection in that they have a long sloping taper from the wide part of the needle to the tip. This is needed to control the flow rate through the very large needle tip (60–80 µm in diameter). Generally, the length of this tapered region in transplantation needles is about 15 mm, versus 3–5 mm for most DNA or RNA injection needles. Good needles can be produced from glass capillaries that are approximately 900 μm (outside diameter), 700 μm (inside diameter), and 78 mm (length), e.g., 30-μl Drummond micropipets (Fisher 21-170J). Glass capillaries from Medical Systems (WGPL1CS, 1.0 mm outside diameter/0.75 mm inside diameter) also work well. These capillaries produce needles with thinner walls that may reduce shear forces on sperm nuclei.

Needles can be made using a variety of commercial pullers (e.g., PN30, Narishige; P-87, Sutter Instruments). Although the setting for each puller varies, a high-heat setting and a low pulling force and velocity are generally desirable. The higher temperature will melt a greater amount of glass so that a slow pull will draw a long, tapered needle. For the Narishige horizontal model (PN30), approximate settings are heat = 55; main magnet = 25; submagnet = 10. The dimensions of a good transplantation needle (using a single pull) are shown in Figure 11.1A. The needle can be clipped with a pair of forceps to produce a beveled tip of 60–80 μm diameter (using the ocular micrometer of a dissecting microscope or a stage micrometer for measurement). The dimension of the tip is important: If it is too narrow, nuclei passing through the needle will be damaged, and if it is too wide, the recipient egg will be damaged. To clip needle tips that have the desired bevel, it often helps to use forceps with slightly unmatched tips, and to pull outward at a 20–30° angle from the needle as the forceps contact the needle.

Alternatively, needles can be made by doing a first pull by hand, in a Bunsen burner flame, and a second pull using a gravity-driven puller. This is accomplished by heating a micropipette (1 mm wide, 80 mm long) in a Bunsen burner flame and pulling it by hand to make the bore of the needle (200–400 μm wide). The drawn pipette should be 10–15 cm in length and the pipette should remain fairly straight when held by one end. To produce a gently sloping needle tip, the upper end of the needle is fixed in the brace of a gravity-driven needle puller. The center of the needle bore of the drawn pipette is placed within a small heating coil and a weight is attached to the lower end of the pipette. The gravity-driven pullers can be custom made (e.g., those used by Kroll and Amaya 1996; Amaya and Kroll 1999), but similar vertical pullers are commercially available from Narishige (e.g., model PB-7). The dimensions of these needles are shown in Figure 11.1B.

Although not essential, treating the inside of needles with Sigmacote (Sigma SL-2) may prevent shearing of sperm nuclei as they pass through the needle. The easiest way to treat the needles is to attach approximately 1-cm Tygon tubing (R-3603 1/32 inch; Fisher 14-169-1A) to the end of a plastic micropipettor tip (200 μl), use the micropipettor to draw up the Sigmacote, and then attach the other end of the tubing to the injection needle (see Figure

A. Needle from single capillary pull:

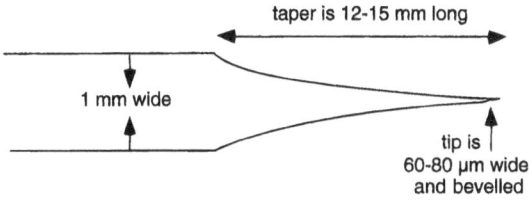

taper is 12-15 mm long

1 mm wide

tip is
60-80 μm wide
and bevelled

B. | Needle from double capillary pull:

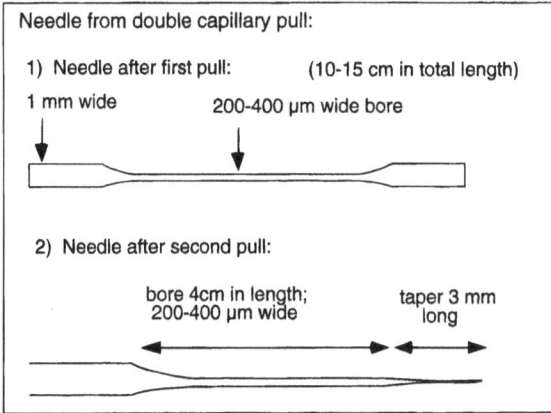

1) Needle after first pull: (10-15 cm in total length)

1 mm wide   200-400 μm wide bore

2) Needle after second pull:

bore 4cm in length;   taper 3 mm
200-400 μm wide   long

C. Backloading needles (see text for explanation):

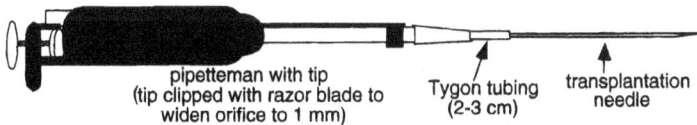

pipetteman with tip   Tygon tubing   transplantation
(tip clipped with razor blade to   (2-3 cm)   needle
widen orifice to 1 mm)

Figure 11.1. Dimensions of nuclear transplantation needles. (*A*) Needle produced using a commercial apparatus and single pull; (*B*) needle produced by a double capillary pull, with the first pull performed by hand over the flame of a Bunsen burner. (Redrawn, with permission, from Amaya and Kroll 1999.) (*C*) Pipettor with tubing attachment, used for backloading needles.

11.1C). Depress the micropipettor plunger to force the Sigmacote through the needle until a few drops emerge from the tip, and then release the plunger of the micropipettor to withdraw most of the solution from the needle. Immediately after coating, rinse the needle with at least 200 μl of water. Make sure that all of the liquid has been removed from the tip because any remaining Sigmacote will block the flow of nuclei into the needle. Needles can be treated several months in advance or as little as 10 minutes before use.

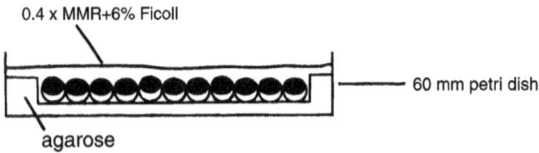

0.4 x MMR+6% Ficoll

60 mm petri dish

agarose

Figure 11.2. Side view of injection dish filled with unfertilized eggs. A depression within an agarose base in a 60-mm petri dish holds eggs ready for injection.

PREPARATION OF AGAROSE-COATED INJECTION DISHES

Injections are most conveniently performed in an agarose-coated petri dish in which a depression has been made to hold the embryos (Figure 11.2). Pour molten agarose (2.5% [w/v] in 0.1x MMR) into several 60-mm petri dishes. Before the agarose solidifies, place small weigh boats on the surface so that as it solidifies, a square depression will be formed. The depression should accommodate approximately 500 eggs (see Figure 11.2). After the agarose has solidified, remove the weigh boats and pour sterile 1x modified Mark's reagent (MMR) into each dish to prevent dehydration. Wrap the dishes in Parafilm and store at 4°C until use. Under these conditions, the injection dishes can be stored for several weeks.

TRANSPLANTATION APPARATUS

Most commercial injection apparatuses used for DNA and RNA injections are unsuitable for nuclear transplantations because of the use of much smaller needle tips. The flow through a 10-μm needle tip used for fluid injection can be controlled even at fairly high pressures, but it is usually not possible to obtain the extremely low positive pressure, or the gentle controlled flow, required to deliver an intact nucleus in a small volume (10–15 nl) through a 60–80-μm needle tip.

Two systems are recommended for performing nuclear injections. One uses an infusion syringe pump (e.g., Cole-Parmer, H-74900, Harvard Apparatus Inc., 55-1111, 55-2222, or PHD-2000) filled with oil, and the other uses an aquarium pump with an adjustable dial.

Infusion pumps have the advantage that they can be set at any desirable flow rate (e.g., 10 nl/second), and they can provide a very accurate, reproducible way of injecting nuclei. A typical infusion pump system (e.g., the Harvard 11 syringe pump; see Figure 11.3) is fitted with a Hamilton gastight 2.5-ml syringe (Fisher 13-684-95) which has a Luer-lok fitting for easy connection. Luer-lok-adapted tubing (Fisher KH90669-0005), made of

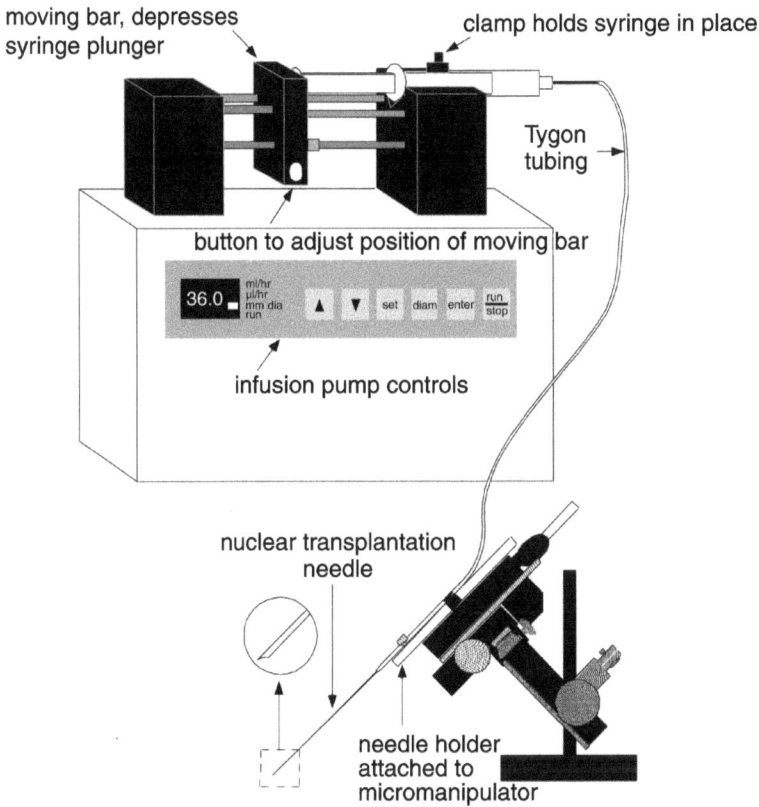

Figure 11.3. Infusion pump set up for nuclear transplantation.

PTFE plastic with Luer-lok connections at each end, is used to connect the Hamilton syringe to a microelectrode holder (model MPH3; WPI, Sarasota, Florida). The tubing and syringe are filled with embryo-tested mineral oil (Sigma M 8410). For the infusion pump set up to work well, all air bubbles must be expelled from the system. The handle of the microelectrode holder is fitted into a micromanipulator, which is mounted on a solid base (e.g., a magnetic base; www.harborfreight.com) adjacent to the stereoscope. Also workable are the Drummond oil-filled injection systems, which are based on a positive displacement mechanism and are therefore not affected by the size of the needle tip. However, they are more difficult to use for nuclear transplantations and are not recommended for this purpose.

The second system for nuclear transplantations, the use of an aquarium pump with an adjustable dial, is a very inexpensive and simple alternative to the infusion pump (see Figure 11.4). It is easy to set up and is capable of

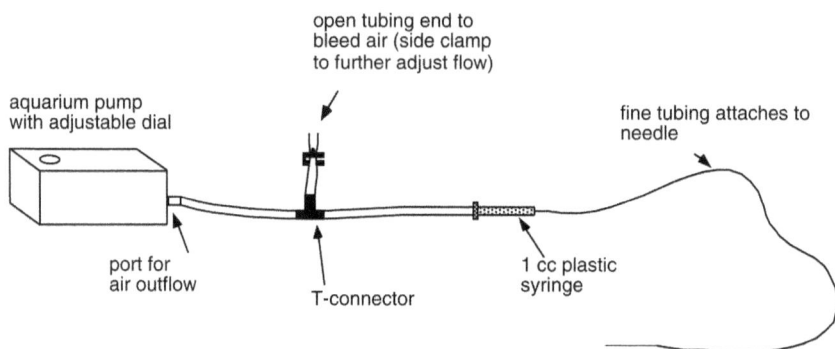

Figure 11.4. Aquarium pump set up for nuclear transplantation. A standard aquarium pump is connected to a l-cc plastic syringe barrel via a length of 3/16-inch bore Tygon tubing. The syringe barrel is connected to a microinjection needle via a 17-gauge needle and a length of capillary Tygon tubing (1/32-inch bore). A side-clamped bleed valve is attached to the wide bore tubing to allow fine adjustment of the flow rate.

delivering a constant, steady flow of fluid through a wide-bore needle tip. The disadvantage of this system is that although the flow rate can be varied, it is not possible to set a specific flow rate. The flow rate can be calibrated by loading the needle with a set volume of fluid and determining how long the needle takes to empty at a particular setting. A pump that delivers an air pressure of about 4 psi and has an adjustable dial to control the flow rate will work well. The aquarium pump set up is shown in Figure 11.4. The adjustable pump is connected to a syringe and hypodermic needle via a series of plastic tubes (the tip of the hypodermic needle has been clipped blunt). A bleed valve is incorporated between the pump and the syringe to allow further adjustment to the rate of flow. The injection needle is attached to the hypodermic needle by a length of fine Tygon tubing (Fisher 14-169-1A) and a low positive pressure is maintained on the fluid in the needle. Although the aquarium pump system is inexpensive and generally gives good results, the oil infusion pump is superior for delivering consistent injection volumes. The infusion pump is recommended for extensive transgenesis work.

PROTOCOL 11.1

High-Speed *Xenopus* Egg Extract Preparation

The procedure for making *Xenopus* egg extract, required for swelling sperm nuclei prior to REMI of DNA, is adapted from Murray (1991). Briefly, a crude cytostatic factor (CSF)-arrested egg extract (cytoplasm arrested in meiotic metaphase) is prepared and calcium is added to allow the extract to progress to interphase. A high-speed centrifugation step is performed to purify the cytoplasmic fraction. Cytochalasin is not included in the protocol, because carryover of cytochalasin into the final extract used for sperm incubations would interfere with normal development of transplant embryos. Using high-speed extracts, rather than crude cytoplasmic extracts, is advantageous because they promote swelling of added sperm nuclei (and some chromatin decondensation) but do not promote DNA replication. DNA replication occurs after the nucleus has been transplanted into the egg. High-speed extract can be stored for several years at –80°C in 25-μl aliquots, which must be thawed just before use. The procedure for preparing extracts takes about 3 hours. An overview of the steps required for this preparation is shown in Figure 11.5.

SOLUTIONS FOR PREPARATIONS OF HIGH-SPEED EGG EXTRACTS

20x Extract Buffer (XB) Salt Stock
 2 M KCl
 20 mM $MgCl_2$
 2 mM $CaCl_2$
 Filter sterilize and store at 4°C.

1x XB Salts (Diluted from 20x XB Salt Stock)
 100 mM KCl
 0.1 mM $CaCl_2$
 1 mM $MgCl_2$

 KCl, $MgCl_2$, $CaCl_2$ (see Appendix for Caution)

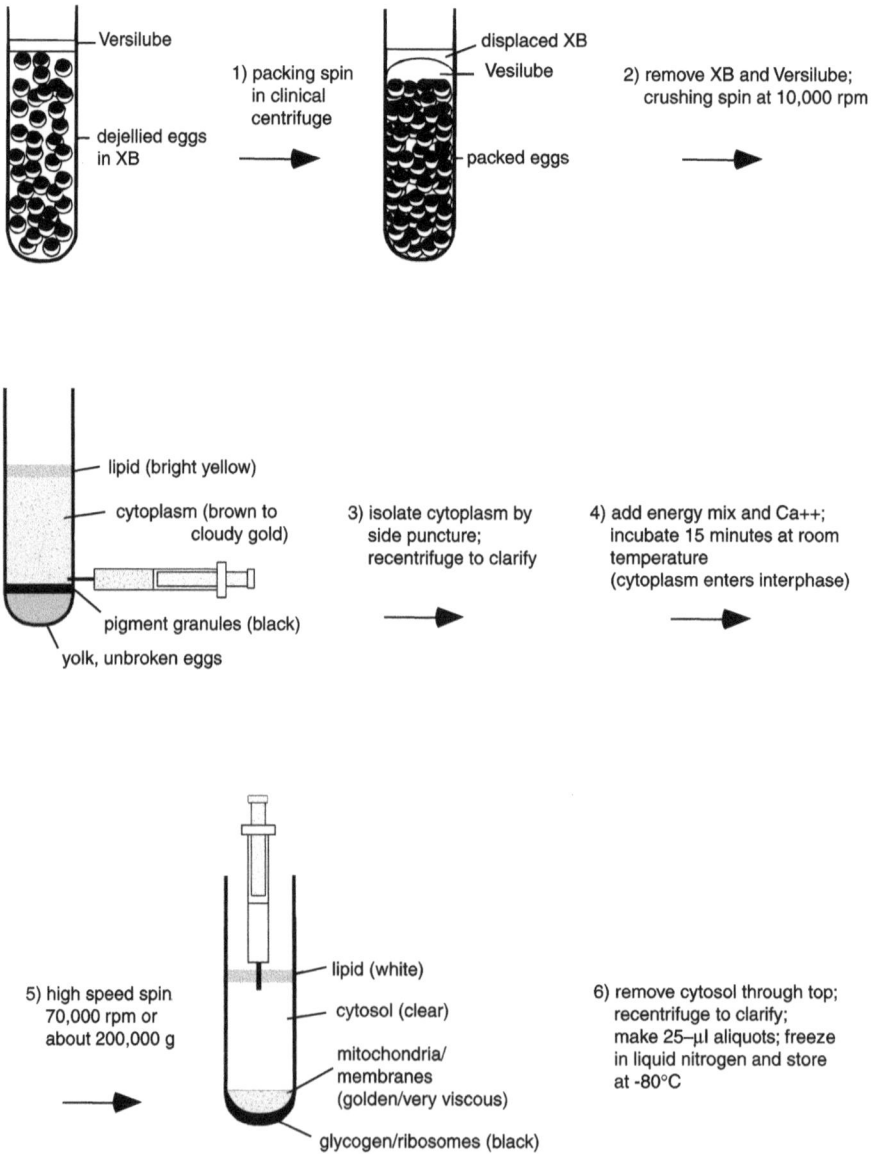

Versilube

1) packing spin in clinical centrifuge

dejellied eggs in XB

displaced XB
Vesilube

2) remove XB and Versilube; crushing spin at 10,000 rpm

packed eggs

lipid (bright yellow)

cytoplasm (brown to cloudy gold)

3) isolate cytoplasm by side puncture; recentrifuge to clarify

4) add energy mix and Ca++; incubate 15 minutes at room temperature (cytoplasm enters interphase)

pigment granules (black)

yolk, unbroken eggs

5) high speed spin 70,000 rpm or about 200,000 g

lipid (white)

cytosol (clear)

mitochondria/ membranes (golden/very viscous)

glycogen/ribosomes (black)

6) remove cytosol through top; recentrifuge to clarify; make 25–µl aliquots; freeze in liquid nitrogen and store at -80°C

Figure 11.5. Preparation of high-speed *Xenopus* egg extract.

1.5 M Sucrose

Filter sterilize and store in aliquots at –20°C.

Extract Buffer (XB)

1x XB salts (diluted from 20x XB salt stock)
50 mM sucrose (diluted from 1.5 M sucrose stock)
10 mM HEPES*

Prepare approximately 400 ml of extract buffer.

*From 1 M stock, titrated with **KOH** such that pH is 7.7 when diluted to 10 mM; should require about 5.5 ml of 10 N KOH for 100 ml; filter sterilize and store in aliquots at –20°C.

KOH (see Appendix for Caution)

2% (w/v) Cysteine (Sigma C 7755) in 1x XB Salts

Prepare within 1 hour of use and titrate to pH 7.8 with **NaOH**. Prepare approximately 600 ml of 2% cysteine.

cysteine, NaOH (see Appendix for Caution)

CSF-XB

1x XB salts (diluted from 20x XB salt stock)
1 mM **MgCl₂***
10 mM potassium HEPES (pH 7.7)
50 mM sucrose
5 mM EGTA (pH 7.7)

Prepare 100 ml of CSF-XB.

*In addition to MgCl₂ present in XB salts; final concentration 2 mM.

MgCl₂ (see Appendix for Caution)

Protease Inhibitors

Prepare a mixture of **leupeptin** (Boehringer Mannheim 1017 101), chymostatin (Boehringer Mannheim 1 004 638), and **pepstatin** (Boehringer Mannheim 600 160), each dissolved to a final concentration of 10 mg/ml (1000x stock solution) in **DMSO**. Store in small aliquots at –20°C.

leupeptin, pepstatin, DMSO (see Appendix for Caution)

1× MMR

> 100 mM NaCl
> 2 mM **KCl**
> 1 mM **MgCl$_2$**
> 2 mM **CaCl$_2$**
> 5 mM HEPES (pH 7.5)

> Prepare a 10× stock and adjust pH to 7.5 with **NaOH**. Sterilize 10× and 1× solutions by autoclaving.

> **KCL, MgCl$_2$, CaCl$_2$, NaOH** (see Appendix for Caution)

1 M CaCl$_2$

> Filter sterilize and store at 4°C.

> **CaCl$_2$** (see Appendix for Caution)

Versilube F-50

> The Versilube name is trademarked by General Electric and is made by them. It can be purchased from Andpak-EMA (1560 Dobbin Drive, San Jose, California 95133; Tel. 408-272-8007). Andpak-EMA refers to Versilube as MIL-S-81087C (F-50). This compound can be omitted from the extract preparation if not available; just draw off as much XB as possible from the buffer layer overlying the eggs after carrying out the packing spin in the clinical centrifuge.

Energy Mix

> 150 mM creatine phosphate (Boehringer Mannheim 127 574)
> 20 mM ATP (Boehringer Mannheim 519 979)
> 20 mM **MgCl$_2$**

> Store in 0.1-ml aliquots at –20°C.

> **MgCl$_2$** (see Appendix for Caution)

Pregnant Mare Serum Gonadotropin (PMSG)

> 100 units/ml PMSG (Calbiochem 367222), prepared in water and stored at –20°C.

Human Chorionic Gonadotropin (hCG)

> 1000 units/ml hCG (Sigma CG-10 from human pregnancy urine), prepared in water and stored at 4°C.

1. To induce ovulation, prime 8–12 female adult *Xenopus* frogs by injecting 50 units of PMSG into the dorsal lymph sac (for instructions on performing this kind of injection, see Protocol 5.2). Maintain the frogs at room temperature for 24 hours.

2. The following evening (i.e., the day before the extract preparation begins), inject each frog with 500–800 units of hCG. Separate the frogs into pairs and maintain in containers with 2 liters of 1x **MMR.**

   Note: *Since one frog with lysing or activating eggs can compromise the whole extract preparation, it is preferable to separate the frogs into pairs for the ovulation.*

   MMR (see Appendix for Caution)

3. Place the frogs at 15–18°C overnight (12–14 hours).

4. The next morning, screen the egg quality in each container and pool healthy batches in 1x MMR. Discard any batch that contains mottled, lysing, or dying cells.

   Notes: *Additional eggs can often be obtained by gently expelling them manually from each frog into a large dish of 1x MMR (see Chapter 5, Manual Egg Collection). Discard any damaged eggs or eggs with uneven pigmentation.*

   *It is **extremely** important to prepare all solutions before beginning the next section of the protocol. Carry the procedure promptly through all steps once it is initiated. Optimally, begin the high-speed spin within 45–60 minutes of dejellying the eggs.*

5. Remove as much 1x MMR as possible from the eggs and dejelly in 1x XB Salts, containing 2% **cysteine.**

   Note: *600 ml of cysteine solution should be ample, even for large volumes of eggs.*

   cysteine (see Appendix for Caution)

6. Add a small amount of cysteine and swirl the eggs. Partially replace with fresh cysteine several times during the dejellying process.

   Note: *Remove any broken eggs with a pipette. Dejelly each batch of eggs separately and discard any batches that show breakage or egg activation. Egg volume after dejellying will be approximately 20–50 ml, depending on how many eggs were laid and how many batches of eggs were discarded.*

7. Wash eggs four times in 100 ml of Extract Buffer and then twice in 50 ml of CSF-XB containing 1x protease inhibitors.

   Note: *Cover the eggs well with solution during each wash.*

8. Use a wide-bore pasteur pipette to transfer eggs to a 14 x 95-mm Beckman ultraclear tube (Beckman 344060) or equivalent. Alternatively, pour eggs from the beaker used for dejellying and washing into the tubes.

   Note: *These tubes hold about 10 ml of solution. If using multiple tubes, try to transfer an equal volume of eggs to each tube. Uneven loading at this stage can make balancing the tubes very difficult later on.*

9. Remove as much CSF-XB as possible and replace with approximately 1 ml of Versilube F-50.

10. Pack the eggs by centrifuging in a clinical centrifuge at room temperature for about 60 seconds at 1000 rpm (150g) and then for an additional 30 seconds at 2000 rpm (600g).

    Note: *Eggs should be packed after this spin but not broken. Versilube is heavier than CSF-XB but lighter than the eggs. After the spin, an inverted meniscus should be clearly visible between the Versilube and displaced CSF-XB.*

11. Remove the excess CSF-XB and Versilube and balance the tubes.

    Note: *At this point, it is not easy to balance the tubes. It is not advisable to add buffer, and it is difficult to remove packed eggs without damaging them.*

12. Centrifuge at 10,000 rpm for 10 minutes at 2°C using rubber adapters in a Sorvall HB-4 swinging bucket rotor (or equivalent).

    Note: *This treatment should crush the eggs, which will then separate into three distinct layers: lipid (top), cytoplasm (center), and yolk (bottom). See Figure 11.5.*

13. Insert an 18-gauge needle through the wall of the centrifuge tube at the base of the cytoplasmic layer and slowly draw out the cytoplasm.

14. Transfer the cytoplasm to a fresh polypropylene tube on ice.

    Note: *If large volumes of darkly pigmented eggs are used, the cytoplasmic layer may be grayish rather than golden at this step. After a second spin to clarify this extract, it should be golden.*

15. Estimate the volume of extracts and add a 1:1000 dilution of protease inhibitor stock solution to the isolated cytoplasm. Do *not* add cytochalasin.

16. Recentrifuge the cytoplasm in polypropylene tubes for an additional 10 minutes at 10,000 rpm to clarify, again using a swinging bucket rotor. Collect the clarified cytoplasm as in step 13.

Note: *Expect to obtain about 0.75–1 ml of cytoplasm per batch of eggs collected from one frog.*

17. Measure the exact volume of the cytoplasm obtained and add 0.05 volume Energy Mix. Transfer the mixture to a 3-ml thick-walled polycarbonate ultracentrifuge tube (e.g., for use with Beckman TL100.3 rotor).

    Note: *Each tube should be at least half full.*

18. Add **CaCl$_2$** to a final concentration of 0.4 mM.

    Note: *This inactivates CSF and pushes the egg extract into interphase.*

    CaCl$_2$ (see Appendix for Caution)

19. Incubate for 15 minutes at room temperature and then, after balancing tubes, centrifuge at approximately 200,000g for 1.5 hours at 4°C (e.g., 70,000 rpm in a Beckman TL100.3 rotor in a TL-100 tabletop ultracentrifuge).

    Note: *The cytoplasm will fractionate into four layers, top to bottom: lipid, cytosol, membranes/mitochondria, and glycogen/ribosomes.*

20. Insert a 17- or 18-gauge needle into the top of the tube, through the lipid layer, and remove the cytosolic layer from each tube (~30–50% total volume). Transfer this fraction to fresh ultracentrifuge tubes and repeat centrifugation at 200,000g for 20 minutes at 4°C.

21. Transfer the final supernatant to 0.5-ml microcentrifuge tubes in 25-μl aliquots. Quick-freeze aliquots in liquid nitrogen and store at –80°C.

    Note: *This protocol typically yields 1–2 ml of extract, the quality of which should be tested as described in the following section. If the interphase extract is active, sperm nuclei should swell visibly (thicken and lengthen) within 10 minutes of being added to extract at room temperature.*

PROTOCOL 11.2

Preparation of Sperm Nuclei

SOLUTIONS FOR PREPARATION OF
SPERM NUCLEI

Bovine Serum Albumin (BSA)
10% (w/v) BSA (Fraction V, Sigma A 7906) Prepare stock in water and titrate to pH 7.6 with **KOH**. Store in 1-ml aliquots at –20°C.

KOH (see Appendix for Caution)

Protease Inhibitors
Leupeptin (Boehringer Mannheim 1 017 101; 10 mg/ml stock in **DMSO**). Store aliquots at –20°C.

Phenylmethylsulfonyl fluoride (PMSF) (Boehringer Mannheim 837 091; 0.3 M stock in ethanol). Store aliquots at –20°C.

leupeptin, DMSO, PMSF (see Appendix for Caution)

1x MMR
100 mM NaCl
2 mM **KCl**
1 mM **MgCl$_2$**
2 mM **CaCl$_2$**
5 mM HEPES (pH 7.5)

Prepare a 10x stock and adjust pH to 7.5 with **NaOH**. Sterilize 10x and 1x solutions by autoclaving.

MMR, KCl, MgCl$_2$, CaCl$_2$, NaOH (see Appendix for Caution)

Hoechst Dye No. 33342 (Sigma B 2261)
10 mg/ml stock in distilled water. Store away from light at –20°C.

Lysolecithin
100 μl of 10 mg/ml L-α-lysophosphatidylcholine (Sigma Type I, L 4129); dissolve in sterile water at room temperature **immediately** before use. Store solid stock at –20°C. Discard the stock powder if it becomes sticky.

2× Nuclear Preparation Buffer (NPB)

Prepare the *stock solutions* listed below, aliquot, and store at −20°C. On the day of preparation, use frozen stocks to prepare 30 ml of this 2× solution.

Volume of stock required (ml)	Stock solution	Comments	Final concentration in 1× NPB solution
10	1.5 M sucrose	filter sterilize	250 mM
0.9	1 M HEPES	pH with **KOH** so that the pH is 7.7 at 15 mM; filter sterilize [a]	15 mM
0.120	0.5 M EDTA (pH 8.0)	sterilize; store at room temperature	1 mM
3.0	10 mM spermidine	Sigma S 2501; filter sterilize	0.5 mM
1.2	10 mM spermine	Sigma S 1141; filter sterilize	0.2 mM
0.608	100 mM **dithiothreitol**	Sigma D 0632; filter sterilize	1 mM

[a] Dilution changes the pH of HEPES, making it impossible to pH the stock directly.

KOH, dithiothreitol (see Appendix for Caution)

Sperm Nuclei Dilution Buffer (SDB)

Store in 0.5-ml aliquots at −20°C. Use stock solutions described above to make SDB. To make 20 ml, use the amounts below and add about 20 µl of 0.1 N NaOH to adjust pH to 7.3–7.5.

Volume of Stock required (ml)	Stock solution	Final concentration in 1× SDB solution
3.34	1.5 M sucrose	250 mM
1.5	1 M **KCl**	75 mM
1.0	10 mM spermidine	0.5 mM
0.4	10 mM spermine	0.2 mM

KCl (see Appendix for Caution)

1. On the day of sperm nuclei preparation, make up 30 ml of 2× NPB solution (from frozen stocks, listed above) and dilute to produce:

   - 40 ml of 1× NPB (used for testis washes and centrifugation)

- 10 ml of 1x NPB containing 3% (w/v) BSA and 1:1000 dilutions of leupeptin and PMSF stocks (see proteinase inhibitors above)
- 5 ml of 1x NPB containing 0.3% (w/v) BSA
- 1 ml of storage buffer
    500 µl 2x NPB
    300 µl glycerol
    170 µl water
    30 µl 10% BSA

Place these solutions on ice.

2. Prepare a fresh stock of lysolecithin (10 mg/ml) and place approximately 40 ml of 1x **MMR** on ice for washes.

   MMR (see Appendix for Caution)

3. Isolate testes from an adult male frog as described in Protocol 5.3. Place testes in a 60-mm tissue culture dish containing cold 1x MMR.

4. Rinse testes three times in ice cold 1x MMR and twice in ice cold 1x NPB. Remove any pieces of fat body or debris with fine forceps.

   Note: *Take care not to release the sperm by puncturing the tissue pouches.*

5. Transfer testes to a dry 35-mm tissue culture dish and macerate well with a clean pair of forceps (until clumps are no longer visible to the naked eye).

   Note: *The thoroughness with which this step is done is a major factor in determining the final yield of nuclei.*

6. Gently resuspend the macerate in 2 ml of 1x NPB by pipetting the solution up and down through a fire-polished, truncated pasteur pipette with an opening of approximately 3 mm in diameter.

7. Filter the macerate through two to four layers of cheesecloth or Nitex mesh (HC-3-110) into a 15-ml round-bottom polypropylene tube (e.g., Fisher 2059).

8. Collect any residual macerate from the forceps and dish by rinsing with 8 ml of 1x NPB.

9. Filter the residue through the same cheesecloth layers and pool the filtrate. Wear gloves to fold the cheesecloth and squeeze any remaining liquid into the 15-ml tube.

   Note: *This constitutes a crude sperm suspension.*

10. Pellet the sperm by centrifuging the filtrate at 3000 rpm for 10 minutes at 4°C (using a Sorvall HB-6 swinging bucket rotor or equivalent). Wash the pellet in 8 ml of NPB and repeat centrifugation.

11. Resuspend the final pellet in 1 ml of 1x NPB, using a cut off micropipette tip, and warm the suspension to room temperature. Add 50 µl of 10 mg/ml lysolecithin (Sigma) and mix by rocking or gentle pipetting. Incubate for 5 minutes at room temperature.

12. Stop the lysolecithin reaction by adding 10 ml of cold 1x NPB, containing 3% (w/v) BSA, 10 µg/ml leupeptin, and 0.3 mM PMSF. Mix gently by inversion, and centrifuge at 3000 rpm as described in step 10. Carefully discard the supernatant.

13. Gently wash the nuclear pellet in 5 ml of ice cold 1x NPB containing 0.3% (w/v) BSA (no protease inhibitors), mix gently by inversion, and repeat centrifugation.

    Note: *To avoid shearing chromosomes, sperm nuclei should not be drawn through small (<100 µm) orifices such as micropipette tips. For all resuspension steps, use a 5- or 10-ml pipette or cut off yellow or blue micropipette tips.*

14. Resuspend the final pellet in 500 µl of 1x NPB containing 0.3% (w/v) BSA and 30% (v/v) glycerol (sperm nuclei storage buffer). Transfer suspension into a 1.5-ml microcentrifuge tube and count the density of sperm nuclei using a hemacytometer (e.g., Fisher 02-671-5). Dilute a small amount of the concentrated nuclei 1:100 in sperm nuclei dilution buffer (see introduction to this protocol) and add 1 µl of 1:10,000 Hoechst dye stock to visualize the nuclei under a fluorescence microscope.

    Note: *A 1:100 primary dilution of Hoechst dye stock can be stored frozen and used for routine assays. This method generally yields a sperm nuclei stock of 75–125 sperm nuclei/nl. If the stock of sperm nuclei is significantly less concentrated (<50 sperm nuclei/nl), it should be repelleted and resuspended in an appropriate volume of sperm nuclei storage buffer.*

15. Store the sperm nuclei at 4°C for up to 3 days and use for transplantations. Alternatively, thoroughly but gently resuspend the nuclei in freezing storage solution (1 ml of freezing storage solution contains 500 µl of 2x NPB, 470 µl of glycerol, and 30 µl of 10% [w/v] BSA) and freeze in single-use aliquots. Store at –80°C for 2–3 months. Thaw just before use.

    Note: *These nuclei will give rise to feeding tadpoles that are indistinguishable from those promoted by fresh nuclei. However, frozen nuclei may not be suitable for experiments where embryos will be reared to adulthood.*

PREPARATION OF LINEARIZED PLASMID DNA

Linearized plasmid for transgenesis should be purified using a silica-based DNA purification resin such as Geneclean (Bio 101, Inc.) and eluted into water. It is essential that all traces of the ethanol wash solution be removed from the pellet of resin-bound DNA prior to elution. Allowing the pellet to dry for about 1 minute before elution may help remove traces of ethanol. The final concentration of DNA should be 200–250 ng/ml.

If necessary, linearized plasmid DNA can be concentrated by standard precipitation with 0.1 volume sodium acetate (3 M stock; pH 5.2) and 2.5 volumes ethanol, followed by a 70% ethanol wash. If another method is used to prepare linearized DNA, it is essential that all traces of contaminants (such as phenol or ethanol) be removed, since large volumes of the plasmid DNA are added directly to sperm nuclei during the REMI reaction and any contamination will adversely affect embryonic development.

USE OF RESTRICTION ENZYME TREATMENT OF SPERM NUCLEI FOR ENHANCING TRANSGENESIS

Most restriction enzymes that function under the moderately high-salt conditions of the egg extract should be suitable for treatment of sperm nuclei. At present, there is no evidence that plasmid linearized with any particular restriction enzyme is more or less effectively utilized in the transgenesis procedure. One enzyme that has been used extensively for this procedure is *Not*I (Boehringer Mannheim or New England Biolabs). It is not known whether using the same enzyme for plasmid linearization and treatment of sperm nuclei improves the frequency of integration, although it is certainly not necessary to use the same enzyme for both. Indeed, integration is unlikely to be mediated by simple ligation of complementary sticky ends of DNA. Instead, integration of the DNA into the genome is more likely to be mediated by chromosome repair mechanisms. *Xba*I, *Not*I, and *Sal*I have been used frequently in transplantation reactions, and plasmids linearized with *Xho*I, *Bam*HI, and *Eag*I (to name but a few) have also been used successfully.

Transgenic embryos containing integrated plasmid can be produced using the REMI procedure in the complete absence of restriction enzyme, but only at a very low frequency. When a high frequency of transgenesis is required, restriction enzyme must be used. However, since addition of restriction enzyme undoubtedly causes some chromosomal damage, it is advisable to omit enzyme, or use lower levels of enzyme, when advanced development of tadpoles (to metamorphosis and beyond) is desired. However, germ-line transmission of transgenes has been observed even when restriction enzymes are used at the levels recommended here for treating sperm nuclei (M. Offield and R. Grainger, unpubl.). The optimum amount of enzyme for each reaction must be determined empirically. Several dilutions of enzyme should be tested (e.g., 1:5, 1:10; 1:20, diluted in SDB, or restriction enzyme buffer, just before setting up a transgenic reaction) to identify a dose that has no apparent deleterious effects on embryo development, but which permits high levels of transgenesis. New batches of *Not*I typically work well at a 1:20 dilution.

PROTOCOL 11.3

Transgenesis by Sperm Nuclear Transplantation into Unfertilized Eggs

An outline of the transgenic procedure is presented in this section. In subsequent sections on monitoring transgenic development and troubleshooting (page 226) and optimizing transgenesis (page 228), detailed information is presented to help investigators optimize the procedure.

REAGENTS FOR TRANSGENESIS

1x **MMR**

2.5% **cysteine** in 1x MMR (titrate to pH 8.0 with **NaOH**) made up fresh each day

Sigmacote (Sigma SL–2)

100 mM **MgCl₂**

0.4x MMR containing 6% (w/v) Ficoll (Sigma Type 400, F 4375)   Sterilize by filtration.

0.1x MMR containing 50 µg/ml gentamycin (a 10 mg/ml stock solution may be purchased from GIBCO/BRL 15710-015). Add 6% (w/v) Ficoll for culturing embryos prior to gastrulation. Culture embryos in 0.1x MMR without Ficoll after gastrulation. Sterilize by filtration.

MMR, cysteine, NaOH, MgCl₂ (see Appendix for Caution)

1. Inject two to four adult female frogs in the dorsal lymph sac with 500–800 units of hCG and incubate at 15°C for 12–16 hours before transplantations (for a description of dorsal lymph sac injection, see Protocol 5.2).

2. Prepare sperm nuclei from testis tissue as described in Protocol 11.2.

   Notes: *If transgenesis is to be carried out in a warm room, chill solutions to 16–18°C and move the petri dish, in which transplantations will be done, to a cool (16–18°C) incubator when transplantations are complete (for further troubleshooting advice, see page 226).*

   *It is **extremely** important to prepare all solutions and to assemble equipment before beginning the transgenic procedure. Once a reaction has been started, promptly proceed through all steps using the timetable described below, since most components of the reaction do not remain stable for more than 30 minutes.*

3. Set up the transplantation apparatus.

   a. For an infusion syringe pump, set the pump to deliver 10 nl/sec or 0.6 µl/min. For an aquarium pump, check that a fluid flow is produced by backloading a needle with water as described above in steps 12 and 13 (see also Figure 11.1C) and attaching the needle to a micromanipulator. Fill a petri dish with 0.4x **MMR** containing 6% (w/v) Ficoll and insert the needle tip into the fluid.

      Note: *For transplantations using an aquarium pump, if flow fails to begin, tap the open end of the bleed tubing or slowly turn up the dial that controls air pressure. A thin line of fluid of a density different from that of the Ficoll solution should be seen flowing from the needle tip (this may be more evident if the dish is moved slowly).*

      MMR (see Appendix for Caution)

   b. Make a number of extra transplantation needles ahead of time as described above in the introduction to allow for breakage and clogging during the procedure.

   c. Prepare agarose injection dishes as described in Protocol 8.4 and fill with a solution of 0.4x MMR containing 6% (w/v) Ficoll (see Figure 11.2).

   d. Other required solutions:
      1x MMR
      1x MMR containing 2.5% cysteine (titrate to pH 8.0 with NaOH); prepare fresh each day
      100 mM **MgCl$_2$**
      plasmid linearized and purified (concentration = 200–250 ng/µl in water) as described on page 218
      aliquots of high-speed extract prepared as described in Protocol 11.1; thaw immediately before adding to the reaction

   e. For needle backfilling, use plastic micropipettor tips (200 µl), clipped to widen the orifice to about 1 mm in diameter, attached to approximately 1-cm length of Tygon tubing (1/32-inch bore; Fisher 12-169-1A) (Figure 11.1C).

      MgCl$_2$ (see Appendix for Caution)

4. Set up a transgenic reaction at room temperature, in a 1.5-ml microcentrifuge tube, by mixing 4 µl of sperm nuclei stock (~4 × 10^5 nuclei) with 5 µl of linearized plasmid (150–250 ng/µl) and incubate for 5 minutes at room temperature.

Note: *To ensure that the stock of sperm nuclei is adequately resuspended, pipette up and down through a cut off micropipettor tip (200 μl) on ice. Some investigators have found that the ratio of nuclei to egg extract is important and so adjust the nuclear concentration, or volume of nuclei, added to keep this ratio fixed. Also, using less DNA may result in less damage to nuclei. As little as 0.5 μl of DNA will yield transgenic animals, although this also appears to reduce the number of copies of active transgenes in recipient embryos, as assessed approximately by the intensity of GFP reporter constructs (M. Offield and R. Grainger, unpubl.).*

5. Mix 0.5 μl of the appropriate restriction enzyme, or dilution thereof, 2.0 μl of 100 mM $MgCl_2$ (to a final concentration of 5 mM), and 25 μl of high-speed egg extract by gentle pipetting and add them to the sperm/nuclei DNA mixture.

   Note: *It has been found that using smaller amounts of extract may result in adequate swelling of nuclei and prevent damage to nuclei by possible over-swelling. Addition of 2 μl of egg extract at this step (and an additional 23 μl of SDB to maintain the volume of the reaction) has yielded larger numbers of normally gastrulating embryos in some laboratories (M. Offield and R. Grainger, unpubl.).*

6. Mix the reaction by gently pipetting it up and down through a cut off micropipette tip (200 μl).

   Note: *Do not push the tip against the bottom of the tube or introduce bubbles to the mixture. Rough handling will damage chromosomes and shear nuclei.*

7. Incubate the reaction for 10 minutes at room temperature.

8. During the 10-minute incubation (step 7) collect unfertilized recipient eggs from individual frogs and dejelly them in 2.5% cysteine in 1x MMR (adjust pH to 8.0 with NaOH). Mix the eggs until the jelly coat is just released.

   Note: *Because dejellying is potentially very damaging to unfertilized eggs, it is imperative to do this very gently. Never allow the eggs to contact the surface (e.g., by overzealous mixing) since this may activate and/or damage them. Reducing the pH to 7.8 and the cysteine concentration to 2% may help reduce the toxicity of the solution during dejellying.*

9. Wash dejellied eggs thoroughly, but gently, in 1x MMR.

   Note: *Eggs collected prior to this step should not be used. Even brief storage periods will be detrimental to embryo development.*

10. Inspect each batch of eggs under the dissecting microscope. Use a wide-bore pasteur pipette to transfer healthy eggs (i.e., those that remain

spherical after dejellying and have even pigmentation) to the square well of an injection dish containing 6% (w/v) Ficoll in 0.4x MMR.

Note: *Fill the square space with eggs so that no space is left between them (~400–500 eggs loaded per 60-mm dish). After about 5 minutes in the injection dish, the eggs will pierce easily.*

11. About 15–20 minutes after beginning the reaction of sperm nuclei with DNA (step 7), dilute the reaction to a final concentration of 1–2 sperm nuclei per 10–15 nl with SDB (this will be a dilution of ~1:25 to 1:100). Add $MgCl_2$ to a final concentration of 5 mM.

Note: *Produce ample amounts (200 μl) of this diluted reaction mix to finish transplantations for the entire dish (allowing for needle breakage or clogging). The diluted reaction is stable for 40–60 minutes at room temperature; the stock reaction appears to be less stable.*

12. Use a piece of Tygon tubing attached to a micropipettor tip (as described for siliconizing needles above in the introduction and Figure 11.1C) to draw up the dilute sperm suspension and mix gently by pipetting up and down. Draw up the dilute sperm nuclei and detach the tip from the micropipettor.

Note: *Try not to create or leave bubbles in the Tygon tubing as these may damage nuclei. Be careful to keep the pipette tip horizontal or the nuclei will dribble out.*

13. Backload an injection needle by attaching it to the Tygon tubing and raising the angle of the pipette tip so that the nuclei flow gently into the needle. As long as no liquid is present at the tip of the needle, the nuclei should flow easily by simple gravity. Once the needle has backfilled completely with nuclei, detach the needle and transfer it to the Tygon tubing filled with mineral oil on the infusion pump (or the tubing attached to the aquarium pump). Place the pipette tip with the remaining nuclei aside (horizontally) for loading other needles.

14. Attach the needle to the micromanipulator and check that the solution is flowing from the needle (as indicated by Schlieren patterns at the needle tip when it enters the liquid in the petri dish).

Note: *Because of the low pressure in the aquarium air pump system, solution will flow out of the needle only when the tip enters the liquid.*

15. Transplant the sperm nuclei into unfertilized eggs. Move the needle fairly rapidly from egg to egg, piercing the plasma membrane of each egg with a single, jabbing motion and then drawing the needle out slightly more slowly.

Note: *For either infusion syringe pump or aquarium sump systems, keep the needle inside each egg for approximately 1 second. For more advice on injection technique, see Helpful Hints.*

16. After injection, incubate the embryos at 16–18°C.

    Note: *At 18°C, cleavage should begin at 1 hour 45 minutes to 2 hours after injection.*

17. When the embryos have reached the 4- to 16-cell stage, separate the normal embryos (those that have been injected with one sperm nucleus) from the abnormal embryos (those having received no sperm nucleus or more than one nucleus). Use a wide-bore pasteur or Spemann pipette to transfer normal embryos to a separate dish of 0.1x MMR containing 6% (w/v) Ficoll and 50 µg/ml gentamycin.

    Note: *The criteria for distinguishing normal and abnormal embryos are discussed below.*

18. Culture embryos in 6-well culture dishes (10 embryos per well) or 24-well dishes (5 embryos per well).

    Note: *Culturing embryos at higher densities can compromise their health, as can the presence of necrotic embryos, which should be removed promptly.*

19. When embryos reach stage 10.5, replace culture medium with 0.1x MMR supplemented with 50 µg/ml gentamycin (no Ficoll).

    Note: *Because of the large needle tip used for transplantations, embryos may develop large blebs at the site of injection. These blebs occur when cells are forced out of the hole left in the vitelline membrane at the injection site, but they generally do not affect development. The blebs usually fall off at the neurula or tailbud stages, but they can be removed manually once the embryos have reached the late blastula stage.*

Helpful Hints

- When injecting unfertilized eggs, hold the needle perpendicular to the surface of the membrane (rather than glancing) to avoid tearing the plasma membrane. Be certain to pierce the membrane by jabbing the needle into the surface. A sharp action must be used, otherwise the membrane will be deformed rather than punctured.

- The rate of flow through the transplantation needle must be fast enough to prevent the solution from being drawn back into the needle and to prevent the needle from clogging with cytoplasm during the injections, and yet slow enough to be manageable. The flow required to deliver one sperm nucleus per injection varies and should be determined empirically; 10 nl/sec is a good starting point. When using an infusion pump, note that the actual volume delivered to each egg appears to be somewhat less than that predicted from the setting on the pump. It may be necessary to set the concentration of nuclei at levels two- or threefold higher than might be predicted in order to deliver one nucleus per egg. Some investigators report better results by reducing the flow rate and injecting smaller volumes (e.g., using a rate of 5 nl/sec). To make this change, be sure to compensate by using a more concentrated nuclear suspension. To ensure a smooth flow of suspension at 5 nl/sec, it may be necessary to replace the 2.5-cc syringe with a smaller model.

- If the needle becomes clogged with cytoplasm, bring the tip to the air-liquid interface of the dish. Sometimes the surface tension at the interface will unblock the needle. If any particulate matter clogs the needle tip during transplantation, change needles (even if solution continues to flow). Debris in the needle will shear sperm nuclei and haploid development will result. Haploid tadpoles have shortened trunks and tails, are thicker than normal throughout the trunk region with a "pigeon-chested" appearance, and often have heads and tails that curl toward the dorsal side; these tadpoles are often edemic and generally die at the beginning of the feeding stage (Gurdon 1960).

- By studying the behavior of the injection site, it is possible to determine whether the correct volume of sperm nuclear suspension has been delivered. Injection of the correct volume produces a hole of approximately the same diameter as the needle tip. This should be visible on the surface of the egg and should remain open for a couple of seconds after injection. It should have clean smooth edges (not jagged) and a small amount of translucent fluid should be visible just over the opaque white cytoplasm in the wound. If the volume injected is too low, the injection wound will close over immediately after the needle is removed. Low injection volume is also indicated by a ring of white cytoplasm that is sometimes drawn from the injection site as the needle is removed. If the flow from the needle is too rapid and excess volume is injected, the surface of the egg near the injection site may ripple, or the injection wound may expand in size significantly.

MONITORING TRANSGENIC DEVELOPMENT
AND TROUBLESHOOTING

Most problems that occur during nuclear transplantations can be diagnosed by carefully observing the developing embryos at first cleavage, blastula, and gastrula stages. Observing the embryos during these early cleavage stages will reveal whether the dilution of the sperm nuclei and the injection volume delivered during the transplantations were appropriate, and normal embryos can be selected. If few of the eggs received a nucleus, the frequency of cleavage will be low. However, pseudocleavage can occur in eggs that have been activated even if they did not receive a nucleus. These eggs undergo an abortive cleavage at around the correct time after injection (often slightly earlier than eggs receiving nuclei), but the cleavage furrow is usually very shallow and never cleaves through the egg completely. Activated eggs continue to undergo false cleavages every 30 minutes or so and can sometimes be mistaken for normal embryos that later die. Note that the cleavage furrows in normal embryos go all the way into the vegetal hemisphere; this does not happen during false cleavage.

In a batch of eggs injected with sperm nuclei at an optimal concentration, one fifth to one third of transplantations result in normally cleaving embryos. At the two-cell stage, each cell of such embryos is the same size, and at the four-cell stage, the new cleavage planes are perpendicular to the first. Because the delivery of nuclei is expected to be a stochastic process, even at an optimal concentration of sperm nuclei, a significant fraction of the eggs would be expected to receive multiple nuclei or no nuclei at all.

Injecting excessive volumes into the eggs may also cause cleavage to fail. Other symptoms caused by overinjection include a mottled or "marbled" appearance in the animal hemisphere and other signs of general unhealthiness. Eggs containing more than one nucleus will divide abnormally at the time of first cleavage, producing three or four (or more) cells with nonorthogonal cleavage planes.

Other morphological features can indicate the health of injected embryos. In embryos injected with an optimal amount of transgenic reaction mix, the injection hole will heal, without leaking, within 1 hour of injection and the egg will not show any discoloration. If a component of the reaction mix was toxic, discoloration may occur around the injection site; otherwise, embryos may appear normal. By 2 hours after injection, a very localized spot of necrosis may become visible and this gradually spreads to the rest of the embryo. Later, the cells in this spot become discolored and many will be dead. If sperm nuclear chromosomes are damaged by handling or during transplantation, the embryo may appear to be normal during the first hour, but may produce abnormal early cleavage patterns, depending on the degree of chromosomal damage. At slightly later stages, cells may still appear healthy (with normal animal pigmentation and good cell-cell contacts), but, if chromosomes have been partially sheared, cleavage may be unsynchronized across the embryos. Uneven cell size can sometimes be observed across the animal hemisphere in blastulae.

At the blastula and gastrula stages, there are again characteristic morphological signs that indicate an optimal set of injections, and characteristic abnormalities resulting from problems with the transgenic procedure. If normally cleaving embryos are selected early, blastulae will be complete (cellularized throughout animal and vegetal pole) and cells will be of normal size and coloration. If eggs are healthy and firm, normal gastrulation and tadpole development can be expected from at least 50% of blastulae. If blastulae looked healthy but a high degree of spina bifida (>50%) occurs during gastrulation, eggs were probably too soft. In this case, the problem may be due to poor egg quality (see below). If embryos gastrulate reasonably well, but give rise to thick, stunted embryos, or embryos with organ problems, this may be due to some subtle toxicity of the reaction mix or to damaged sperm chromosomes. Subtle toxicity may also result in embryos with localized regions of dead cells. In this case, the entire embryo will die eventually; these embryos usually do not gastrulate. If a toxic component in the transgenic reaction mix is suspected, nuclei can be transplanted in dilution buffer, with no plasmid or extract. Excluded components of the reaction mix can be reintroduced one by one to track down the toxic ingredient. If embryos receiving nuclei alone are still unhealthy, the needle may not be sufficiently tapered, or the membrane may have been damaged at the injection site due to poor transplant technique (e.g., using a glancing rather than a perpendicular angle of injection).

If there is extensive chromosome damage to sperm nuclei, gastrulation occurs normally and haploid development results. If chromosomes are less extensively damaged, gastrulation may arrest or the blastopore may fail to close. In the case of very subtle chromosome damage, embryos do gastrulate but will have organ problems later in development. To avoid these problems, make sure that no tissue obstructs or constricts the needle tip during transplantations. Use a fresh needle for each transplantation reaction, since debris attached to the sides of the needle may shear nuclei passing through it. Store nuclei on ice, but do not place nuclei/plasmid reaction or diluted nuclei on ice. Handle nuclei in suspension very gently with cut off micropipette tips.

If too few nuclei are delivered to a batch of eggs, there will be little normal cleavage in the batch and few or no blastulae or gastrulae will develop. In this case, the concentration of sperm nuclei must be checked; if there are less than 1×10^8 sperm nuclei/ml, the suspension must be concentrated by centrifuging and resuspending in an appropriately smaller amount of buffer. The suspension must be mixed just before the transgenic reaction is set up, just before the nuclei are diluted into transfer buffer, and just before the injection needles are loaded. Needles must *not* be loaded before the egg dish is prepared, and they must be used within 20 minutes of loading.

If too many nuclei are delivered to the eggs, there will be multiple cleavages at early stages, as noted above, but many blastulae with healthy-looking cells will be formed. However, regions of the embryo will then begin to divide asynchronously, resulting in a range of cell sizes across the embryo. Cells may arrest and die in a large region of the embryo prior to or during gastrulation, with the remainder of the embryo continuing the process of gastrulation, resulting in the formation of partial embryos. Embryos receiving many nuclei do not initiate gastrulation at all, whereas those receiving three to four nuclei undergo abnormal gastrulation. In these embryos, blastopore closure is typically incomplete, resulting in the formation of two wings of somites and neural tissue on each side of the exposed yolky tissue lying in the center of the trunk. Reducing the concentration of nuclei in the nuclei/plasmid reaction (step 4 of Protocol 11.3) will solve this particular problem. Early selection of normally cleaving embryos will remove embryos containing multiple nuclei. At late cleavage or blastula stages, it is often impossible to distinguish these two classes.

OPTIMIZING TRANSPLANTATION-BASED TRANSGENESIS

Egg quality contributes significantly to the level of postgastrula development obtained from sperm nuclear transplantation. To obtain good postgastrula development, eggs must be healthy. In particular, they should have even pigmentation and be firm enough to hold their shape after dejellying. In addition, it is important that they do not become activated before they are injected with nuclei. When egg quality is poor, a fraction of the embryos show morphogenetic defects resulting in incomplete blastopore closure during gastrulation. This problem is often compounded by high levels of foreign gene expression during the gastrula stages. Embryos expressing genes from the CMV promoter, which is active from the mid blastula stage onward, are more likely to show nonspecific gastrulation defects than embryos expressing genes from strong promoters activated after gastrulation. If suboptimal egg quality is thought to be a problem, it may be advisable to use young frogs, not previously ovulated, as a source of recipient eggs for transgenesis.

Nuclear transplantation and posttransplant incubation should be carried out at temperatures no higher than 22°C. Embryo survival rates can be enhanced by incubating at 16–18°C through the early stages. Incubation at elevated temperatures (24–25°C) lowers survival rate and the frequency with which plasmids are expressed in resulting embryos. Embryos raised at elevated temperatures exhibit a high frequency of mosaic transgene expression; in many cases, only 25–50% of expected cells show expression. It may be that at high temperatures, the accelerated first cell cycle is too short for plasmids to integrate into the genomic DNA, a prerequisite for stable transgene expression.

The transgenesis procedure described in this chapter is very efficient and workable, but it involves a multitude of steps, all of which are critical for its success. When learning the technique, it is advisable to master each step in turn, rather than all at once. For example, first learn to isolate sperm nuclei and transplant them into eggs. Once this can be done successfully, resulting in normal development, then determine whether sperm nuclei, swollen in extracts, give normal development. If swelling of sperm in extract has no adverse effects on the level of development obtained after transplantation, plasmid and enzyme can be added to the reaction, thus reconstituting the whole transgenesis procedure.

The techniques described here are an evolving set of procedures now being utilized, and modified, in a number of labs. An alteration in the sperm nuclear preparation involving a Percoll gradient step and the use of digitonin to permeabilized sperm (Huang et al. 1999) provide an effective complement to the procedure described here. The use of heated embryo extract for swelling sperm nuclei (M. Wu, L. Browder, and J. Gerhart, unpubl.; Web site: http://www.ucalgary.ca/UofC/eduweb/virtualembryo/frogs2.html) may help the efficiency of transgenesis. These alterations and several others (e.g., partial dejellying of eggs) have been used to make transgenic *X. tropicalis* embryos at high efficiency and to generate permanent transgenic frog lines (M. Offield, L. Zimmerman, and R. Grainger, unpubl.). Many of these methods are summarized on the *X. tropicalis* Web site at http://minerva.acc.Virginia.EDU/~develbio/trop/.

FUTURE PROSPECTS: GENETICS

The advantages of the frog system over other developmental models are numerous, but one major disadvantage is that *Xenopus laevis* cannot be exploited at the genetic level because it is allotetraploid and has a long generation time. *Xenopus tropicalis*, which is diploid and has a generation time of approximately 4–6 months (Tymowska and Fischberg 1973), may be a tractable organism for such studies (Amaya et al. 1998). The transgenic technology described here works very well in *X. tropicalis*, and germ-line transmission of several transgenes has been demonstrated (M. Offield and R. Grainger, unpubl.). The prospects for performing mutagenesis in *X. tropicalis* are therefore quite good. The transgenic technique can be readily adapted for an insertional mutagenesis scheme using gene trap approaches

(O. Bronchain and E. Amaya, unpubl.). For similar reasons, *X. tropicalis* will also be the species of choice for carrying out targeted mutations.

REFERENCES

Amaya E. and Kroll K.L. 1999. A method for generating transgenic frog embryos. *Methods Mol. Biol.* **97:** 393–414.

Amaya E., Offield M.F., and Grainger R.M. 1998. Frog genetics: *Xenopus tropicalis* jumps into the future. *Trends Genet.* **14:** 253–255.

Gurdon J.B. 1960. The effects of ultraviolet irradiation of the uncleaved eggs of *Xenopus laevis. Q.J. Microsc. Sci.* **101:** 299–312.

Huang H., Marsh-Armstrong N., and Brown D.D. 1999. Metamorphosis is inhibited in transgenic *Xenopus laevis* tadpoles that overexpress type II deiodinase. *Proc. Natl. Acad. Sci.* **96:** 962–967.

Knox B.E., Schlueter C., Sanger B.M., Green C.B., and Besharse J.C. 1998. Transgene expression in *Xenopus* rods. *FEBS Lett.* **423:** 117–121.

Kroll K.L. and Amaya E. 1996. Transgenic *Xenopus* embryos from sperm nuclear transplantations reveal FGF signaling requirements during gastrulation. *Development* **122:** 3173–3183.

Kroll K.L., Salic A.N., Evans L.M, and Kirschner M.W. 1998. Geminin, a neuralizing molecule that demarcates the future neural plate at the onset of gastrulation. *Development* **125:** 3247–3258.

Latinkic B.V., Umbhauer M., Neal K.A., Lerchner W., Smith J.C., and Cunliffe V. 1997. The *Xenopus* Brachyury promoter is activated by FGF and low concentrations of activin and suppressed by high concentrations of activin and by paired-type homeodomain proteins. *Genes Dev.* **11:** 3265–3276.

Mohun T.J., Garrett N., and Gurdon J.B. 1986. Upstream sequences required for tissue-specific activation of the cardiac actin gene in *Xenopus laevis* embryos. *EMBO J.* **5:** 3185–3193.

Murray A.W. 1991. Cell cycle extracts. *Methods Cell Biol.* **36:** 581–605.

Pownall M.E., Isaacs H.V., and Slack J.M. 1998. Two phases of Hox gene regulation during early *Xenopus* development. *Curr. Biol.* **8:** 673–676.

Tymowska J. and Fischberg M. 1973. Chromosome complements of the genus *Xenopus. Chromosoma* **44:** 335–342.

Tape 3 of the video series "Manipulating the Early Embryo of Xenopus laevis*" illustrates several aspects of the transgenic procedure, including sperm nuclear isolation, needle preparation, sperm nuclear injection techniques, and how to distinguish normal and abnormal embryos after injection with sperm nuclei.*

CHAPTER 12

Immunohistochemistry

Staining embryos with antibodies is extremely useful because it provides spatial information on gene expression. This chapter discusses methods for staining embryos in whole mount. The procedure is sufficiently robust to be used as a routine assay for experiments, being limited only by the abundance of the antigen and the quality of the antibodies used. A condensed protocol is presented at the end of this chapter.

Antibody staining and in situ hybridization provide valuable alternatives to histological examination of sectioned and stained embryos. Conventionally stained sections provide information about the arrangements and types of many tissues in embryos or embryonic explants, but the use of antibodies provides a less ambiguous assay for individual tissues. For example, animal caps treated with activin develop cells with the morphology of the primitive red cell lineage. However, closer analysis shows that despite their morphological similarity to primitive red cells, they do not express globin antigens, a characteristic feature of this type of cell (Green et al. 1990).

The whole-mount immunohistochemical staining protocols developed for *Drosophila* embryos (Mitchison and Sedat 1983) have been adapted for staining the much larger embryos of *Xenopus* (Dent et al. 1989; Hemmati-Brivanlou and Harland 1989). In short, fixed embryos are incubated with a primary antibody, raised against the antigen of interest, washed thoroughly, and then incubated with a secondary antibody, coupled to an enzyme (often horseradish peroxidase, HRP). The embryos are then washed and cleared to reveal stained regions corresponding to enzyme activity and hence the antigen of interest. The development of a good clearing medium by Andrew Murray was an essential feature in the development of whole-mount methods: The mounting medium matches the refractive index of yolk platelets and therefore renders *Xenopus* embryos transparent.

Various antibodies are now available for many embryonic tissues, including notochord (MZ15, Smith and Watt 1985; Tor70, Bolce et al. 1992), muscle (12/101, Kintner and Brockes 1984), neural tube (6F11, Lamb et al. 1993; Xen1, Ruiz and Altaba 1992), differentiated neurons (3A10, Serafini et al. 1996; Tor25, Hemmati-Brivanlou and Melton 1994), and epidermis

(London et al. 1988). Many antibodies raised against antigens from other species cross-react well with *Xenopus* antigens, although few of these have been documented in the literature. A useful database of antibody reactivity is maintained at http://vize222.zo.utexas.edu/. Antibodies can be obtained from individual laboratories or from the Developmental Studies Hybridoma Bank (http://www.uiowa.edu/~dshbwww/index.html). The preparation of antibodies is beyond the scope of this manual; for details, see Harlow and Lane (1988).

In general, monoclonal antibodies stain with very low background in embryos, whereas polyclonal antibodies must be purified prior to use. Most polyclonal sera contain contaminating antibodies that react with the epidermis. These antibodies cause a high level of background staining, which obscures any deeper staining. High-titer polyclonal antibodies can be cleaned up initially by selection against the antigen and then by preabsorption with embryos. This treatment removes the epidermis-binding components. Similarly, monoclonal antibodies that have been grown in ascites can be preabsorbed to remove background binding activities from serum. Because the specific binding antibodies are usually at much higher titer than nonspecific binding components, preabsorption against embryos usually removes unwanted binding activities without affecting the specific binding activity; thus, antibodies often give very clean signal-to-noise ratios when recycled for a second use.

Immunohistochemistry requires that the antigen be fixed in place in the embryo. The two most common methods used for fixing *Xenopus* antigens are formaldehyde cross-linking, which tends to give better results for nuclear antigens, and alcohol precipitation, which is often superior for membrane and cytoskeletal antigens. Different antigens respond variably to fixation, so it is worth testing alternative procedures. For many antigens, little difference exists in staining efficiency, but for some, there is a dramatic difference.

Penetration of reagents, and removal of excess reagents, can take some time with larger embryos. Therefore, one of the principal differences between this procedure and others for use with smaller embryos is an increase in incubation times. It is advantageous to use baskets when staining large numbers of *Xenopus* embryos (see Figure 12.1). Not only do these baskets reduce the labor of changing solutions, but they can also be placed in a large excess of buffer, thus increasing the efficiency of the washes.

The stain in embryos is easily obscured by the embryo's natural pigment. This problem can be overcome by using albino embryos or by bleaching normal embryos. Bleaching embryos with hydrogen peroxide is straightforward, and most antigens are resistant to the bleaching conditions. Assuming that this is the case, a whole batch of embryos can be bleached before staining. The advantage of prebleaching is that it allows the subsequent staining reaction to be monitored against minimal background pigmentation, thus enabling the optimum staining period to be determined with greater accuracy. However, it is also possible to bleach embryos after staining, since diaminobenzidine precipitates are resistant to the treatment. Note that some of the color enhancements, e.g., nickel enhancement, are not stable in bleach. If it is necessary to bleach after staining, arrange to have a stock of fixed albino embryos so that one or two albino controls can be added to the staining reaction to monitor color development.

APPARATUS FOR IMMUNOHISTOCHEMISTRY

Agitators

Listed below are three kinds of agitators used in staining procedures.

- A rotator that provides end-over-end rotation of vials is useful when the vial is full, since it keeps the embryos constantly in suspension. Various low-speed rotators are available (e.g., Labquake shaker, Labindustries, Berkeley, CA, which rotates at a constant speed of 12 rpm).

- A Nutator (Adams Nutator, Clay Adams, model 1105) is valuable when reagents are used in small volumes. This device provides a circular tipping movement that agitates embryos without subjecting them to excessive turbulence.

- An orbital shaker with adjustable speed (e.g., Gyrotory shaker G2, New Brunswick) can be used in place of the Nutator, although it is often difficult to find an ideal setting that agitates embryos gently. An orbital shaker is essential to agitate embryos in baskets (see below).

Containers

Glass Vials

For a small number of samples, use 3-ml screw-cap glass vials (e.g., Fisher 06-408B). Embryos can be kept in the same vial for all steps of the procedure (fixation, antibody incubations, washes, enzyme reaction).

Baskets

For larger experiments, changing solutions in vials becomes tedious and baskets should be used. Commercially available baskets, e.g., 15-mm Netwell baskets (Costar 3481), fit into 12-well tissue culture plates. Although they are expensive, they can with extensive washing be reused. However, these baskets are not readily adaptable to large-scale use, and their relatively shallow wells make cross-contamination between wells a real danger. Homemade baskets, made from plastic microcentrifuge tubes and nylon mesh, are more adaptable to large-scale experiments (Stachel et al. 1993). The baskets are narrow and deep and thus can be placed in a rack at high density. To change solutions, individual tubes or whole racks of tubes can be moved from one bath of solution to another. Since many tubes can be manipulated simultaneously, there is an enormous saving in work. It is also quite difficult to lose embryos in baskets; embryos in vials stand a good chance of being sucked into the aspirator during solution changes and lost. Follow the procedure below to make the baskets.

1. Remove the bottom of the required number of polypropylene microcentrifuge tubes with a single-edged razor blade. Make sure that all the tubes are cut at the same level.

   Note: *Polypropylene tubes vary in their ability to be cut and in the thickness of the lip or flange at the mouth of the tube. Sarstedt (72.693) 2-ml tubes are relatively easy to cut, but one side of the flange must also be removed to allow the tubes to pack closely.*

2. Adjust a hot plate to a temperature that melts polypropylene but not nylon.

3. Place a small piece of aluminum foil on the hot plate and then place a small square of nylon screen over the foil.

   Note: *Depending on the size of the embryo or explant, different sizes of screen may be useful. For all but the smallest explants, a 100-µm mesh (e.g., Tetko Nitex 3-100-47 or Spectramesh 100-µm nylon) will suffice.*

4. Press the cut side of the tube onto the mesh until a good seal is made. Allow the tube to cool and remove the foil.

   Note: *Once the tube has cooled, the foil will detach readily.*

5. Trim excess Nitex with scissors and a large-bore pencil sharpener (dedicated to this purpose).

   Note: *Cleanly trimmed edges are only important for parallel-sided 2-ml tubes in order to fit them into racks.*

Arranging Baskets in Racks

Baskets should be arranged in a rack that fits snugly into a transparent container, e.g., a 45 × 70 × 32-mm glass staining dish (Wheaton Inc. 900170) that holds 20 tubes or a 105 × 70 × 85-mm dish (Wheaton Inc. 900203) that holds 30 tubes. The washes are carried out in standing racks on an orbital shaker (do not use a rotator or a Nutator!). The wave action cycles the buffer up and down through the tubes, moving the embryos and ensuring good washes.

Standing racks are preferable to floating racks. Improvised racks can be made from fluorescent light diffuser panels (egg crate type, 1/2-inch-square cells from a lighting store or McMaster Carr) or the top of a Gilson fraction collector rack/polypropylene storage rack (Fisher 05-5641; the bottom of the rack must be cut off with a band saw). Both alternatives are inexpensive. Syringe needles or plastic pipette pieces can be melted and fused to the rack to serve as legs. Arrange the legs such that the mesh of the baskets sits close to the bottom of the dish, so that excessive volumes of wash solution are not needed.

For most steps, the baskets can be processed en masse. The volumes of solution used depend on their value. For washes in simple buffers, the rack can be moved to a large dish and washed in an excess of buffer. Where buffers are more precious, smaller volumes can be used, and the rack should be transferred to a glass staining dish (e.g., Wheaton Inc. 900170 or 900203). Approximately 25 ml of solution is required for a 20-tube Wheaton dish, and 50 ml of solution is needed for the 30-tube dish. These relatively large volumes may appear wasteful, but they are quite reasonable if the solution can be recycled.

Baskets can also be manipulated individually if, for example, different antibody treatments are to be used. For individual treatment, the basket can be transferred with forceps to a 15-ml polypropylene tube (e.g., 2059 tube, Sarstedt). Approximately 250–500 μl of solution will cover the embryos, and the solutions are removed using a fine plastic tube attached to an aspirator as shown in Figure 12.1A.

If all of the samples receive the same treatment, the embryos can be kept in the rack throughout the procedure. If samples are treated differently, then the appropriate baskets can be moved to a 15-ml tube or the embyros themselves can be transferred from the baskets to glass vials at any step. To transfer embryos from the basket, it is most straightforward to immerse the basket in a petri dish of buffer, and then tip the embryos out into the dish. They can then be moved from the dish with a pipette.

A

B

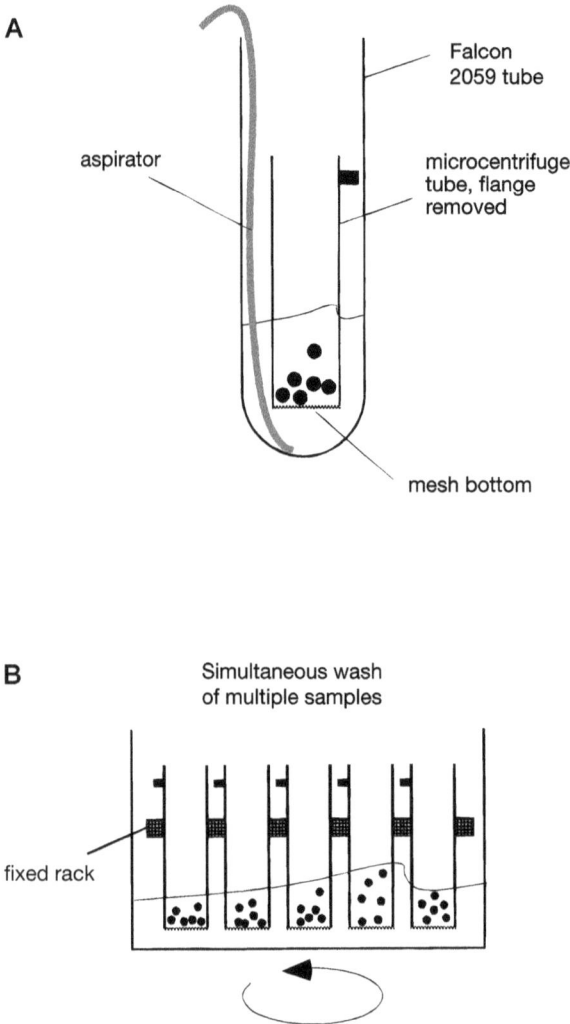

Figure 12.1. Baskets for staining embryos. (*A*) An individual basket in a 15-ml tube, where solutions can be exchanged by aspiration and replacement. (*B*) A rack of tubes in a staining dish, where all tubes are exposed to the same solution. The solution is agitated by placing the dish on a rotating platform. Solutions are changed by moving the rack from one dish to another. (Courtesy of Mark Curtis.)

EMBRYO PREPARATION AND FIXATION

For optimal staining of early embryos, remove the vitelline membrane. For neurula and tailbud tadpole stages, it is generally better to remove the membrane well before fixation, since this allows the embryo to develop clear surface morphology; embryos that develop inside a vitelline membrane are often constrained into a sphere and are therefore difficult to stage. However, early-stage embryos are delicate and removal of the vitelline membrane may cause the embryo to flatten or large blastomeres to burst. As an alternative to early devitellination, remove the membranes after a few minutes in formaldehyde fixative (e.g., MEMFA). This treatment causes the embryo to shrink away from the membrane, but does not make the membrane brittle. However, formaldehyde fumes are very unpleasant and this treatment must be carried out with good ventilation. A more convenient option is to treat the embryos briefly in a protease solution, whereupon the membrane softens and expands over time. The embryos must be monitored carefully, since batches vary in their susceptibility to the treatment. Early-stage embryos may flatten in a loose vitelline membrane, but they will assume a spherical shape during fixation, provided they stay in suspension in an end-over-end rotator. A 5–10-minute treatment in 10 μg/ml proteinase K or 0.25 mg/ml pronase is sufficient to soften the membrane. If any of the embryos break out of their membrane, stop the treatment by rinsing with a buffer supplemented with 1 mg/ml bovine serum albumin or by smashing one or two of the embryos to release soluble proteins.

A final option for devitellination is to remove the membrane after fixation with a cocktail of proteases, as suggested in some in situ hybridization procedures (Islam and Moss 1996). Embryos are treated for 10 minutes at room temperature in phosphate-buffered saline containing 10 μg/ml proteinase K, 2 mg/ml collagenase A, and 20 units/ml hyaluronidase. Obviously, it is difficult to control the rate of protease attack, and excess treatment can destroy antigens; thus, this treatment is only recommended for antigens that are protease-resistant.

Vital Staining of Albino Embryos

Since it is difficult to see detail in albino embryos, particularly at early stages, these embryos can be lightly stained after dejellying with a vital dye such as Nile blue. Nile blue can be added to a final concentration of 0.01%, from a 10% stock solution. Note that Nile blue precipitates in most buffers over time and staining may be uneven. For an even stain, prepare the Nile blue in 50 mM sodium phosphate (pH 7.8). Addition of 2% Ficoll collapses the vitelline space and provides the most even staining, but for most purposes, the staining afforded by simple addition of Nile blue to embryos is adequate. After a few minutes in 0.01% Nile blue, the embryos are rinsed, and incubation is continued. The cleavage furrows, blastopore, and neural folds show up very clearly after such a stain. Nile blue is soluble in alcohols, and thus it is removed later in the procedure by methanol.

Fixation of Embryos

Two fixing methods are commonly used, depending on the epitope of interest. Formaldehyde fixation (e.g., MEMFA) is suitable for detection of most antigens, particularly nuclear antigens. However, alcohol fixative (e.g., Dent's) is superior for some antigens, including many cytoskeletal and membrane antigens. For a new antigen, these two fixatives provide a useful starting point, although if staining efficiency is to be optimized, other fixatives should be tested (see, e.g., Larabell et al. 1997).

PROTOCOL 12.1

Formaldehyde Fixation

1. Wash the embryos (see above for advice on devitellination) in water, or a buffer that does not contain free amines, by allowing them to sink through the water to the bottom of a glass vial.

   Note: *Free amines (as are present in Tris buffers) will compete for fixative.*

2. When the embryos are all at the bottom of the vial, remove most of the water and replace with MEMFA.

   Note: *Take care to prevent contact between devitellinized embryos and the surface since unfixed embryos will explode at an air/water interface.*

3. Immediately place the vial on a rotator, to maintain the embryos in suspension, and allow flattened embryos to return to a more spherical shape.

   Note: *Formaldehyde is highly toxic to embryos, so it is important not to transfer formaldehyde solution back to a dish of live embryos. Use different pipettes for transferring live embryos, and for pipetting the MEMFA fixative.*

   formaldehyde (see Appendix for Caution)

4. Incubate the embryos in MEMFA for 1–16 hours.

   Notes: *The time of fixation is not critical; fixation times from 1 hour to overnight give similar results. Use shorter fixation times (20–30 minutes) if the immunohistochemistry is to follow a stain for enzyme activity such as β-galactosidase. However, once the enzyme reaction is complete, refix the embryos to prevent disintegration during subsequent steps.*

   Even if the procedure is to be continued immediately, do not omit this dehydration step since it appears to be essential for optimal penetration of reagents into the embryos.

   Following fixation, transfer the embryos to **methanol** or to Dent's fixative for storage.

   Methanol-DMSO Fixation (Dent's Fixative)

   Dent's fixative is applied in a fashion similar to that of MEMFA; incubate the embryos for 2 hours at room temperature on a rotator and then leave overnight at –20°C.

   methanol, DMSO (see Appendix for Caution)

PROTOCOL 12.2

Bleaching of Pigment

The main reason to bleach early in the procedure is to allow the development of stain to be seen easily and to destroy any endogenous peroxidase activity, which may interfere with the enzymatic staining reaction if HRP is used. Bleaching can also be carried out after staining. The benzidine precipitate produced in the reaction of HRP with DAB (diaminobenzidine) is not affected by bleaching, but some stains, such as nickel intensification of DAB staining, will be reversed by the process. Bleaching also eliminates endogenous peroxidase activity from globin-containing cells of older embryos.

1. Transfer the fixed embryos to bleaching solution (see Appendix 1).

2. Place the dish of embryos on a Nutator over an aluminum foil reflective backing and under a fluorescent light.

   Note: *A light box can be inverted and propped over the Nutator, or a fluorescent reading light can be placed close to the vials.*

3. Incubate under these conditions for about 1 hour, or until the embryos are adequately bleached.

   Note: *Different bleaching solutions that vary dramatically in their effectiveness have been recommended. All bleaching solutions contain hydrogen peroxide, but a bleaching solution that contains **formamide** and is buffered by SSC (Mayor et al. 1995) is much more effective than a higher concentration of hydrogen peroxide in Dent's fixative, which can take several days to bleach embryos effectively (Klymkowsky and Hanken 1991).*

   formamide (see Appendix for Caution)

4. After fixation and bleaching, store the embryos at –20°C in 100% **methanol** or dehydrate in methanol and continue the processing for immunostaining.

   Note: *Fixed embryos can be shipped in methanol at room temperature. Use overnight shipping and fill the vials to the brim with methanol to avoid turbulence.*

   methanol (see Appendix for Caution)

PROTOCOL 12.3

Embryo Rehydration

Embryos can be transferred from aqueous solutions straight to methanol, but the reverse should be done gradually to avoid excessive deformation. Prior to antibody incubation, rehydrate the embryos by passing them through a graded series of alcohols over a period of 10–15 minutes. Suggested steps are 80% methanol/20% water, 50% methanol/50% PBS, 20% methanol/80% PBS, and finally PBS for 10 minutes. Embryos should be agitated during these solution changes.

If embryos are kept in glass vials, the solutions must be removed carefully with an aspirator, and replaced as needed. If baskets are used, the volumes must be adjusted to keep the embryos immersed.

Prior to incubation in the primary antibody, nonspecific binding sites on the embryos must be saturated and blocked with serum. Goat serum is often used because most secondary antibodies are raised in goats; however, the blocking serum should correspond to whatever secondary antibody is used, since the secondary antibody should have no reaction against a blocking reagent from the same species.

When using different antibodies, take care to avoid cross-contamination. Inappropriate staining can result when embryos are exposed even for a few seconds to an incorrect antibody. Embryos that are being stained with different antibodies should not be washed together.

1. Incubate embryos in glass vials for 15 minutes at room temperature in PBT (PBS with 2 mg/ml BSA and 0.1% Triton X-100) with continuous rotation (end-over-end, 12 rpm).

2. Replace the PBT solution with 500 μl of fresh PBT containing 10% goat serum. Incubate for 1 hour at room temperature on a Nutator.

3. Replace this solution with 500 μl of fresh PBT and the appropriate dilution of the primary antibody (without serum).

Note: *The appropriate dilution of the primary antibody is dependent on the source, titer, and specificity of the immunoglobulin used. Dilutions vary enormously and the optimal dilution must be determined empirically. Dilutions in the 1:100 to 1:10,000 range are generally used for rabbit serum, affinity-puri-*

fied IgG from immunized rabbits, affinity-purified monoclonal immunoglobulin, and ascites from immunized mice. However, a hybridoma supernatant containing monoclonal antibody is usually used at a 1:1 to 1:10 dilution.

4. Incubate the embryos overnight at 4°C, or for 4 hours at room temperature on a Nutator.

5. Remove excess antibody with a brief wash in PBT. Then wash the embryos, again in PBT, at room temperature three to five times, for 2 hours each wash.

   Note: *The longer the wash, the lower the background. Washes in excess of 10 hours should be carried out at 4°C.*

6. Incubate embryos for 4–6 hours at room temperature, or overnight at 4°C with PBT and the appropriate dilution of secondary antibody.

   Note: *Dilutions in the range of 1:200 to 1:1000 are usual for seconardy antibodies, but it is necessary to determine the optimal dilution empirically. In each experiment, it is advisable to include a control sample that has not been exposed to primary antibody, to determine the levels of background staining due to secondary antibody alone.*

7. Agitate the embryos gently on a Nutator.

8. Wash embryos as described in step 5.

PROTOCOL 12.4

Immunodetection and HRP Staining

Various proprietary reagents are available for HRP staining. However, many of these react so quickly with the embryos that the staining does not penetrate below the surface. If the reaction with HRP is rapid, then a crust of stain (often just background staining) may collect at the surface and deep staining will not be obtained. This problem can be avoided by carrying out a slow and controlled HRP reaction in the cold.

1. Prepare a solution of **DAB** (1 mg/ml in PBT).

   Note: *Nonspecific precipitation of DAB can occur on particulate material, so pass the DAB through a 0.2-μm syringe filter before use.*

   DAB (see Appendix for Caution)

2. Wash the embryos and resuspend them in 0.5 ml of PBT. Add 0.5 ml of DAB (1 mg/ml in PBT) and incubate with vigorous shaking on an orbital shaker for 10 minutes at 4°C.

   Note: *This treatment allows the DAB to penetrate the embryo before the hydrogen peroxide is added, thus ensuring homogeneous staining.*

3. Initiate staining by adding 1 μl of a freshly prepared 1:1 mixture of 30% hydrogen peroxide and PBT. Shake vigorously (using a horizontal shaker) at 4°C.

4. Observe the embryos under a dissecting microscope during the staining process. When adequate staining has been achieved, remove the DAB and replace it with PBT.

   Notes: *The time of staining varies depending on the abundance of antigen and quality of the antibodies. It can be as short as a few minutes or as long as several hours. If staining is slow, warm the embryos to room temperature and add fresh hydrogen peroxide. However, staining does not appear to intensify after 6 hours, presumably because the HRP activity is destroyed by the reaction with peroxide.*

   Unlike other enzymatic reactions, HRP staining appears to reach a saturated endpoint. It is therefore very difficult to use HRP staining for quantitative analysis. However, lightly stained embryos will give a more accurate impression of the relative abundance of antigens than will darkly stained embryos.

5. Wash the embryos twice in PBT for 5 minutes per wash.

6. Dehydrate embryos in two changes of 100% **methanol**, over the course of 5 minutes.

   methanol (see Appendix for Caution)

7. For detailed observation, tip the embryos into a petri dish containing methanol. From here they can be transferred to depression slides or a glass spot test plate and cleared.

   Note: *Even though spot plates are optically poor under the compound microscope, they are sufficiently good for most bright-field photomicrography. For superior optics, Sylgard slides should be made for mounting embryos in **benzyl benzoate/benzyl alcohol (BB/BA)** (see Chapter 13).*

   BB/BA (see Appendix for Caution)

8. Remove most of the methanol and clear the embryos by adding approximately 250 µl of a 2:1 mixture of BB/BA (or Murray's clearing medium).

9. As the methanol diffuses out of embryos, they will sink through the dense medium. Draw off the BB/BA and replace with fresh.

   Note: *Embryos can be observed after a few minutes. BB/BA is not volatile, and embryos can be stored in these plates for several days without loss of signal. Over the course of weeks on the benchtop, staining will fade, so if the embryos are to be stored, they should be replaced in glass vials and the vials topped up with methanol.*

10. If the embryos are to be sectioned after examination as whole mounts, immediately transfer them to the wax clearing medium (e.g., **xylenes** or xylene substitutes such as histosol) and follow directions given in Protocol 14.1, starting at step 6).

    xylene (see Appendix for Caution)

Helpful Hints

- If the clearing solution becomes milky, the dehydration step was insufficient. Remove BB/BA and replace with fresh 100% methanol until dehydration is complete. If water contacts embryos after they have been cleared, they must be dehydrated in methanol before reclearing.

- Methanol is volatile and care must be taken to ensure that the embryos are submerged at all times. Rehydration of an air-dried embryo leads to distortion and bubbles in the embryo. If this does occur, transfer the embryos to a large volume of methanol and place them in a desiccator under partial vacuum for a brief period. They will still be distorted, but the bubbles should disappear. Embryos should only be left in large volumes of methanol, or in tightly capped vials.

CONDENSED PROTOCOL

1. Remove the vitelline membranes from embryos.

2. Allow the embryos to sink through distilled water, and replace most of the water with MEMFA. Fix for 1 hour. Alternatively, fix in Dent's solution for 1 hour.

3. Transfer to **methanol** or Dent's to dehydrate (embryos can be stored at this point).

   methanol (see Appendix for Caution)

4. Bleach embryos for 1 hour or more under strong fluorescent light.

5. Rehydrate over the course of 10 minutes in changes of

   80% methanol/20% water

   50% methanol/50% PBS

   20% methanol/80% PBS

6. Incubate in PBT for 10 minutes.

7. Replace PBT with 500 µl of PBT plus 10% goat serum and incubate for 1 hour on the Nutator.

8. Replace with PBT-containing diluted antibody and incubate for 4 hours at room temperature or overnight at 4°C.

9. After one brief wash in PBT, wash three to five times, at least 2 hours each wash (or overnight at 4°C).

10. Incubate with 500 µl of PBT-containing secondary antibody for 4 hours at room temperature or overnight at 4°C on the Nutator.

11. Wash embryos as described previously in three to five changes of PBT.

12. Add 0.5 ml of **DAB** (1 mg/ml in PBT). Incubate with shaking at 4°C for 10 minutes. Add 1 µl of a 1:1 mixture of 30% H_2O_2 mixed with an equal volume of PBT.

    DAB (see Appendix for Caution)

13. When embryos are sufficiently stained, remove the DAB solution and replace with PBT.

14. Wash twice in PBT for 5 minutes.

15. Dehydrate with two changes of 100% methanol over the course of 5 minutes. Clear in two changes of **BB/BA**.

    BB/BA (see Appendix for Caution)

REFERENCES

Bolce M.E., Hemmati-Brivanlou A., Kushner P.D., and Harland R.M. 1992. Ventral ectoderm of *Xenopus* forms neural tissue, including hindbrain, in response to activin. *Development* **115**: 681–689.

Dent J.A., Polson A.G., and Klymkowsky M.W. 1989. A whole-mount immunocytochemical analysis of the expression of the intermediate filament protein vimentin in *Xenopus*. *Development* **105**: 61–74.

Green J.B., Howes G., Symes K., Cooke J., and Smith J.C. 1990. The biological effects of XTC-MIF: Quantitative comparison with *Xenopus* bFGF. *Development* **108**: 173–83.

Harlow E. and Lane D. 1988. *Antibodies: A laboratory manual.* Cold Spring Harbor Laboratory Press, Cold Spring Harbor, New York.

Hemmati-Brivanlou A. and Harland R.M. 1989. Expression of an engrailed-related protein is induced in the anterior neural ectoderm of early *Xenopus* embryos. *Development* **106**: 611–617.

Hemmati-Brivanlou A. and Melton D.A. 1994. Inhibition of activin receptor signaling promotes neuralization in *Xenopus*. *Cell* **77**: 273–281.

Islam N. and Moss T. 1996. Enzymatic removal of vitelline membrane and other protocol modifications for whole mount in situ hybridization of *Xenopus* embryos. *Trends Genet.* **12**: 459.

Kintner C.R. and Brockes J.P. 1984. Monoclonal antibodies identify blastema cells derived from differentiating muscle in newt limb regeneration. *Nature* **308**: 67–69.

Klymkowsky M.W. and Hanken J. 1991. Whole-mount staining of *Xenopus* and other vertebrates. *Methods Cell Biol.* **36**: 419–441.

Lamb T.M., Knecht A.K., Smith W.C., Stachel S.E., Economides A.N., Stahl N., Yancopolous G.D., and Harland R.M. 1993. Neural induction by the secreted polypeptide noggin. *Science* **262**: 713–718.

Larabell C.A., Torres M., Rowning B.A., Yost C., Miller J.R., Wu M., Kimelman D., and Moon R.T. 1997. Establishment of the dorso-ventral axis in *Xenopus* is presaged by early asymmetries in β-catenin that are modulated by the Wnt signaling pathway. *J. Cell. Biol.* **136**: 1123–1136.

London C., Akers R., and Phillips C. 1988. Expression of Epi 1, an epidermis-specific marker in *Xenopus laevis* embryos, is specified prior to gastrulation. *Dev. Biol.* **129**: 380–389.

Mayor R., Morgan R., and Sargent M.G. 1995. Induction of the prospective neural crest of *Xenopus*. *Development* **121**: 767–777.

Mitchison T.J. and Sedat J. 1983. Localization of antigenic determinants in whole *Drosophila* embryos. *Dev. Biol.* **99**: 261–264.

Ruiz i Altaba A. 1992. Planar and vertical signals in the induction and patterning of the *Xenopus* nervous system. *Development* **116**: 67–80.

Serafini T., Colamarino S.A., Leonardo E.D., Wang H., Beddington R., Skarnes W.C., and Tessier-Lavigne M. 1996. Netrin-1 is required for commissural axon guidance in the developing vertebrate nervous system. *Cell* **87:** 1001–1014.

Smith J.C. and Watt F.M. 1985. Biochemical specificity of *Xenopus* notochord. *Differentiation* **29:** 109–115.

Stachel S.E., Grunwald D.J., and Myers P.Z. 1993. Lithium perturbation and goosecoid expression identify a dorsal specification pathway in the pregastrula zebra fish. *Development* **117:** 1261–1274.

CHAPTER 13

Whole-mount In Situ Hybridization

Whole-mount in situ hybridization is the most versatile method for determining where embryonic transcripts are expressed. The information gained using this technique is similar to that obtained from immunohistochemistry and includes not only the location of expression, but a semiquantitative estimate of the relative levels of gene expression in different parts of the embryo. Since most genes are first analyzed as genomic or cDNA clones, it is straightforward to derive specific probes from these sequences. In contrast, it can take considerable time and effort to derive antibody probes to gene products, and in many cases, the antigen is so rare that antibodies are ultimately not useful for localization. Whole-mount in situ hybridization is particularly versatile since many embryos or explants can be processed in one vial; therefore, information about stage-specific differences in gene expression is obtained by processing one sample of mixed embryos. Although a detailed analysis of gene expression often requires analysis of sectioned material, the whole-mount approach is invariably the first method used to localize gene expression. *Xenopus* embryos are rather large for whole-mount in situ hybridization, and reagents tend not to detect transcripts very effectively in the deep yolky endoderm; however, the method is extremely effective for mesodermal and ectodermal gene expression.

Whole-mount in situ hybridization is used to describe when and where newly isolated genes are expressed, but it has also become an invaluable tool for analyzing experimentally manipulated embryos and explants. In whole embryos, experimental manipulation or mRNA injection may not alter the overall levels of gene expression very much, but if the induced gene expression is ectopic, then whole-mount analysis makes new expression very obvious. For example, induction of an ectopic organizer only increases organizer-specific gene expression twofold, but the induced gene expression is prominent because it occurs in a region that normally has no expression (see, e.g., Christian and Moon 1993; Steinbeisser et al. 1993). In addition, the normal location of gene expression provides a built-in control in each embryo. In situ hybridization has been very useful for analyzing manipulated embryos at early stages, since region-specific transcripts can be detected early. In contrast, most useful antibodies used in immunohistochemical analysis only react with anti-gens that accumulate in later development, as cells differentiate.

When analyzing explants, in situ hybridization is an invaluable adjunct to methods such as RNase protection or reverse transcription–polymerase chain reaction (RT-PCR). These methods are extremely sensitive, but they analyze an average response of explants. In situ hybridization can be used to detect variations between explants (see, e.g., Bolce et al. 1992) and therefore provides clues about the contribution of prepatterns in the explant to the heterogeneous response. This technique is particularly useful for determining whether induction of gene expression is a uniform response to a treatment or the result of an occasional atypical response. An example of such an atypical response is occasionally seen in animal caps that are neuralized by noggin treatment (Lamb et al. 1993). Normally, these caps only express rostral markers, but if animal caps are cut large, then noggin may induce some mesoderm that, in turn, is sufficient to induce additional more posterior markers (Lamb et al. 1993; T.M. Lamb and R.M. Harland, unpubl.).

One disadvantage of in situ hybridization analysis is that only one or two markers can be examined in each explant. Therefore, it is important in many experiments to use both whole RNA analysis (to examine several markers) and in situ hybridization (to examine selected markers) in individual explants.

OVERVIEW OF THE METHOD

Whole-mount in situ hybridization relies on nonradioactive methods for probe labeling and detection (Tautz and Pfeifle 1989). The method was first used extensively with *Drosophila* embryos, and the initial methods were then adapted to the much larger embryos of *Xenopus* (Hemmati-Brivanlou et al. 1990; Frank and Harland 1991; Harland 1991).

RNAs are fixed in place with an aldehyde cross-linking reagent, e.g., MEMFA (see Appendix 1). Following permeabilization of the embryo, the fixed RNAs are hybridized with a labeled probe. The most common labeling method for the probe is to transcribe RNA in the presence of a digoxigenin-substituted nucleotide. Following the hybridization step, unhybridized probe is washed away from the embryos, and the digoxigenin is then detected with alkaline-phosphatase-linked antibodies that are highly specific for the digoxigenin moiety. The antibodies are coupled to alkaline phosphatase so that a chromogenic reaction can be used to reveal the enzyme, and therefore (indirectly) the hybridized probe.

The procedure can be varied to include different methods of labeling and therefore achieve detection of multiple RNAs. Other variations include different wash stringencies, with or without ribonuclease; such variations change signal and background levels and also affect how specifically the probe recognizes one mRNA sequence versus a family of related sequences. Finally, if a target mRNA is fairly abundant, and the probe gives robust results, many steps in the procedure can be omitted to save time and work.

Because of the various steps in the in situ hybridization procedure that destroy or inhibit RNases, it is not necessary to take special precautions to remove ribonuclease contaminants. If baskets are used, then both these and their corresponding dishes can be reused. Apparently, the extensive washes and fixation steps are sufficient to remove or destroy RNases that are used in the stringent wash steps. However, it is recommended that a dedicated or disposable bath (such as the top of a blue tip box) be used for the RNase incubation. If RNase remains a concern, solutions can be treated with diethyl pyrocarbonate (DEPC). A DEPC bottle must be opened behind the screen of a chemical fume hood, since accidental exposure to water can cause high pressure to build up in the bottle. One drop of DEPC is used per 100 ml of solution. The solution is shaken, vented, and then autoclaved. DEPC breaks down into ethanol and CO_2. Many solutions (e.g., PTw) can be autoclaved, but autoclaving is dangerous or damaging for others (such as formaldehyde or hybridization solution). No autoclaving is necessary if distilled water is used. If desired, glassware may be baked, and polypropylene plasticware washed in 1 M NaOH before rinsing in distilled water and autoclaving.

Large-scale In Situ Hybridization Screens: Use of Baskets

The in situ hybridization procedure is fairly laborious because it involves multiple solution changes and wash steps. These steps can be greatly simplified if the embryos or explants are transferred to baskets (Stachel et al. 1993). A solution change then only involves moving a rack of baskets from one bath to another. The use of baskets is strongly recommended, and these are discussed in Chapter 12. The amount of the various wash solutions required depends on the configuration of tube racks and baths, and thus the volumes given in these protocols may need to be adjusted. The rest of the procedure is written assuming that the embryos are processed in 5-ml glass vials, but again the use of baskets is strongly recommended. If each sample is to be hybridized to a different probe, single baskets can be inserted into individual tubes for hybridization steps (see Figure 12.1B); then, following the preliminary washes, the baskets are transferred back to racks for the subsequent washes and antibody steps. Solutions such as the hybridization buffer used in prehybridizations can be reused many times.

Preparation of Digoxigenin-labeled RNA Probes

Bacteriophage polymerases are highly selective for their own promoter, so that a single strand of the sequence inserted downstream from the promoter is transcribed. In vitro transcription in the direction opposite to in vivo transcription of the mRNA makes an antisense RNA that is complementary to the mRNA.

The RNA probe is synthesized in the presence of a digoxigenin-substituted nucleotide, dig-UTP. The ratio of dig-UTP to UTP has been optimized to give efficient synthesis and efficient detection of the probe. The three commercially available bacteriophage polymerases (SP6, T7, and T3) all incorporate the substituted nucleotide with equivalent efficiencies (Hemmati-Brivanlou et al. 1990), and thus there is a wide choice of plasmid vectors to prepare the template. In general, probes complementary to large regions of the target mRNA yield higher signal than do short probes. However, very occasionally and unpredictably, long probes cause high background; in which case, several shorter probes may need to be tested, to eliminate the sequence that is causing the problem.

Some cloned cDNAs, which may contain AU-rich 3 -untranslated regions (3 UTR), and a long poly(A) tract, can also be problematic. Polymerases often terminate prematurely in such sequences, and since the 3 UTR is the first sequence encountered in an antisense transcript, such templates can be very inefficient. The problems are usually not severe, especially since poly(A) tracts are unstable in plasmids and tend to shrink to inconsequential length. However, for those plasmids that do transcribe poorly, the cDNA may need to be recloned to eliminate some of the 3 UTR.

Template Preparation

The DNA template is linearized with a restriction enzyme, extracted with phenol, and precipitated prior to use. Occasionally, there is no convenient restriction site available to linearize the template just distal to the *Xenopus* DNA insert. In such cases, the template can be linearized using a site in the plasmid sequence—small stretches of prokaryotic transcript do not interfere or cause excessive background hybridization. Enzymes such as *Hae*II, which contain a CpG dinucleotide in their recognition sequence, are useful for template linearization since they cut frequently in plasmids, but infrequently in *Xenopus* DNA.

When preparing templates, it is necessary to consider what constitutes a good control for nonspecific hybridization. The standard control is a sense probe (i.e., of the same sense as the mRNA). However, background levels caused by different probes vary inexplicably, and some sense probes cause anomalously high background, much higher than the corresponding anti-sense probe. Therefore, acceptable control probes can include irrelevant plasmid sequences or even another specific probe that gives a distinctive pattern of hybridization.

PROTOCOL 13.1

Probe Synthesis

A standard RNA synthesis (Melton et al. 1985) using SP6, T7, or T3 RNA poly-merase is carried out incorporating a digoxigenin-substituted ribonucleotide.

The mixed components can precipitate if cooled, due to the interaction of spermidine with DNA, and thus the reaction should be assembled in a microcentrifuge tube at room temperature. Stocks of sensitive components, such as the enzyme, can still be kept on ice. The reaction recommended here is carried out in a slightly smaller volume than has been recommended else-where, since efficient polymerization of long transcripts is sensitive to nucleotide concentration. The addition of [^{32}P]CTP is optional but very use-ful for estimating the yield of probe.

1. Add the following reagents to a microcentrifuge tube. Mix well and incubate for 2 hours at 37°C.

5x SP6 buffer	5 µl
1 M **dithiothreitol (DTT)**	0.5 µl
2.5 mM dig-NTP stock	10 µl
water (final volume)	to 25 µl
linear template DNA (1µg/µl)	2.5 µl
[^{32}P]CTP (*optional*)	1 µl
RNasIn (20 µ/µl)	0.5 µl
RNA polymerase	90 units

   radioactive materials, DTT (see Appendix for Caution)

2. To the RNA synthesis reaction, add 1 µl of RNase-free DNase I (1 mg/ml) and incubate for a further 10 minutes at 37°C

3. Stop the reaction by adding 100 µl of embryo homogenization buffer (1% **SDS**, 20 mM EDTA, 20 mM Tris at pH 7.5, 100 mM NaCl).

   SDS (see Appendix for Caution)

4. To calculate the yield of probe, remove a 1-µl aliquot from the reaction and determine the total available counts.

   Note: It is most convenient to determine Cerenkov counts (i.e., without added scintillant). The sample can be counted in a microcentrifuge tube in most scintillation counters.

5. Remove unincorporated nucleotides by purifying the probe on a 1-ml Sephadex G-50 spun column (see Appendix 1 and Protocol 13.2).

6. Collect the eluate from the column in a microcentrifuge tube and precipitate the RNA by adding 10 µg of carrier (Torula RNA or glycogen) and 2 volumes of **ethanol**. Collect the precipitate by centrifuging at maximum speed for 15 minutes at 4°C.

   Notes: *The yield should be sufficient for precipitation without the use of carrier, but addition of carrier ensures efficient precipitation and a visible pellet*
   Protocol 13.2 assumes that the Sephadex is prepared in 0.3 M sodium acetate. If not, then salt must be added to ensure precipitation.

   ethanol (see Appendix for Caution)

7. Carefully remove the ethanol and resuspend the RNA in 50 µl of water, *or* partially fragment the RNA by resuspending in 50 µl of 40 mM sodium bicarbonate/60 mM **sodium carbonate** and incubating at 60°C for 35–50 minutes.

   Note: *Under these conditions, the probe should be fragmented to an average size of 300 nucleotides. This can be verified by formaldehyde agarose gel electrophoresis, if necessary. Fragmentation of the probe is optional. It is unnecessary for probes less than 500 nucleotides in length, and even for very long probes fragmentation, it is not essential. However, a side-by-side comparison of intact and fragmented probes shows that fragmentation does improve the hybridization signal.*

   sodium carbonate (see Appendix for Caution)

8. Measure the incorporated counts by counting Cerenkov radiation of the entire sample.

   Note: *If a microcentrifuge tube cannot be placed in the scintillation counter, transfer an aliquot of the reaction to a suitable tube and measure the radiation. Remember to correct for sampling.*

9. Determine the yield of the probe using the following equation.

$$\text{yield in } \mu g = \frac{\text{Incorporated CPM} \times 33}{\text{Total available CPM (obtained in step 4 above)}}$$

   Note: *Incorporation of all the available nucleotide would correspond to 33 µg of RNA in the standard reaction, i.e., $(2.5 \times 10^{-3}$ molar nucleotide$) \times (10 \times 10^{-6}$ liters of nucleotide$) \times 330$ (g/mole nucleotide$) \times 4$ (since only CTP is labeled) $= 33 \times 10^{-6}$ g.*

10. Precipitate the probe be adding 200 µl of water, 25 µl of **sodium acetate** (3 M, pH 5.5), and 600 µl of ethanol. Almost all the counts should precipitate.

    sodium acetate (see Appendix for Caution)

11. Carefully remove the supernatant and resuspend the probe in hybridization buffer to a concentration of 10 µg/ml. Digoxigenin probes are stable at –20°C for more than 8 years and can be diluted as needed for use.

PROTOCOL 13.2

Probe Purification Using a Sephadex Column

1. Prepare a general-use Sephadex for RNA by suspending it in water and allowing it to settle. Remove excess water and resuspend the slurry in 0.3 M **sodium acetate** (pH 5.5) containing 0.1% **SDS**.

   sodium acetate, SDS (see Appendix for Caution)

2. Repeat the procedure to remove soluble dextrans from the Sephadex, and finally resuspend the slurry in 0.3 M sodium acetate (pH 5.5) containing 0.1% SDS. Treat with **diethyl pyrocarbonate** and autoclave.

   Note: *For simple removal of nucleotides to make a probe, washing is not necessary. However, the washing is essential if the same Sephadex is to be used to prepare RNA for injection.*

   diethyl pyrocarbonate (see Appendix for Caution)

3. To prepare the Sephadex column, remove the plunger and protective tip from a 1-ml syringe and place it in a Falcon 2059 tube in a rack.

4. Use two pairs of forceps (sterilized in a flame) to stuff a plug of glass wool into the syringe and tamp it down with a long pasteur pipette.

5. Fill the syringe with Sephadex slurry, from step 2, and centrifuge at low speed in a benchtop clinical centrifuge (e.g., 1000 rpm for 5 minutes in a Sorvall RT6000). Remove the column and discard the eluate from the bottom of the Falcon tube.

6. Remove the cap from a microcentrifuge tube and place the tube inside the Falcon tube. Stand the column inside the microcentrifuge and apply the probe reaction to the surface of the Sephadex. Centrifuge as described in step 5.

7. Transfer the eluate to a fresh tube and precipitate in 2 volumes of **ethanol**.

   ethanol (see Appendix for Caution)

PROTOCOL 13.3

Preparation of Embryos

Embryos are prepared for in situ hybridization by removal of the vitelline membrane, fixation, and dehydration. Albino embryos are the preferred subjects for this technique because histochemical color development can be easily monitored. However, the advantages of albino embryos must be balanced against the advantages of pigmented embryos; the latter are easier to stage, easier to inject in specific locations, and, most importantly, generally more robust. Although the pigment in these embryos obscures in situ hybridization stains almost completely, a rapid bleaching can be carried out after staining with little loss of signal. It is still extremely useful to keep a stock of albino embryos; these albinos can be included in the experiment and used to monitor the color reaction.

The lack of pigment in albino embryos makes identification of cleavage planes and surface features, such as the blastopore and neural folds, very difficult, which makes stage determination very difficult. To assist developmental stage determination, albino embryos can be stained, after dejellying, with Nile blue (0.01% Nile blue in 50 mm phosphate buffer at pH 7.8).

1. Remove the vitelline membranes after proteinase K treatment or manually (see Protocol 6.2).

   Notes: *For large numbers of embryos, an enzymatic pretreatment with proteinase K (5 μg/ml for ~5 minutes) is recommended. Monitor digestion under the dissecting microscope so that when the membrane begins to lift from the surface of the embryo, or if any membrane ruptures, the embryos can be washed immediately in 0.3x Ringer's solution. Softening the vitelline membrane at blastula stages allows easy removal of the membrane at various stages thereafter. This early protease treatment is particularly recommended for the subsequent harvesting of neurula and tailbud stages, since the embryo develops detailed surface morphology, assisting stage determination.*

   Loosening of the vitelline membrane can inhibit closing of the blastopore in some embryos. This problem can be prevented by immersing embryos in 0.3x Ringer's solution containing 2% (w/v) Ficoll during gastrulation. Transfer the embryos to Ficoll-free medium following gastrulation.

   The greatest control over subsequent steps is provided by removal of the membrane before fixation. However, if this step is omitted, the proteinase K step prior to hybridization (Protocol 13.4, step 4) will remove the membrane from almost all the embryos. Alternatively, remove the membrane before hybridization with a cocktail of proteases (embryos can be treated for 10 minutes at room temperature in PBS containing 10 μg/ml proteinase K, 2 mg/ml collagenase A, and 20 units/ml hyaluronidase [Islam and Moss 1996]).

The membrane softens and lifts from the surface of the embryo, making it easy to grasp with forceps.

2. For embryos at blastula through tailbud stages, use an eyebrow knife or needle to pierce the blastocoel and archenteron to prevent the development of background staining in these cavities.

Note: *This step is not necessary when using probes that give robust signals, but it is necessary when using weaker probes. Tadpoles can develop background staining in the pharynx, so this too can be punctured.*

3. Transfer the embryos to a 5-ml screw-cap glass vial (e.g., Fisher 3338B), prefilled with distilled water.

Notes: *To prevent lysis of the embryos, avoid contact with the air. Label the vial using a diamond pencil (pen marks and tape are easily lost during the subsequent steps of the protocol).*

4. When the embryos have settled to the bottom of the vial, remove most but not all of the water (the embryos will break if all of the water is removed). Refill the vial almost completely with the **MEMFA** fixative.

MEMFA (see Appendix for Caution)

5. Rotate the vial, end over end, for 1–2 hours at room temperature (e.g., 8 rpm in a Labquake rotator; Labindustries Inc.).

Notes: *Make sure that the embryos remain in suspension. Early-stage embryos flatten considerably unless kept in suspension during the early part of the fixation.*

When fixing a large number of embryos, use a larger vial (e.g., a large scintillation vial) to prevent the embryos from clumping.

6. Replace the MEMFA with absolute **ethanol** and continue the rotation for a further few minutes at room temperature.

Note: *Earlier protocols call for replacement of MEMFA with methanol. This is no longer recommended since methanol (unlike ethanol) does not dissolve lipid; remaining lipid tends to accumulate in the blastocoel where it accumulates stain and contributes severely to background staining.*

ethanol (See Appendix for Caution)

7. Replace the ethanol with fresh ethanol and store the fixed embryos at
   –20°C.

   Note: *After fixation, embryos are quite robust, and in all subsequent steps, all
   of the liquid can be removed when changing solutions. Embryos can be stored
   for several months, but give the most intense signals if used in the first few
   days after fixation.*

PROTOCOL 13.4

Hybridization

For hybridization, fixed embryos are removed from storage, rehydrated, and then subjected to protease treatment, which increases sensitivity to the probe; this is followed by incubation in prehybridization solutions, which helps to prevent nonspecific sticking of probe. The hybridization conditions then allow efficient duplex formation between the fixed mRNAs in the sample and the digoxigenin-labeled probe.

In many experiments, embryos of different stages are taken from storage and combined so that a single hybridization reaction can be carried out on multiple stages. For example, when investigating the expression of a newly isolated gene, it is important to use a comprehensive series of embryonic stages and to use enough embryos to determine whether a particular result is reproducible. It is useful to set up duplicate vials, so that the chromogenic reaction can be carried out for a short time (to detect abundant expression), as well as a longer period of time (to detect weak sites of gene expression). Approximately 30 embryos can be processed in a single vial.

It is also important to set up a control hybridization with a sense probe, particularly to judge the rate of appearance of background staining in the chromogenic reaction. In addition, it is useful to set up a probe with a known expression pattern as a positive control. For tadpoles, En-2 (Hemmati-Brivanlou et al. 1991) is a useful positive control, since it is expressed locally, has two very different intensities of expression (at the midbrain-hindbrain boundary and in the mandibular arch), and has very low background elsewhere in the embryo. Robust expression in the brain should be evident in the chromogenic reaction after a few minutes; however, successful detection of weaker expression will provide a more stringent test of the procedure.

1. Transfer fixed embryos from their storage vials to a dish of **ethanol**.

    Note: *It may be necessary to first tap the vial to dislodge embryos that have stuck to the glass.*

    ethanol (see Appendix for Caution)

2. Use a wide-bore plastic pipette to select good quality embryos and transfer them to a hybridization vial.

   Note: *Plastic pipettes are favored over glass since embryos in ethanol are less likely to stick to the walls.*

3. Rehydrate the embryos by washing in each of the following solutions (2–5 minutes per wash):

   75% **methanol,** 25% water
   50% methanol, 50% water
   25% methanol, 75% PTw
   100% PTw

   Carry out a final wash in PTw for 5 minutes.

   methanol (see Appendix for Caution)

4. Permeabilize the embryos by incubating them for approximately 15 minutes at room temperature in 1 ml of 10 µg/ml proteinase K. This is best achieved using a nutator. *Do not* subject embryos in small volumes to end-over-end rotation or horizontal rocking.

   Notes: *Proteinase K treatment is not always necessary but it does increase the embryos' sensitivity to staining; e.g., En-2 transcripts in the brain are easily detected without treatment, but good visualization of the branchial arch transcripts is only possible after proteinase K treatment. Unfortunately, excessive treatment does not become evident until the end of the hybridization procedure, when over-proteolyzed embryos break up. The epidermis often peels away from the embryo if protease use is heavy; thus, if transcripts are expected in the superficial layer, it is advisable to omit protease treatment.*

   Incubation for 15 minutes is given as a conservative guideline for gastrula stages. A shorter incubation of 5 minutes should be used for explants. If proteinase K is to be omitted, continue with step 5 below.

5. Wash the embryos twice for 2–5 minutes in 5 ml of 0.1 M **triethanolamine** (pH 7.0–8.0; Sigma T 1502) with horizontal rocking on the nutator.

   triethanolamine (see Appendix for Caution)

6. Add 12.5 µl of acetic anhydride (Sigma) and incubate for a further 5 minutes.

   Notes: *This treatment acetylates and neutralizes free amines and helps to prevent elecrostatic interaction between the probe and basic proteins in the embryo.*

Acetic anhydride is an oil that dissolves gradually in the aqueous layer. It breaks down rapidly when exposed to water. Breakdown is evident if the added acetic anhydride mixes immediately with the aqueous phase. Addition of poorly stored acetic anhydride is equivalent to a pointless addition of acetic acid.

7. Repeat step 6.

8. If proteinase K was used, refix embryos for 20 minutes in PTw containing 4% **paraformaldehyde**.

   paraformaldehyde (see Appendix for Caution)

9. Wash embryos twice for 3 minutes in Ptw.

10. Remove all but approximately 1 ml of PTw from the vial and add 250 µl of hybridization buffer (without mixing).

11. Once embryos have settled through the dense layer of buffer, remove all the buffer and replace with 0.5 ml of fresh hybridization buffer per vial. Incubate for 10 minutes at 60°C in a shaking water bath.

12. Replace hybridization buffer and prehybridize for 4–6 hours at 60°C. Buffer used for prehybridization steps can be reused.

13. Remove hybridization buffer and replace with fresh hybridization buffer containing 0.5 µg/ml probe. Hybridize overnight at 60°C.

14. Remove solution containing probe.

    Note: *Used probe can be stored at 20°C for reuse (probes can be used at least three times).*

PROTOCOL 13.5

Washing the Embryos

1. Rinse the embryos for 10 minutes at 60°C in fresh hybridization buffer and then wash three times in 2x SSC for 20 minutes at 60°C.

2. Replace wash solution with 2x SSC containing 20 µg/ml RNase A and 10 µg/ml RNase T1. Incubate vials for 30 minutes at 37°C.

   Notes: *If using baskets in a glass dish, carry out this step in a separate box such as a plastic food storage container reserved for RNase treatment.*

   This step ensures that the hybridization is extremely stringent. Cross-reaction of probes, even between members of closely related gene families, is rare. However, the stringent wash does reduce the signal somewhat, and the RNase step is often omitted in order to boost signal. Under these conditions, cross-hybridization between members of a gene family can occur.

3. Remove excess RNase by washing the embryos once for 10 minutes at room temperature in 5 ml of 2x SSC, and then twice for 30 minutes at 60°C in 5 ml of 0.2x SSC.

4. Wash twice for a total of 10 minutes at room temperature in 5 ml of **MAB**.

   MAB (see Appendix for Caution)

PROTOCOL 13.6

Antibody Incubation

1. Replace **MAB** with 1 ml of MAB containing 2% BMB blocking reagent (Boehringer Mannheim 1096 176). Place vials vertically on the nutator to agitate and wash for 1 hour at room temperature.

   Note: *Blocking reagent dissolves readily on heating but the solution remains cloudy. The inclusion of MAB/BMB blocking reagent was developed by Tabitha Doniach and significantly reduces background staining in prolonged incubations (Lamb et al. 1993; Doniach and Musci 1995).*

   MAB (see Appendix for Caution)

2. Replace solution with fresh MAB containing 2% BMB blocking reagent and a 1/2000 dilution of the affinity-purified antidigoxigenin antibody coupled to alkaline phosphatase (Boehringer Mannheim 1093 274). Rock vertically overnight at 4°C, or for 4 hours at room temperature.

   Note: *The antibody solution can be preincubated overnight at 4°C with fixed embryos to remove nonspecific binding components. In practice, there is very little background from such nonspecific interactions, so such a preincubation is generally unnecessary. The antibody working solution can be recycled at least three times.*

3. Remove excess antibody by washing several times in MAB (e.g., eight 20-minute washes or five 1-hour washes) at room temperature. One of the washes can be overnight at 4°C.

   Note: *It is very difficult to overwash the samples, so the wash schedule can be determined by the worker's degree of impatience. The antibody complex is extremely stable and is not significantly removed even by washing for 3 days at 37°C.*

4. For the chromogenic reaction, first wash embryos twice for 5 minutes at room temperature in alkaline phosphatase buffer (see Appendix 1)

5. Replace the last wash with BM purple (Boehringer Mannheim 1442 074) or with alkaline phosphatase buffer containing 4.5 μl/ml **NBT** and 3.5 μl/ml **BCIP**. Use 0.5–2 ml of buffer per vial and incubate at room temperature until staining becomes apparent. This can take from 5 minutes to 24 hours.

Note: *If expecting deep staining, incubate at a lower temperature (4–15°C) to improve the penetration of stain (see Chapter 12; Sasai et al. 1996). For incubations in excess of 1 hour, cover the rack with foil.*

NBT, BCIP (see Appendix for Caution)

6. Monitor the staining process periodically by tipping the vial and placing it under the dissecting microscope. Accelerate or decelerate the staining by altering the temperature from 4°C to 37°C. It is worth staining the embryos particularly intensely if they are to be sectioned.

   Note: *BM purple can give two shades of blue stain, particularly if there are sites of very intense gene expression. Therefore, for detecting transcripts that are very strongly expressed, the NBT/BCIP stain is preferable. A modified buffer for NBT/BCIP uses polyvinyl alcohol and reduces background. However, the solution is viscous and messy and is thus only recommended for use with weaker probes. For PVA alkaline phosphatase buffer, see Appendix 1.*

7. When staining becomes apparent, tip the embryos into a dish of alkaline phosphatase buffer and examine under the dissecting microscope. Return any poorly stained embryos to the staining solution. If the staining is sufficiently clear, carry out examination through the glass vial.

8. Rinse the embryos, replace the solution with fresh MEMFA, or Bouin's Fix, and incubate at room temperature overnight. If embryos are to be bleached, Bouin's must be used.

   Note: *This refixing treatment is essential if the embryos are to be mounted in **BB/BA**. Unfixed stains dissolve on clearing. Bouin's recipe fixes the stain more satisfactorily than MEMFA and tends to remove background staining more efficiently. However, it gives the embryos a yellow counterstain, and unless this fix is washed out completely, embryos tend to remain somewhat opaque in clearing solution. Omitting the picric acid from Bouin's prevents the yellow stain, but retains the advantage of efficient removal of chromogens.*

   BB/BA (see Appendix for Caution).

9. Remove background and Bouin's fix with several washes of buffered 70% **ethanol** (prepare the ethanol with TE or PTw as the aqueous component).

   Note: *Several washes are essential to remove chromogenic components before bleaching. Residual components can cause the embryos to turn black.*

   ethanol (see Appendix for Caution)

10. If pigmented embryos were used, transfer well-washed embryos to vials containing bleaching solution (1% H_2O_2, **5% formamide**, 0.5x

SSC [Mayor et al. 1995]). Place vials on a background of aluminum foil, as close as possible to a fluorescent light, and incubate at room temperature, with rocking, for 1 hour or until adequate bleaching is obtained. If albino embryos were used, proceed with step 11.

Note: *For superficial stains, the embryos may be best examined and photographed in aqueous media such as PBS. However, for deep stains, the embryos must be cleared, as described in steps 11 and 12.*

H_2O_2, **formamide** (see Appendix for Caution)

11. Dehydrate the embryos by incubating in two changes of **methanol** (5 minutes each) and transfer to depression slides, spot plates, or Sylgard slides (construction of Sylgard slides is described in Protocol 13.7).

    methanol (see Appendix for Caution)

12. Clear the embryos by removing most of the methanol and replacing it with 250 µl of 2:1 BB/BA. After embryos settle through the BB/BA, replace with fresh BB/BA.

    Note: *Polystyrene plates dissolve in BB/BA. To avoid contamination with BB/BA, use it sparingly and clean any spills promptly with wipes and 95% ethanol. It is useful to have an aspirator nearby to draw up any droplets of BB/BA that creep out of the wells.*

Helpful Hints

- BB/BA is a very effective clearing agent, and it can be difficult to see features other than the stain. Therefore, if some contrast is needed, e.g., to see the blastopore lip of an early gastrula, mount the embryos in 100% benzyl benzoate and view with Nomarski optics.

- BB/BA is not volatile, and refixed embryos can be stored in these plates for several hours without loss of signal. However, the stain is sensitive to light and air, so for long-term storage, embryos should be transferred back to vials or sealed in Sylgard slides and placed in a dark box or drawer. The stain is stable for several months if the embryos are sealed and kept in the dark. For long-term storage, transfer to vials containing 70% ethanol or 70% methanol and seal well to prevent drying out.

- Alternatively, wash the embryos in methanol and rehydrate. If storing in aqueous media, it is advisable to add EDTA to 1 mM to prevent fungal growth. The stain is very stable in 80% glycerol, and there is no danger of the embryos drying out, even if the vial is not tightly capped.

PROTOCOL 13.7

Preparation of Sylgard slides

Suitable mounting chambers can be made in the laboratory using Sylgard 184 elastomer (Dow Corning Corp.). The chambers use a microscope slide as a base, and therefore provide a good optical mount for photography. The mounting well itself is composed of a layer of Sylgard elastomer, which is resistant to benzyl benzoate. A chamber is cut into the layer of Sylgard with a razor blade. Coverslips seal well onto the tacky surface of Sylgard, making a temporary chamber that is ideal for microscopy.

1. Arrange the appropriate number of microscope slides in the bottom of an alluminium foil tray, as shown in Figure 13.1.

    Note: *The sides of the tray should be approximately 1 cm deep to prevent loss of the elastomer.*

2. If possible, place the tray on a microscope slide warmer at 60°C.

3. Prepare approximately 40 ml of the Sylgard elastomer (sufficient for ~165 slides), according to the manufacturers instructions, and pour it evenly onto the slides. Cure overnight at room temperature, or for 2 hours at 60°C.

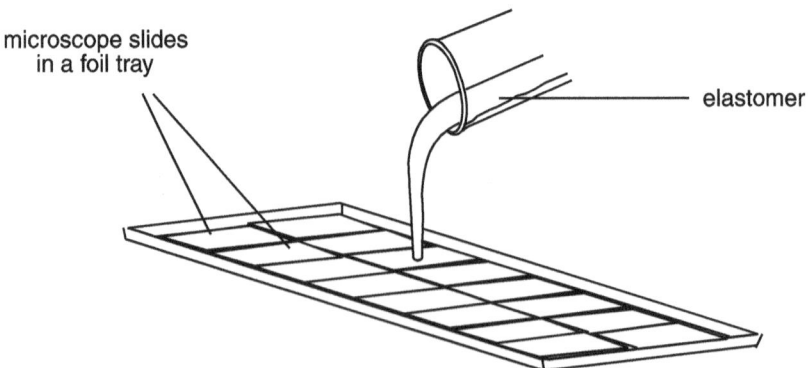

Figure 13.1. Set up for pouring Sylgard slides. Microscope slides are set on aluminum foil, and the edges of the aluminum are folded up. The Sylgard is poured over the slides to make a layer that polymerizes over the entire surface. The slides are then cut apart.

Note: *The volume of elastomer can be adjusted to produce slides of different thickness; 1-mm depth is a good starting point for gastrula stages, but tadpole stages can be mounted in much thinner slides.*

4. Cut the slides apart with a sharp scalpel or single-edge razor blade and cut a rectangular well in the center of each slide (see Figure 13.2).

   Note: *At least 5 mm of elastomer should be left around the edges of the well to prevent leakage of BB/BA. Sylgard can be cut away from the end of the slide to leave a space for labeling. The elastomer is more easily removed while fresh.*

5. Transfer the embryos from **BB/BA** to the well, remove excess BB/BA to leave only a small convex meniscus. Add a coverslip and use a diamond pencil to label the slides.

   Note: *BB/BA in which the specimen is mounted removes ink.*

   BB/BA (see Appendix for Caution)

6. Tip the slide and blot excess BB/BA from the edge of the coverslip.

   Note: *The coverslip will stay in place, but after several days, the BB/BA may need to be replaced. Take care with BB/BA; it is an irritant and has been reported to dissolve microscope glue. BB/BA can be cleaned up with 95%* ***ethanol.***

   ethanol (see Appendix for Caution)

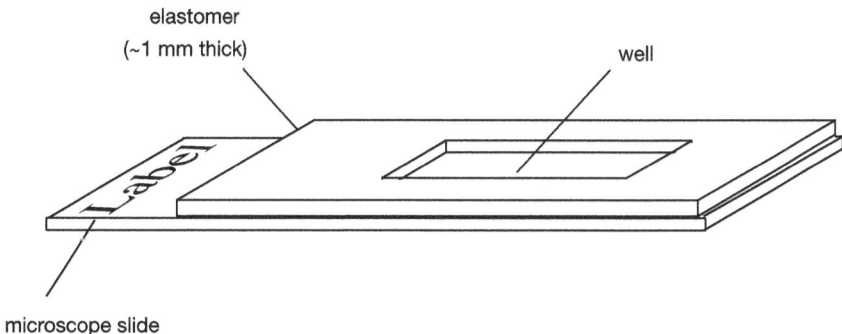

Figure 13.2: Sylgard slide. The end of the Sylgard is trimmed to make a space for the label, and a well is cut into the Sylgard to receive the embryos. (Courtesy of Mark Curtis.)

DOUBLE STAINING IN SITU PROCEDURE FOR WHOLE-MOUNT IN SITU HYBRIDIZATION

Protocols 13.4 to 13.6 can be used with fairly minor modifications to detect two different mRNAs in the same embryos. The two probes must be labeled with different nucleotide analogs, such as digoxigenin-UTP and fluorescein-UTP. The embryos are then incubated sequentially with alkaline-phosphatase-coupled antibodies against digoxigenin and fluorescein, with different chromogenic reactions for the two steps. The modified technique has the obvious advantage that the same embryo can yield information on the relative distribution of two mRNAs, which is difficult to do with a side-by-side comparison of singly stained embryos or explants. The method is most effective when the two mRNA are expressed in distinct regions, since the overlap of two colored products is not always apparent.

The method has been used mainly with digoxigenin probes in combination with fluorescein probes, but other substitutions are possible and have been exploited, e.g., dinitrophenyl (DNP)-substituted probes, bringing the possibility of triple staining (Jowett and Lettice 1994; Doniach and Musci 1995; Knecht et al. 1995).

The Probes

For double staining in situ, one probe is synthesized with digoxigenin-substituted UTP, and the other probe is synthesized with fluorescein-substituted UTP (Boehringer Mannheim 1427 857), using the same concentrations and procedures previously described in Protocol 13.1. If required, a third probe can be labeled with DNP-UTP (Molecular Probes); however, this label appears to be less efficient than the digoxigenin and fluorescein labels.

The Hybridization Procedure

The hybridization procedure for double staining is carried out exactly as described for single staining in Protocol 13.4. At step 13 of Protocol 13.4, add the two (or three) probes simultaneously to the hybridization solution, at a final concentration of 0.5 μg/ml each.

Antibody Incubation

Although single staining with digoxigenin and fluorescein probes gives comparable results, in double staining, the second probe usually stains less efficiently. For this reason, it is advisable to stain the less intensely expressed RNA first. To reduce background staining with fluoresceinated probes, use the anti-fluorescein-AP-Fab fragment (Boehringer Mannheim 1426 338) at 1:10,000 rather than at the 1:2,000 dilution recommended by the manufacturer. For DNP detection, a primary anti-DNP followed by a secondary enzyme-linked antibody must be used, since no alkaline-phosphatase-linked antibody to DNP is currently available (Doniach and Musci 1995).

Inactivating the Alkaline Phosphatase

The procedure for double staining in situ is the same as that for single staining except that the first-round staining reaction must be stopped and the alkaline phosphatase inactivated before the second antibody incubation and staining reaction can be carried out.

When staining reaches the desired level, wash samples several times in MAB to remove the alkaline phosphatase substrate. There are several methods to destroy alkaline phosphatase. The usual procedure is to use two methods in series: (1) incubate embryos for 10 minutes at 65°C, in MAB containing 10 mM EDTA and (2) dehydrate embryos in methanol.

At this point, the samples can be cleared in BB/BA and examined. Otherwise, they should be rehydrated gradually in MAB. After the 1-hour blocking reaction with MAB containing 2% BMB blocking reagent, the antibody against the second probe is added. The washes are done as described above, and the second alkaline phosphatase substrate is used. Finally, after all staining is done, the embryos are refixed with either MEMFA or Bouin's fix. Bouin's is recomended for fixing unstable stains, such as BCIP. Excess picric acid should be removed by extensive washing with 70% ethanol (in TE or PBS at neutral pH). MEMFA supplemented with 0.2% glutaraldehyde also stabilizes stains better than MEMFA alone.

An alternative procedure uses low pH to inactivate alkaline phosphatase. This is essential when using Fast Red or INT/BCIP as substrates (see below), as the color leaches out on heating or in organics. After the first alkaline phosphatase reaction, wash three times (5 minutes each) in PTw, and then once in 0.1 M glycine-HCl (pH 2) and 1% Tween for 40 minutes. Wash four times in MAB for 10 minutes each; then proceed with second antibody incubation and color reaction.

Substrates for Alkaline Phosphatase

A growing list of alkaline phosphatase substrates and kits is available. For single stains, NBT/BCIP and BMB purple are very effective. However, in double stains, BMB purple can react to give stains of different colors, a dark purple stain and a turquoise stain, very similar to the product of BCIP alone. This has only been observed with abundantly expressed RNAs, but it can be confusing in double stains, so BMB purple should be reserved for weakly expressed mRNAs and not used in conjunction with BCIP alone. The following reagents and kits give good results in double-staining procedures.

BCIP Alone

BCIP stains turquoise and is usually somewhat diffuse. After several hours in BB/BA, this stain fades away completely in embryos refixed with MEMFA. Therefore, embryos should be refixed with Bouin's (see above) or with MEMFA containing 0.2% glutaraldehyde (Stachel et al. 1993).

Vector Black

Vector black is obtained from Vector Laboratories (AP Substrate Kit II SK-5200). This dye stains dark brown.

Magenta-phos

Magenta-phos is obtained from Biosynth (B-755). This reagent stains magenta. The staining reaction with this reagent is very slow, but it gives low background. Magenta-phos should be made up at 25 mg/ml in DMF and used at 7 µl/ml of buffer. Be aware that the pinkish purple darkens to a bluish purple during very long stainings, especially at 37°C. Incubations can be carried out for up to 24 hours at 37°C, but then should be continued at room temperature to preserve the color. With weaker signals, several days staining may be required, but fortunately there is little or no background from magenta-phos reagent. Since this product has such low background, it is good for relatively rare RNAs and provides excellent color contrast with the product of BCIP.

Interactions occur between some other substrates and magenta-phos. For example, BMB purple and NBT/BCIP appear to alter the magenta color significantly, making it difficult to distinguish the two colors. The following stain combinations work well:

- NBT/BCIP first, followed by Vectastain Kit II (Vector labs)
- BCIP alone, then NBT/BCIP
- Magenta-phos and BCIP

Fast Red and INT (Iodonitrotetrazolium Voilet)/BCIP

Although all of the above stains are suitable for embryos that will be cleared, other stains such as Fast Red and INT/BCIP can also be used. However, they are not stable to clearing agents and thus should only be used for small explants or superficial stains.

For abundant RNAs, Fast Red gives very good color contrast with the product of NBT/BCIP. Fast Red tablets (Boehringer Mannheim 1486 549) are dissolved in 2 ml of staining solution. Because the Fast Red substrate is only active for approximately 4 hours, the solution must be prepared just before use. When additional staining proves to be necessary, a fresh solution must be prepared. If Fast Red is used first in a double-stain reaction, the staining must be stopped by acidification. Heating in MAB plus EDTA or methanol fixation will cause the red color to leach out.

The product of INT (Molecular Probes; used in combination with BCIP) is also soluble in alcohols, but for specimens that will be viewed in aqueous solution, it provides a good orange color contrast with the violet from NBT/BCIP.

REFERENCES

Bolce M.E., Hemmati-Brivanlou A., Kushner P.D., and Harland R.M. 1992. Ventral ectoderm of *Xenopus* forms neural tissue, including hindbrain, in response to activin. *Development* **115:** 681–688.

Christian J.L. and Moon R.T. 1993. Interactions between Xwnt-8 and Spemann organizer signaling pathways generate dorsoventral pattern in the embryonic mesoderm of *Xenopus. Genes Dev.* **7:** 13–28.

Doniach T. and Musci T.J. 1995. Induction of anteroposterior neural pattern in *Xenopus*: Evidence for a quantitative mechanism. *Mech. Dev.* **53:** 403–413.

Frank D. and Harland R.M. 1991. Transient expression of XMyoD in non-somitic mesoderm of *Xenopus* gastrulae. *Development* **113:** 1387–1393.

Harland R.M. 1991. In situ hybridization: An improved whole-mount method for *Xenopus* embryos. *Methods Cell Biol.* **36:** 685–695.

Hemmati-Brivanlou A., Frank D., Bolce M.E., Brown B.D., Sive H.L., and Harland, R.M. 1990. Localization of specific mRNAs in *Xenopus* embryos by whole-mount in situ hybridization. *Development* **110:** 325–330.

Hemmati Brivanlou A., de la Torré J.R., Holt C., and Harland R.M. 1991. Cephalic expression and molecular characterization of *Xenopus* En-2 Development **111:** 715–724

Islam N. and Moss T. 1996. Enzymatic removal of vitelline membrane and other protocol modifications for whole mount in situ hybridization of *Xenopus* embryos. *Trends Genet.* **12:** 459.

Jowett T. and Lettice L. 1994. Whole-mount in situ hybridizations on zebrafish embryos using a mixture of digoxigenin- and fluorescein-labelled probes. *Trends Genet.* **10:** 73–74.

Knecht A.K., Good P.J., Dawid I.B., and Harland R.M. 1995. Dorsal-ventral patterning and differentiation of noggin-induced neural tissue in the absence of mesoderm. *Development* **121:** 1927–1935.

Lamb T.M., Knecht A.K., Smith W.C., Stachel S.E., Economides A.N., Stahl N., Yancopolous G.D., and Harland R.M. 1993. Neural induction by the secreted polypeptide noggin. *Science* **262:** 713–718.

Mayor R., Morgan R., and Sargent M.G. 1995. Induction of the prospective neural crest of *Xenopus*. *Development* **121:** 767–777.

Melton D.A., Kreig P.A., Rebagliati M.R., Maniatis T., Zinn K., and Green M.R. 1985. Efficient in vitro synthesis of biologically active RNA and RNA hybridization probes from plasmids containing a bacteriophage SP6 promoter. *Nucleic Acids Res.* **12:** 7035–7056.

Sasai Y., Lu B., Piccolo S., and De Robertis E.M. 1996. Endoderm induction by the organizer-secreted factors chordin and noggin in *Xenopus* animal caps. *Embo J.* **15:** 4547–55.

Stachel S.E., Grunwald D.J., and Myers P.Z. 1993. Lithium perturbation and goosecoid expression identify a dorsal specification pathway in the pregastrula zebra fish. *Development* **117:** 1261–1274.

Steinbeisser H., De Robertis E.M., Ku M., Kessler D.S., and Melton, D.A. 1993. *Xenopus* axis formation: Induction of goosecoid by injected Xwnt-8 and activin mRNAs. *Development* **118:** 499–507.

Tautz D. and Pfeifle C. 1989. A non-radioactive in situ hybridization method for the localization of specific RNAs in *Drosophila* embryos reveals translational control of the segmentation gene hunchback. *Chromosoma* **98:** 81–85.

CHAPTER 14

Histology

Histology is a very useful tool for scoring the effects of various treatments or the results of misexpressing a gene on explants or embryos. In combination with in situ hybridization, it can be used to obtain accurate information on the location of RNA or protein expression.

Early *Xenopus* embryos can be difficult to section because the cells are large and have a high yolk content. However, with good fixation and embedding, excellent results can be obtained. Embryos to be sectioned after staining in whole mount should be deliberately overstained to compensate for the relatively small amount of stain in single sections.

Two protocols are presented here, a paraffin-embedding method (Protocol 14.1) and a plastic-embedding method (Protocol 14.2). The paraffin method gives good results for general light microscopy and is relatively simple. The plastic method is rather more complicated but yields superior preservation of tissue and cell morphology.

Materials for embedding and sectioning are available from most major catalogs. For small numbers of samples, glass vials are used for embedding. For larger numbers, the "baskets" described in Chapter 12, can be used.

PROTOCOL 14.1

Embedding in Paraffin

Various paraffin waxes are available for histology. Paraplast, or a 1:1 mixture of Paraplast with Tissue-prep 2 (Fisher), is commonly used. To orient the specimen in wax, the following equipment is required:

- Warming plate (a heating block with the tube holder inverted works well, e.g., Baxter/SP H2029) preheated to melt the wax, slightly above 60°C.

- Dissecting microscope

- Old forceps/mounted needle

- Molten Paraplast. It is useful to maintain several hundred milliliters of molten paraffin wax in a backup oven. When embedding specimens, the oven can cool excessively because of repeated door opening.

- Embedding rings and base molds

- Small Bunsen burner or alcohol burner

- Microtome

- Paint brushes

- Slide warmer

- Subbed slides. Soak microscope slides briefly in an adhesive such as Gatenby and Cowdry, drain well, and allow to air dry. Subbed slides can be accumulated and stored in microscope slide boxes.

- Gatenby and Cowdry adhesive. Melt 15 g of gelatin in 500 ml of hot water, and dissolve 1 g of chromic potassium sulfate (chrome alum) in 120 ml of cold water. When the gelatin solution has cooled, mix with the chrome alum solution and add to 70 ml of **acetic acid** and 300 ml of **methanol.**

  acetic acid, methanol (see Appendix for Caution)

1. Prepare phosphate buffer (0.4 M stock and 0.1 M working solution).

   For 1 liter:

   $NaH_2PO_4 \cdot H_2O$ 10.6 g   (or 9.22 g NaH_2PO_4)

   K_2HPO_4 56 g   (or 73 g $K_2KPO_4 \cdot 3H_2O$)

   Titrate with concentrated HCl to pH 7.4.

   Dilute one part 0.4 M stock with three parts water to make 0.1 M phosphate buffer.

   NaH_2PO_4 (see Appendix for Caution)

2. Prepare 4% **paraformaldehyde** in 0.1 M phosphate buffer.

   For 100 ml:

   a. Heat 60 ml of water to 65°C.

   b. Add 4 g of paraformaldehyde.

   c. Dissolve. If cloudy, add a few drops of 6 M **NaOH.**

   d. At room temperature, add 25 ml of 0.4 M phosphate buffer.

   e. Add water to 100 ml.

   paraformaldehyde, NaOH (see Appendix for Caution)

3. Fix the specimens by incubating at room temperature for 30–60 minutes in 4% paraformaldehyde in 0.1 M phosphate buffer.

4. Remove fixation buffer and wash the embryos at room temperature for 1 hour in 0.1 M phosphate buffer containing 5% sucrose.

5. Dehydrate the embryos in a series of 10-minute **ethanol** washes, beginning with 70% ethanol, increasing to 90% ethanol for the second wash, and to absolute ethanol for the final two washes.

   ethanol (see Appendix for Caution)

6. Clear the specimens by incubating them in a **xylene** substitute, such as Hemo-de (Fisher) or **amyl acetate**, at room temperature overnight in a chemical fume hood.

   Note: *Samples should not be left for extended periods in clearing solvents or heated paraffin since they can become brittle.*

   xylene, amyl acetate (see Appendix for Caution)

7. Infiltrate samples with paraffin wax by incubating in a series of solvent/paraffin mixtures at 60°C for 20 minutes, beginning with 1:1 solvent/paraffin and progressing through three changes of paraffin wax for 20 minutes each.

   Note: *It is helpful to carry out the final infiltration at reduced pressure (in a vacuum oven) to ensure good penetration of wax and removal of solvent, although this is more important for specimens larger than* Xenopus *embryos.*

8. Assemble equipment required for embedding and sectioning, as described in the introduction to this protocol.

9. Place the base mold on the heated plate and fill partially with molten Paraplast. Tip the embryos from the vial into a small petri dish containing molten paraplast.

10. Use a heated wide-bore pipette to transfer the infiltrated embryo(s) into the mold (if a solid scum forms on the surface of the wax, it can be remelted with heated forceps).

    Note: *When learning, it is best to orient just one embryo per block; with practice, up to four can be included in a single block. Embryos and paraffin should not be left on the block for long periods, since the elevated temperature will damage the paraffin wax.*

11. Move the mold to the side of the heating block so that the base of the paraplast begins to solidify. As a carpet of solid Paraplast forms, nudge the embryos into place with heated forceps.

12. Apply the embedding ring and fill the ring to the top with Paraplast. Before topping up, remelt any solid Paraplast that has formed at the surface (to prevent formation of a brittle interface in the Paraplast). Cool to solidify.

13. When cool, the block should separate easily from the base mold. Using a dissecting microscope to view the specimen, trim the block with a scalpel or single-edge razor blade to within about 1 mm of the embryos. Ensure that the top and bottom surfaces of the block are parallel. This will prevent the ribbon from curling during sectioning.

    Note: *To avoid cracking the specimen from the mounting ring, trim the wax by shaving the sides of the block. The block should be slightly trapezoidal, with the wider surface closer to the knife when mounted on the microtome.*

14. Sectioning is difficult to describe, and it is best to watch a demonstration of the technique. Paraplast sections make fine ribbons of 8–20 μm thickness, which can be collected and manipulated with paint brushes. Collect the ribbons in a box for later mounting. To relax the wrinkles in the sections, warm the sections by floating them on the surface of a water bath or by floating them on a drop of water on the microscope slide.

15. Mount ribbons on slides that have been subbed as described in the introduction to this protocol. To mount the sections, put one or two drops of water on the slide, and "paint" the water out into a rectangle that leaves space on all sides. Cut ribbons into lengths shorter than the drop (they will expand) and place as many ribbons as possible onto this area of the slide.

16. Place the slides on a slide warmer at 45°C to relax the sections, and then allow them to dry overnight.

17. Dewax the samples in xylene or one of its substitutes. Two changes of clearing agent are necessary to remove all of the wax.

18. After they have been dewaxed, the slides can be stained, or if they have already been stained in whole mount, they can be mounted immediately in Permount (Fisher).

   Note: *Do not allow the slides to dry, since small bubbles can be trapped in the section. After mounting the sample and adding a coverslip, the slides can be kept indefinitely in a slide storage box. The sections clear as the solvent evaporates from the Permount, which is facilitated by using a slide warmer.*

PROTOCOL 14.2

Embedding in Plastic

Numerous plastic resins are available for this protocol. The resin and other embedding equipment chosen here are all available from Polysciences.

1. For analysis of in situ patterns, fix embryos in MEMFA or **paraformaldehyde** (as described in steps 3 and 4 of Protocol 14.1). For optimal preservation of cell morphology, fix samples at room temperature for 5 hours in freshly prepared 0.05 M phosphate buffer, containing 2% paraformaldehyde and 1% **glutaraldehyde**, with gentle rotation (Hausen and Riebesell 1991).

   Note: *The Hausen and Riebesell fixative gives significantly better preservation than paraformaldehyde alone. However, this extensive fixation protocol is not compatible with optimal in situ hybridization signals.*

   paraformaldehyde, glutaraldehyde (see Appendix for Caution)

2. Rinse the embryos twice in 70% phosphate-buffered saline (PBS) for 10 minutes per rinse.

3. Assemble the following equipment for embedding:

   JB-4 embedding kit (Polysciences 00226)
   plastic block holders (Polysciences 15899)
   polyethylene molding cup trays (Polysciences 17177A)

4. Prepare catalyzed JB-4 according to the manufacturers instructions and dehydrate the fixed embryos in a series of 10-minute washes in **ethanol**:PBS. Begin with 30:70 (ethanol:PBS) and increase through 50:50, 70:30, and 95:5 for successive washes. Use absolute ethanol for the final two washes.

   ethanol (see Appendix for Caution)

5. Remove the ethanol from the samples and rinse with catalyzed JB-4 to ensure that no residue remains (ethanol in combination with JB-4 can leach out in situ staining).

6. Infiltrate the embryos in three to four changes (1–1.5 hours each) of catalyzed JB-4 solution at 4°C with gentle rocking. For maximum preservation, extend the final incubation period to overnight.

7. Mix the embedding solutions according to the manufacturer's instructions and store on ice to slow polymerization. Fill block holder well with this solution, and drop the embryo(s) into the well.

8. Orient embryo(s) as polymerization proceeds. The solution usually begins to solidify around the embryo after approximately 20–30 minutes.

   Note: *If a number of blocks are being embedded, this process can be staggered. Contact with air inhibits polymerization, so the resin will harden at the bottom of the well first.*

9. When the embryos have been correctly orientated, add more resin and place the block holder on top of the block. Allow the solution to polymerize completely for 1–2 hours at room temperature.

10. Remove the polymerized block from the holder and trim to produce a rectangular block containing all of the sample tissue.

11. For visualization of in situ hybridization signals, cut 10-μm sections with a glass knife (e.g., Leica 2065 Supercut). For visualization of cell morphology, cut 1-μm sections with a standard electron microscopy microtome.

12. For sections taken after in situ hybridization, counterstaining is not usually required. For visualizing morphology, stain samples with Toluidine Blue (1% [w/v] aqueous, filtered; available from Rowley Biochemical Institute). Cover sections in staining solution for 30–60 seconds at 50–60°C (low setting on a hot plate).

    Note: *The length of the staining period will vary depending on the tissue type and thickness of the sample.*

13. When staining is complete (the edges of the staining solution start to dry up), rinse the samples with distilled water and dry on a hot plate for 1–2 minutes.

14. Mount the samples using Crystal/Mount (Biomedia Corp.) or Permount (Fisher), according to the manufacturer's instructions.

REFERENCES

Hausen P. and Riebesell M., eds. 1991. *The early development of* Xenopus laevis: *An atlas of the histology*. Springer-Verlag, Berlin.

Culture Media and Solutions

CULTURE MEDIA

Numerous media exist in which to culture *Xenopus* embryos. The chief differences among these media are the concentration of monovalent cations and the buffering agent. The original culture media devised by Holtfreter (1931) used bicarbonate as buffer, which is not an efficient buffer and has generally been superseded by Tris and HEPES-based buffers. However, bicarbonate buffer is still superior in some applications, e.g., progesterone-induced oocyte maturation. Solutions should be stored at room temperature unless otherwise noted.

GENERAL MEDIA

Modified Barth's Saline (MBS)

Prepare two solutions: 0.1 M **CaCl$_2$** and 10x **MBS** salts (**NaCl/KCl/MgSO$_4$/HEPES/NaHCO$_3$**)

0.1 M CaCl$_2$
　　11.1 g/liter

　　Autoclave and store in aliquots at –20°C or 4°C.

10x MBS salts
　　880 mM NaCl
　　10 mM KCl
　　10 mM MgSO$_4$
　　50 mM HEPES (pH 7.8)
　　25 mM NaHCO$_3$

　　Adjust final pH to 7.8 with NaOH; autoclave.

Prepare the final MBS solution by mixing 100 ml of 10x salt solution with 7 ml of 0.1 M $CaCl_2$; adjust the volume up to 1 liter with distilled water. The following are the final concentrations.

88 mM NaCl
1 mM KCl
0.7 mM $CaCl_2$
1 mM $MgSO_4$
5 mM HEPES (pH 7.8)
2.5 mM $NaHCO_3$

Note: *Omit HEPES if MBS is to be used for oocyte maturation.*

$CaCl_2$, MBS, NaCl, KCl, $MgSO_4$ (see Appendix for Caution)

Antibiotics

Antibiotics can be added to MBS. The following are the most commonly used antibiotics for embryos and explants.

Gentamycin (50 µg/ml)
Make a 1000x stock solution, 50 mg/ml, in water, filter sterilize, and refrigerate.

Oxytetracycline (12.5 µg/ml)
Make a 1000x stock solution, 12.5 mg/ml, in water. Store frozen.

Penicillin (commonly used for oocyte culture at 100 units/ml)
Make a 1000x stock solution, 100,000 units/ml, in water. Store frozen.

Streptomycin Sulfate (Life Technologies)
Make a 1000x stock solution, 100 mg/ml, in water. Store frozen.

High-salt MBS

0.1 M **$CaCl_2$**	7 ml
10x **MBS** salts	100 ml
5 M NaCl	4 ml
water	888 ml

$CaCl_2$, MBS (see Appendix for Caution)

Amphibian Ringers

Modified Frog Ringers (MR)

0.1 M NaCl
1.8 mM **KCl**
2.0 mM **CaCl$_2$**
1.0 mM **MgCl$_2$**
5.0 mM HEPES-**NaOH** (pH 7.6) or 300 mg/liter NaHCO$_3$

KCl, CaCl$_2$, MgCl$_2$, NaOH (see Appendix for Caution)

Marc's Modified Ringers (MMR) (Ubbels et al. 1983)

0.1 M NaCl
2.0 mM **KCl**
1 mM **MgSO$_4$**
2 mM **CaCl$_2$**
5 mM HEPES (pH 7.8)
0.1 mM EDTA

Note: *Most current formulations of MMR omit EDTA and are adjusted to pH 7.4.*

KCl, MgSO$_4$, CaCl$_2$ (see Appendix for Caution)

Normal Amphibian Medium (NAM) (Slack and Forman 1980)

110 mM NaCl
2 mM **KCl**
1 mM **Ca(NO$_3$)$_2$**
1 mM **MgSO$_4$**
0.1 mM EDTA
1 mM NaHCO$_3$
2 mM sodium phosphate pH 7.4

Note: *Assemble this solution from autoclaved stocks of 50x phosphate buffer (0.1 M sodium phosphate, pII 7.5) and a 10x stock of remaining salts excluding sodium bicarbonate. A 100x stock of sodium bicarbonate (0.1 M) should be filter-sterilized.*

KCl, Ca(NO$_3$)$_2$, MgSO$_4$ (see Appendix for Caution)

Holtfreter's Solution (Holtfreter 1931)

Prepare a stock of 25 mM $NaHCO_3$ and 10x Holtfreter's salts. Autoclave the salts, and filter-sterilize the $NaHCO_3$.

25 mM $NaHCO_3$
2.1 g/liter

10x Holtfreter's salts

600 mM NaCl
6 mM **KCl**
9 mM **$CaCl_2$**

Prepare the final solution by mixing 100 ml of 25 mM $NaHCO_3$ with 100 ml of 10x Holtfreter's salts and adjust the volume to 1 liter with distilled water.

KCl, $CaCl_2$ (see Appendix for Caution)

SPECIALIZED MEDIA

Danilchik's Blastocoel Buffer (Keller et al. 1985)

Prepare 500 ml of solution as follows.

Stock solution	Volume required
4.0 M NaCl	6.63 ml
0.8 M $NaHCO_3$	9.38 ml
0.45 M K-gluconate (Aldrich)	5.0 ml
2 M $CaCl_2$	0.25 ml
0.83 M $MgSO_4$	0.60 ml
bicine (Mann)	0.408 g
distilled water	450 ml

Adjust the pH to 8.3 with a measured volume of 1.0 M Na_2CO_3 (1.8 ml), calculate total Na^+ concentration, and adjust to 95 mM with Na-isethionate (2-hydroxy ethanesulfonic acid; Sigma) (add ~1.73 g). Adjust the volume to 500 ml with distilled water and store frozen.

Notes: *A small amount of precipitation occurs after 2–3 weeks so fresh stock should be made up weekly.*

This medium depresses apical-surface healing responses interpreted as evidence that internal cells are content in this medium. Do not use this medium for rearing intact embryos.

The ion concentrations for this buffer, shown below, are taken from the mean free intercellular activities as measured by J. Gillespie (1983) for Xenopus.

Na^+	*95 mM*	HCO_3^-/CO_3^{--}	*18–19 mM*
K^+	*4.5 mM*	*bicine*	*5.0 mM*
Ca^{++}	*1.0 mM*	SO_4^{--}	*1.0 mM*
Mg^{++}	*1.0 mM*	*isethionate*	*23–24 mM*
Cl^-	*55.0 mM*	*gluconate*	*4.5 mM*

$CaCl_2$, $MgCO_4$, Na_2CO_3 (see Appendix for Caution)

Sater's Modified Blastocoel Buffer

4 M NaCl	12.38 ml
gluconic acid, sodium salt	7.95 g
Na_2CO_3	0.53 g
2 M KCl	2.25 ml
1 M $CaCl_2$	1 ml
1 M $MgSO_4$	1 ml

Adjust the pH to 8.1 with HEPES (use ~6 mM) and add 50 µg/ml gentamycin sulfate. If desired, standard amounts of penicillin and streptomycin and 1 mg/ml BSA can also be added. Filter sterilize and store frozen in 50-ml aliquots.

Na_2CO_3, **KCl**, $CaCl_2$, $MgSO_4$ (see Appendix for Caution)

Low-calcium Magnesium Ringer's (LCMR) (Stewart and Gerhart 1990)

Various recipes of this solution have been used to slow healing of dissected embryos. Embryo LCMR is recommended for embryo dissection, but long-term survival of explants is better in explant LCMR (Hemmati-Brivanlou et al. 1990).

Embryo LCMR

66 mM NaCl
1.33 mM **KCl**
0.33 mM **$CaCl_2$**
0.17 mM **$MgCl_2$**
5 mM HEPES (pH 7.1; adjust pH after making)
50 µg/ml gentamycin

Explant LCMR

43 mM NaCl
0.85 mM KCl
0.37 mM $CaCl_2$
0.19 mM $MgCl_2$
5 mM HEPES
50 µg/ml gentamycin
0.5% BSA if growth factors are added

KCl, CaCl₂, MgCl₂ (see Appendix for Caution)

Calcium Magnesium-free Medium (CMFM)

This medium is used for dissociating embryonic tissue. It will dissociate the inner, but not the outer, layer of an animal cap (Sargent et al. 1986).

88 mM NaCl
1 mM **KCl**
2.4 mM $NaHCO_3$
7.5 mM Tris (pH 7.6)

Prepare from sterile stocks and store at room temperature.

KCl (see Appendix for Caution)

PhoNaK Buffer

This buffer is used for dissociating embryos. It is much more vigorous than CMFM and will completely dissociate animal caps (Godsave and Slack 1989)

50 mM **NaH_2PO_4**
35 mM NaCl
1 mM **KCl**

NaH₂PO₄, KCl (see Appendix for Caution)

OOCYTE CULTURE MEDIA

O-R2 (Wallace et al. 1973)

Prepare two stock solutions, A and B.

Stock Solution A

NaCl	48.221 g
KCl	1.864 g
CaCl$_2$-H$_2$O	1.470 g
MgCl$_2$-6H$_2$O	2.030 g
HEPES	11.915 g
NaOH	1.520 g
water	to 1 liter

Stock Solution B

Na$_2$HPO$_4$	1.420 g
water	to 1 liter

Mix 1 volume of Stock Solution A, 1 volume of Stock Solution B, and 8 volumes of distilled water. This should yield a solution with a pH of 7.8 and the following final salt concentrations.

82.5 mM NaCl
2.5 mM **KCl**
1 mM **CaCl$_2$**
1 mM **MgCl$_2$**
1 mM **Na$_2$HPO$_4$**
5 mM HEPES
3.8 mM **NaOH**

KCl, CaCl$_2$, MgCl$_2$, Na$_2$HPO$_4$, NaOH (see Appendix for Caution)

Oocyte Culture Medium (Modified from Opresko et al. 1980)

50% L-15 + glutamine (Life Technologies; Leibovitz stock)
40% HEPES/insulin stock (see note below)
10% fetal calf serum or frog serum with 50–200 mg/ml VTG serum or VTG (vitellogenin)
100 µg/ml gentamycin
0.5% fungizone/penicillin/streptomycin (100x stock; Life Technologies)

Note: *100 ml of HEPES/insulin stock is 37.5 mM HEPES (pH 7.8), 2.5 µg/ml Sigma insulin stock (22.5 IU/mg; 5 mg/ml insulin slurry in distilled water, solubilized with 50 µl of 0.5 M EDTA, pH 7.7)*

SOLUTIONS FOR IMMUNOHISTOCHEMISTRY

MEMFA (MOPS/EGTA/Magnesium Sulfate/Formaldehyde Buffer)

0.1 M **MOPS** (pH 7.4)
2 mM EGTA
1 mM **MgSO$_4$**
3.7% **formaldehyde**

A 10x solution of the salts, without formaldehyde, can be prepared and stored. This solution turns yellow if autoclaved or aged.

MOPS, MgSO$_4$, formaldehyde (see Appendix for Caution)

Dents Fix (Methanol/DMSO)

4 volumes **methanol**
1 volume **DMSO**

methanol, DMSO (see Appendix for Caution)

Bleaching Solution (Mayor et al. 1995)

1% **H$_2$O$_2$**
5% **formamide**
0.5x SSC (standard saline citrate)

H$_2$O$_2$, formamide (see Appendix for Caution)

20x PBS (Phosphate-buffered Saline)

NaH$_2$PO$_4$·H$_2$O	5.12 g
NaHPO$_4$·H$_2$O	23.88 g
NaCl	204.4 g
distilled water	800 ml

Adjust pH to 7.4 and volume to 1 liter.

NaH$_2$PO$_4$, NaHPO$_4$ (see Appendix for Caution)

Goat Serum (GS) (Life Technologies; Mycoplasma-tested) 500 ml

Heat deactivate complement for 30 minutes at 56°C. Store as 1-ml aliquots at –20°C.

PBT

PBS (phosphate-buffered saline)
2 mg/ml BSA (bovine serum albumin) (Fraction 5, reagent grade)
0.1% Triton X-100

DAB (Diaminobenzidine; Polyscience)

1 mg/ml in PBT

Filter through a 0.2-µm filter and store as 1-ml aliquots at –70°C.

DAB (see Appendix for Caution)

H_2O_2 (Hydrogen peroxide; Sigma; 30% stock solution)

Store at 4°C. Dilute 1/2x in PBT just before use.

Secondary Antibodies Coupled to Horseradish Peroxidase (e.g., Bio-Rad)

Store as 100-µl aliquots at –20°C.

Murray's Clearing Medium (Benzyl Benzoate/Benzyl Alcohol, BB/BA)

2 volumes of benzyl benzoate
1 volume of benzyl alcohol

BB/BA (see Appendix for Caution)

SOLUTIONS FOR RNA AND DNA PREPARATION (See Appendix 3)

10x mRNA Transcription Buffer

120 mM $MgCl_2$
800 mM HEPES-Cl (pH 7.5)
20 mM spermidine-HCl

$MgCl_2$, **HCl** (see Appendix for Caution)

2× mRNA Nucleotide Triphosphate Mix

6 mM ATP
6 mM CTP
6 mM UTP
3 mM GTP
9 mM m7(5′)Gppp(5′)G, cap analog (e.g., New England Biolabs 1404 or Ambion 8050)*

*Alternatively, use unmethylated GpppG (e.g., New England Biolabs 1407), which is less expensive. It is reported not to be as good as the methylated analog in reticulocyte translation, but in oocytes and embryos, this reaction yields RNA that has a potency similar to that of Ambion's mMessage Machine (Ambion).

TE

10 mM Tris (pH 7.5)
1 mM EDTA

Homogenization Buffer

1% **SDS**
10 mM EDTA
20 mM Tris (pH 7.5)
100 mM NaCl

SDS (see Appendix for Caution)

Reticulocyte Standard Buffer (RSB)

10 mM NaCl
10 mM Tris-**HCl** (pH 8)
5 mM **MgCl$_2$**

HCl, MgCl$_2$ (see Appendix for Caution)

Nuclei Freezing Buffer

50% glycerol
0.15 M NaCl
5 mM **MgCl$_2$**
10 mM Tris-**HCl** (pH 8)
0.1 mM EDTA
1 mM **DTT** (dithiothreitol)

MgCl$_2$, HCL, DTT (see Appendix for Caution)

REAGENTS FOR IN SITU HYBRIDIZATION

Nucleotide Stocks

2.5 mM nucleotide mix with digoxigenin-11 UTP (Boehringer Mannheim 1209 256; 25 nmoles)

10 mM CTP	10 µl
10 mM GTP	10 µl
10 mM ATP	10 µl
10 mM UTP	6.5 µl
10 mM dig-11 UTP	3.5 µl

Alternatively, prepare the whole tube of dig-UTP (25 nmoles; Boehringer Mannheim 1209 256) at one time to make a 2.5 mM stock.

100 mM CTP	7.1 µl
100 mM GTP	7.1 µl
7100 mM ATP	1 µl
100 mM UTP	4.6 µl
10 mM dig-UTP	25 µl
DEPC-treated water	234 µl

DEPC (see Appendix for Caution)

5× SP6 Buffer

200 mM Tris (pH 7.5)
30 mM **MgCl$_2$**
20 mM spermidine

Autoclave and store frozen in aliquots.

MgCl$_2$ (see Appendix for Caution)

MEMFA

0.1 M **MOPS** (pH 7.4)
2 mM EGTA
1 mM **MgSO$_4$**
3.7% **formaldehyde**

Makeup a stock of 10x MEMFA salts (omitting formaldehyde) in advance; add fresh formaldehyde (1/10 volume of a standard 37% stock) just prior to use. It is normal for this solution to turn yellow, especially upon autoclaving. Paraformaldehyde can also be used (MEMPFA), especially if fresh formaldehyde is not available. Preparation of a paraformaldehyde stock is described below. Use of formaldehyde gives slightly better signals, perhaps because formaldehyde stocks are stabilized with 10% methanol and the fixative penetrates the embryos more rapidly.

MOPS, MgSO₄, formaldehyde (see Appendix for Caution)

PTw

1x PBS (phosphate-buffered saline)
0.1% Tween-20

This solution is used for numerous washes so make up liter quantities.

Triethanolamine, 0.1 M (pH 7.0–8.0; Sigma T 1502)

Make up 0.1 M **triethanolamine** in **DEPC**-treated water, and correct the pH with 1 M NaOH. Monitor the pH by spotting small samples onto a pH paper. Do not expose the solution to a pH meter of dubious cleanliness.

triethanolamine, DEPC (see Appendix for Caution)

PTw Containing 4% Paraformaldehyde

1 volume of 20% **paraformaldehyde**
4 volumes of Ptw

Note: *To make 20% paraformaldehyde in water, neutralize the cloudy solution with NaOH and then heat at 65°C, with shaking, until clear. Store at –20°C.*

paraformaldehyde (see Appendix for Caution)

Hybridization Buffer

50% **formamide**, redistilled (BRL 5515UB)
5x SSC
1 mg/ml Torula RNA* (Type IX, Sigma R 3629)
100 µg/ml heparin (Sigma H 3125)
1x Denhart's solution
0.1% Tween 20 (Sigma P 1379)
0.1% CHAPS (Sigma C 3023)
10 mM EDTA

*Prepare Torula RNA in DEPC-treated water. Store in 1-ml aliquots at –20°C. It is normal for this solution to be brown and thus it needs no further processing.

formamide (see Appendix for Caution)

20× SSC, Stock Solution

175.3 g NaCl
88.2 g sodium citrate
Dissolve in 800 ml of water, adjust pH to 7.0 with **NaOH**, and adjust volume to 1 liter.

NaOH (see Appendix for Caution)

100× Denhart's Solution

2% BSA (ICN 810661)
2% **polyvinylpyrrolidone** (PVP-40 Sigma)
2% Ficoll 400 (Pharmacia)

Add a small amount of water to the dry ingredients, make a slurry, and then dilute.

polyvinylpyrrolidone (see Appendix for Caution)

RNase A (Sigma R 5000)

Dissolve 10 mg/ml in TE (pH 7.8), and boil for 10 minutes before use. Store RNases in aliquots at –20ºC.

RNase T1 (Sigma R 8251)

Dissolve 10,000 units/ml in 0.1 M **sodium acetate** at pH 5.5. Boil for 10 minutes before use. Store RNases in aliquots at –20ºC. While in use, keep at 4ºC. Avoid repeated freezing and thawing.

sodium acetate (see Appendix for Caution)

Maleic Acid Buffer (MAB)

100 mM maleic acid (Sigma M O375)
150 mM NaCl (pH 7.5)

This solution is used for numerous washes so make up liter quantities.

Alkaline Phosphatase Buffer

100 mM Tris (pH 9.5)
50 mM **MgCl$_2$**
100 mM NaCl
0.1% Tween 20 (Sigma P 1379)
2 mM levamisol (Sigma)

Add fresh levamisol just before use. This reagent inhibits endogenous phosphatase. Alkaline phosphatase buffer can be stored, but magnesium hydroxide tends to precipitate over time. A slight misty precipitate has no effect on the reaction.

MgCl$_2$ (see Appendix for Caution)

NBT/BCIP

Nitro blue tetrazolium (NBT): 75 mg/ml in 70% dimethylformamide
5-bromo-4-chloro-3-indolyl-phosphate (BCIP): 50 mg/ml in 100% dimethyl-
    formamide

NBT/BCIP (see Appendix for Caution)

PVA Alkaline Phosphatase Buffer

Alkaline phosphatase buffer (see above) containing 10% **polyvinyl alcohol** (Aldrich 36,313-8).

Dissolve PVA (98–99% hydrolyzed, average m.w. 31,000–50,000) to a final concentration of 10% in alkaline phosphatase buffer. Heat the solution to help dissolve the PVA. PVA tends to precipitate during storage, so do not store this buffer for more than a few days. The final buffer is quite viscous, and thus the NBT, BCIP, and levamisole must be mixed thoroughly before adding it to embryos. The embryos must also be washed well after staining, to prevent precipitation of PVA in methanol.

polyvinyl alcohol (see Appendix for Caution)

Bouin's Fix

To make up 100 ml of Boudin's fix, dissolve 1 g of **picric acid** in 70 ml of water and then add 25 ml of 37% **formaldehyde** and 5 ml of glacial acetic acid.

picric acid, formaldehyde (see Appendix for Caution)

REFERENCES

Gillespie J.I. 1983. The distribution of small ions during the early development of *Xenopus laevis* and *Ambystoma mexicanum* embryos. *J. Physiol.* **344:** 359–377.

Godsave S.F. and Slack J.M. 1989. Clonal analysis of mesoderm induction in *Xenopus laevis. Dev. Biol.* **134:** 486–490.

Hemmati-Brivanlou A., Stewart R.M., and Harland R.M. 1990. Region-specific neural induction of an engrailed protein by anterior notochord in *Xenopus. Science* **250:** 800–802.

Holtfreter J. 1931. Über die Aufzucht isolierter Teile des Amphibienkeimes. II Züchtung von Keimen und Keimteilen in Salzlösung. *Wilhelm Roux Arch. Entwicklungsmech. Org.* **124:** 405–464.

Keller R.E., Danilchik M., Gimlich R., and Shih J. 1985. The function and mechanism of convergent extension during gastrulation of *Xenopus laevis. J. Embryol. Exp. Morphol.* **89:** 185–209.

Mayor R., Morgan R., and Sargent M.G. 1995. Induction of the prospective neural crest of *Xenopus. Development* **122:** 767–777.

Opresko L., Wiley H.S., and Wallace R.A. 1980. Differential postendocytotic compartmentation in *Xenopus* oocytes is mediated by a specifically bound ligand. *Cell* **22:** 47–57.

Sargent T.D., Jamrich M., and Dawid I.B. 1986. Cell interactions and the control of gene activity during early development of *Xenopus laevis. Dev. Biol.* **114:** 238–246.

Slack J.M. and Forman D. 1980. An interaction between dorsal and ventral regions of the marginal zone in early amphibian embryos. *J. Embryol. Exp. Morphol.* **56:** 283–299.

Stewart R.M. and Gerhart J.C. 1990. The anterior extent of dorsal development of the *Xenopus* embryonic axis depends on the quantity of organizer in the late blastula. *Development* **109:** 363–372.

Ubbels G.A., Hara K., Koster C.H., and Kirschner M.W. 1983. Evidence for a functional role of the cytoskeleton in determination of the dorsoventral axis in *Xenopus laevis* eggs. *J. Embryol. Exp. Morphol.* **77:** 15–37.

Wallace R.A., Jared D.W., Dumont J.N., and Sega M.W. 1973. Protein incorporation by isolated amphibian oocytes. 3. Optimum incubation conditions. *J. Exp. Zool.* **184:** 321–333.

Timing of Development of *Xenopus laevis* Embryos and Temperature Dependence

For a thorough discussion of the timing of development of *Xenopus laevis* embryos, see Nieuwkoop and Faber (*Normal table of* Xenopus laevis. Garland Publishing, New York [1994]). This appendix presents a brief summary table of the number of hours required to reach particular stages and also lists key features found at these stages. The times listed in the table are for embryos raised at 22°C, the maximum temperature compatible with healthy embryos during early stages. It is often useful to raise embryos at lower temperatures to prolong the time required to reach a particular stage. For example, investigators might wish to generate embryos at different stages at one time or to produce embryos at a given stage at different times during a particular day. If the rate of development at 22°C is taken to be 1x, then the rate at 20°C will be approximately 3/4x, at 16°C, the rate will be approximately 1/2x, and at 14°C, it will be approximately 1/3x. These are only estimates, and the fractional values are not completely linear at all developmental stages, but they are useful in planning experiments.

Abbreviated Table of *Xenopus laevis* Embryonic Development

Stage	Hours at 22°C	Description
1	0	1 cell
2	1 1/2	2 cells
3	2	4 cells
4	2 1/4	8 cells
5	2 3/4	16 cells
6	3	32 cells
6 1/2	31/2	about 48 cells
7	4	large-cell blastula; tangential cleavage
8	5	medium-cell blastula
9	7	fine-cell blastula
10	9	initial gastrula; pigment concentration at blastopore
10 1/4	10	early gastrula; dorsal blastopore groove
10 1/2	11	crescent-shaped blastopore
11	11 3/4	horse-shoe-shaped blastopore,
12	13 1/4	medium yolk plug stage
13	14 3/4	slit-blastopore stage
14	16 1/4	neural plate stage
15	17 1/2	early neural fold stage
16	18 1/4	mid neural fold stage
18	19 3/4	neural groove stage
19	20 3/4	initial neural tube stage
20	21 3/4	neural folds fused; suture still present
24	26 1/4	initial motor reactions to external stimulation
25	27 1/2	beginning of fin formation
29/30	35	tail bud distinct
33/4	44 1/2	melanophores appearing dorsally on the head
35/6	50	melanophores appearing on back
40	66	mouth broken through; outlines of proctodeum and tail myotomes forming angle of 90°
42	80	torsion of intestine about 90°
44	92	coiling part of intestine showing s-shaped loop
46	106	hindlimb bud becomes visible

Nucleic Acid Methods

The methods presented here have been adapted for use on *Xenopus* embryos. The more generally applicable methods are not provided. Although many of the protocols listed call for treatment with DEPC (diethylpyrocarbonate), this is generally not necessary when starting with good quality sterile distilled water. The recipes for solutions listed in this appendix are given in Appendix 1.

RNA ISOLATION

Xenopus embryos contain a large amount of yolk, which can make RNA preparation from large numbers of embryos rather messy. Although only a single method is presented here, alternative methods work well as long as the volumes are increased to compensate for the yolk. This method is suitable for large numbers of embryos and can be adapted for smaller numbers by scaling the protocol appropriately. Additionally, a method for preparing RNA by in vitro transcription is given that yields RNA suitable for microinjection.

Method 1: Preparation of RNA Using Proteinase K/LiCl

1. Dejelly embryos as described in Protocol 6.1.

2. Allow the embryos to settle and resuspend in 4 volumes of premixed proteinase K mix.

   > 20 mM Tris (pH 7.6)
   > 100 mM NaCl
   > 30 mM EDTA
   > 1% **SDS**
   > 0.5 mg/ml proteinase K
   >
   > **SDS** (see Appendix for Caution)

3. Homogenize the embryos by douncing with a "B" pestle. For smaller volumes (<1 ml), homogenize by vortexing.

   Note: *The vortexing method is inefficient for embryos beyond stage 20. They will not dissolve completely.*

4. Incubate the proteinase K reaction for 1.5 hours at 37°C.

5. Divide the reaction into two 50-ml tubes (e.g., Corning) and add 20 ml of a 1:1 mix of **phenol:chloroform** to each tube.

   phenol, chloroform (see Appendix for Caution)

6. Vortex the tubes and then centrifuge at 12,000*g* for 5 minutes. Transfer the aqueous phases to fresh tubes.

7. Back-extract the phenol:chloroform by adding 5 ml of TE (see Appendix 1) to each tube, vortexing, and centrifuging at 12,000*g* for 5

minutes. Pool the aqueous phases with those obtained at step 5 above, and discard the organic phase.

8. Repeat phenol:chloroform extraction twice (without back extractions), or until no interface is visible between layers, and extract once with 1 volume of chloroform.

9. Transfer the aqueous layers to fresh tubes and add 2.5 volumes of **ethanol**. Store for 1–2 hours at –20°C.

ethanol (see Appendix for Caution)

10. Collect the precipitated nucleic acids by centrifuging at 12,000g for 20 minutes. Carefully decant the ethanol and resuspend the pellet in 360 μl of TE and 40 μl of 10x NEB4 (New England Biolabs).

11. Add 1 μl of RNase-free DNase I (Ambion, 2 units/μl) and incubate for 20 minutes at room temperature.

12. Stop the DNase reaction by extracting with 800 μl of a 1:1 mix of phenol:chloroform and then with 400 μl of chloroform. Transfer the aqueous phase to a clean tube and add 135 μl of 10 M **LiCl** (autoclaved). Store at 4°C for several hours, or overnight. LiCl will highly preferentially precipitate RNA and not DNA oligonucleotides remaining after DNase treatment.

LiCl (see Appendix for Caution)

13. Centrifuge for 20 minutes at 12,000g. Carefully discard the supernatant and resuspend the pellet in 200–500 μl of distilled water.

14. Add 2.5 volumes of ethanol and 0.1 volume of **sodium acetate** (3 M, pH 4.8). Collect the precipitated RNA by centrifuging for 20 minutes at 12,000g.

sodium acetate (see Appendix for Caution)

15. Redissolve the RNA in 500 μl of distilled water. Determine the concentration and quality of the solution by measuring the $OD_{260/280}$. Store the RNA as an aqueous solution at –80°C.

Note: *Embryos contain approximately 5 μg of total RNA. This protocol routinely produces yields of 90%.*

Helpful Hints

- This method can be used for smaller numbers of embryos or explants (one to two explants), but the volumes must be scaled down. Use about 50 µl of proteinase K buffer per embryo, or 10 µl per animal cap. Incubate the reaction for 1 hour at 37°C. Extract once with phenol:chloroform and treat with 1 unit of DNase I, as described above. Precipitate with LiCl (add 34 µl of 10 M LiCl for each 100 µl of RNA solution) and then with 2.5 volumes of ethanol and 0.1 volume of sodium acetate (pH 5.5).

Method 2: Preparation of RNA (In Vitro Transcription)

RNAs produced by in vitro transcription can be introduced into embryos by microinjection and used to test the effects of overexpression, misexpression, or expression of dominant-negative constructs. This method (adapted from Green et al. 1983) assumes that the coding region of interest is already inserted behind an appropriate bacteriophage promoter. A cap analog, GpppG, is included in the nucleotide mix to result in capped RNA.

1. For a 20-ml reaction volume, assemble the following ingredients, in a prewarmed tube, in the order indicated. If the nucleotide precipitates before synthesis, dilute and warm the reaction.

linear template DNA (0.5–1 mg/ml)	2 µl
2x mRNA nucleotide triphosphate mix	10 µl
(see Appendix 1)	
[α-^{32}P]UTP (if desired)	0.1 µl (trace)
DTT (200 mM)	2 µl
BSA (1 mg/ml)	2 µl
10x mRNA transcription buffer (see Appendix 1)	2 µl
RNase inhibitor (20 units/ml)	0.5 µl
bacteriophage RNA polymerase (20 units/ml)	1.5 µl

   Note: *A precipitate invariably forms during the reaction, but the yields and bioactivity are good.*

   radioactive substances, DTT (see Appendix for Caution)

2. Mix gently and incubate the reaction for 2 hours at 37°C.

3. Add 2 units of RNase-free DNase I and incubate for a further 10 minutes at 37°C. Cerenkov count (i.e., with no scintillation fluid) the tube to obtain the total available cpm.

4. Dilute the reaction to 100 µl with a buffer containing 100 mM NaCl, 30 mM EDTA, 20 mM Tris (pH 7.5), and 1% **SDS**.

SDS (see Appendix for Caution)

5. Purify the RNA through a spin column (Qiagen or Sephadex G-50).

   Note: *The column removes unincorporated nucleotides and cap analog, which would otherwise aggregate at the interphase of a phenol/chloroform extraction and necessitate multiple extractions. The cap analog is extremely toxic to embryos, so must be removed efficiently. Removal of the unincorporated nucleotides allows the efficiency of the reaction to be measured. If the reaction has worked well, most of the counts will be associated with the eluate. If the reaction has not worked well, the counts will remain in the column.*

6. Extract the column eluate with **phenol:chloroform** and check the recovery of counts from the interface and, if necessary, extract again. Precipitate the aqueous phase with 2.5 volumes of **ethanol** and 0.1 volume of 3 M **sodium acetate** (pH 5.5).

   phenol, chloroform, ethanol, sodium acetate (see Appendix for Caution)

7. Collect the precipitate by centrifuging for 15 minutes at maximum speed in a microcentrifuge at 4°C. Redissolve the pellet in 100 μl of **DEPC**-treated water and take a 1-μl aliquot for scintillation counting. Calculate the total incorporated cpm to obtain the percentage incorporation.

   Note: *The expected yield is 10–20%, or 8–16 μg of RNA.*

   DEPC (see Appendix for Caution)

8. (*Optional*) Reprecipitate the RNA in 2.5 volumes of ethanol and 0.1 volume of 3 M sodium acetate (pH 5.5). Wash the pellet in 70% ethanol. Resuspend at desired concentration.

9. Load 1 μl of the transcription mix onto a denaturing agarose gel (see Note below) to assess the transcript yield and its integrity. Check the final resuspension of RNA by recovery of counts, if radioactivity has been used.

   Note: *A simple denaturing agarose gel uses a standard Tris-acetate buffer but denatures the sample prior to loading.*

The sample should be heated to 65°C for 10 minutes in

   50% **formamide**
   2.2 M (6.7%) **formaldehyde** from a 37% stock
   1x **MOPS** buffer
   0.05% **bromophenol blue**
   100 μg/ml **ethidium bromide**

This is conveniently done using a stock of "FFM"

FFM
    10 ml formamide
    3.5 ml formaldehyde
    2.5 ml 10x MOPS buffer
    bromophenol blue/xylene cyanol (0.05% each)
    ethidium bromide (1 mg/ml)

10x MOPS

    0.2 M **MOPS** (pH 7.0)
    50 mM sodium acetate
    5 mM EDTA

After electrophoresis, the gel can be photographed directly.

formamide, formaldehyde, MOPS, bromophenol blue, ethidium bromide (see Appendix for Caution)

Helpful Hints

- *Problems with Yield and Translation*
  Poor yield is usually due to poor template, although occasionally commercial reagents can be at fault. Some batches of GpppG inhibit transcription, or worse, they promote synthesis of untranslatable RNA. It can be useful to keep stocks of known good quality reagents as positive controls, which will provide a basis for complaints to the manufacturer.

- *Calculation of Yield*
  If the labeled nucleotide is UTP, then the reaction above contains 10 μl of 6 mM UTP. In an average RNA, for every mole of UTP incorporated, there will be one mole of each of the other three nucleotides incorporated. The average molecular weight of nucleotide monophosphate is assumed to be 330 g/mole. If all the UTP were incorporated, the yield would be: $(10 \times 10^{-6}$ liters$) \times (6 \times 10^{-3}$ moles/liter$) \times 4 \times 330$ g/mole $=$ 79.2 μg, i.e., (volume of label) x (concentration of label) x (number of nucleotides) x (average molecular weight of each nucleotide).

  For the purposes of calculating mass, each 1% of yield corresponds to 0.8 μg of RNA. For synthesis of large numbers of RNAs, e.g., as in expression cloning, half-scale transcription reactions can be performed. For a half-scale reaction, each 1% incorporation of UTP corresponds to 0.4 μg of RNA.

mRNAs of higher specific activity can be synthesized without excessive radiolytic degradation. Radioactive nucleotide (10 µCi) can be added to the standard reaction. The RNA will retain approximately 25% of its original activity after storage for 2 years at –80°C in DEPC-treated water.

- *Choice of Vector*
  See Chapter 3. All polymerases add a dinucleotide cap with the same efficiency (based on stability of RNA in oocytes). However, some polylinkers contain sequences that appear to inhibit translation (Kuo et al. 1996). The effect is observed when cDNAs containing 5 dC tracts are cloned downstream from T7 or T3 polymerase promoters. The effect is thought to be due to formation of secondary structure between the initial GGG triplet of T3 and T7 transcripts, and the oligo(dC) tracts. Since SP6 transcripts initiate with AAA, and are not capable of forming this secondary structure, their translation is unaffected. In general, the less polylinker present the better.

  DNAs for translation are often cloned into plasmids that contain globin 5 UTR (untranslated region) and 3 UTR, e.g., pSP64 or derivatives. Although no systematic analysis has been published, in some cases, the UTR sequences enhance translation manyfold; in other cases, they appear to have no effect. These plasmids may be particularly useful for the over-expression of maternal mRNAs. Regulation of maternal RNA expression is mainly translational, and regulatory regions are often contained in the 3 UTR and 5 UTR stretches. Removing these regions and substituting the strong globin UTRs, or even the AUG Kozak consensus sequence, may help significantly.

  The CS series of vectors (Turner and Weintraub 1994) incorporate a cytoplasmic polyadenylation signal in the primary transcript. Cytoplasmic polyadenylation (Richter 1991) enhances translation tenfold, or more, compared to that achieved by equivalent transcripts from pSP64T. Conversely, for oocyte injections, where translation occurs during long-term incubation, pSP64T transcripts, which are polyadenylated in vitro, are translated more efficiently than the equivalent CS transcripts polyadenylated in vivo (after injection into the oocyte nucleus). In vitro polyadenylation by poly(A) tracts encoded by the vector provides a useful means by which to select full-length transcripts, using oligo(dT) cellulose chromatography if desired. However, poly(dA) tracts encoded into the 3 polylinker of a plasmid can be unstable in plasmids, more than 50 As are seldom maintained.

DNA ISOLATION

Genomic DNA is used for Southern blotting, for determining gene structure, and for detecting the presence or absence of genes of interest.

Method 1: Extraction of DNA from Single Embryos

1. Homogenize the embryo using a micropestle in a microcentrifuge tube, containing 0.5 ml of homogenization buffer (see Appendix 1).

2. Store the homogenate at –20ºC until ready to process.

3. Thaw the homogenate and add 2.5 µl of proteinase K (20 mg/ml).

4. Mix well and incubate overnight at 55ºC.

5. Extract with 1 volume of aqueous **phenol**.

   Note: *If the DNA solution is viscous and seems to drag phenol with it, it may be easier to remove the lower phenol layer rather than removing the upper aqueous layer.*

   phenol (see Appendix for Caution)

6. Extract once with a 1:1 mix of phenol:**chloroform**, and then once with chloroform.

   Note: *Embryos appear to contain compounds that inhibit restriction diges-tion. It is essential to thoroughly extract the DNA in order to remove these inhibitors.*

   chloroform (see Appendix for Caution)

7. Transfer the aqueous phase to a fresh tube and precipitate the DNA by adding **ammonium acetate** to 2 M and 0.6 volumes of **isopropanol**.

   ammonium acetate, isopropanol (see Appendix for Caution)

8. Mix gently but thoroughly by inverting the tube.

   Note: *The precipitate may be viscous and mixing can take some time.*

9. Hold the tube on ice until a stringy DNA precipitate appears (~30 min-utes). This may not be apparent from a single embryo before stage 20–30.

   Note: *If DNA fails to precipitate on ice, store at –20ºC overnight.*

10. Recover precipitate by centrifuging at 12,000g for 5 minutes.

11. Wash the pellet in 70% **ethanol**. Redissolve in 100 µl of TE.

    ethanol (see Appendix for Caution)

12. Add RNase A to 10 µg/ml and RNase T1 to 10 µg/ml. Incubate for 30 minutes at room temperature.

13. Precipitate the reaction by adding ammonium acetate to 2 M and 0.6 volumes of isopropanol.

14. Centrifuge at 12,000g for 5 minutes and resuspend the pellet in 20 µl of TE.

   Note: *The older the embryo, the greater the number of nuclei and the greater the amount of genomic DNA isolated.*

Method 2: Isolation of DNA from Red Blood Cells

Unlike the mammalian equivalent, frog red blood cells contain nuclei from which genomic DNA can be extracted.

Isolation of Red Blood Cell Nuclei

1. Anesthetize the frog as described in Protocol 8.1, and place the animal belly up on a piece of plastic-coated bench paper, plastic side up.

2. Introduce a 1-cm^2 opening in the upper ventral thorax to expose the heart (which should still be beating).

   Note: *Keep the opening small so that the liver does not pop out.*

3. Use heavy scissors to cut and remove the shield-shaped bone connecting the shoulders.

   Note: *The heart should visible through the silvery pericardium.*

4. Expose the heart by carefully lifting and cutting open the pericardium.

5. Inject the heart with 0.5 ml of heparin (2 mg/ml) and allow blood to flow freely from the wound.

6. Collect the blood in a beaker containing 0.85x SSC (see Appendix 1). Drape the frog over the top of the beaker such that the heart is actually in the solution.

   Note: *Approximately 2–3 ml of blood will be pumped into the beaker over a period of 10 minutes. If the blood flow stops, a second incision can be made in the heart. Massage the frog's legs to help maintain blood flow.*

7. When the procedure is complete, wrap the frog in a plastic bag and place overnight at –20°C before discarding.

8. Collect the cells by centrifuging in a benchtop centrifuge at about 2000 rpm. Pour off the supernatant, and place the tube on ice.

   Note: *For the remainder of the protocol, keep the cells on ice and use ice-cold buffers at all times.*

9. Partially resuspend the pellet in the residual supernatant and lyse the cells in RSB containing 0.05% Nonidet P-40 (NP-40). Start with a small volume (~5 ml), resuspend cells, and then increase volume to approximately 50 ml.

   Note: *NP-40 is a nonionic detergent that breaks down cell membranes, but not nuclear membranes.*

10. Hold on ice for 1 minute and collect the nuclei by centrifuging in a benchtop centrifuge at about 2000 rpm.

11. Pour off the supernatant gently and resuspend the loose pellet by vortexing. When the nuclei are thoroughly resuspended, adjust the volume to 50 ml with RSB containing 0.05% NP-40.

12. Collect the nuclei by centrifuging in a benchtop centrifuge at about 2000 rpm. Repeat steps 11 and 12 until the nuclear pellet is white.

    Note: *If the nuclei aggregate, dounce gently with a "B" pestle or remove clots by filtering through 2–4 layers of cheescloth.*

13. Store the nuclei at –20°C in nuclei freezing buffer at a concentration of 1 ng/ml. Proceed with preparation of genomic DNA as described below.

Preparation of Genomic DNA from Red Blood Cells

The preparation of genomic DNA from red blood cell nuclei requires gentle treatment to avoid shearing the DNA. Avoid vortexing and violent phenol extraction where possible. The following procedure should yield several milligrams of DNA, more than 200 kb in size, suitable for cosmid cloning and genomic Southern blots.

1. Thaw nuclei prepared in the protocol above, and transfer to a 50-ml tube (Corning). Dilute to 100–200 µg/ml in RSB and add proteinase K to a final concentration of 200 µg/ml. Mix gently—*do not vortex.*

2. Add one volume of 0.6 M NaCl, 20 mM Tris-**HCl** (pH 7.4), 20 mM EDTA, and 1% **SDS** and mix gently, but thoroughly.

   Note: *The solution will become very viscous as the nuclei lyse.*

   HCl, SDS (see Appendix for Caution)

3. Precipitate the DNA by adding 0.25 volumes of 10 M **ammonium acetate** and mixing thoroughly.

   Note: *The DNA should precipitate immediately.*

   ammonium acetate (see Appendix for Caution)

4. Use a thin glass rod, or a heat-sealed pasteur pipette, to spool the DNA out of the mixture. Discard the remaining solution and replace with 20 ml of 70% **ethanol**. Use a second pasteur pipette to return the spooled DNA to the tube.

   ethanol (see Appendix for Caution)

5. Wash the DNA by gently inverting the tube, until it is completely white and rather stringy looking.

6. Carefully draw off the ethanol with a pasteur pipette, avoiding the DNA.

7. Resuspend the DNA at an estimated concentration of 200 µg/ml.

REFERENCES

Green M.R., Maniatis T., and Melton D.A. 1983. Human β-globin pre-mRNA synthesized in vitro is accurately spliced in *Xenopus* oocyte nuclei. *Cell* **32:** 681–694.

Kuo J.S., Veale R., Maxwell B., and Sive H. 1996. Translational inhibition by 5 - polycytidine tracts in *Xenopus* embryos and in vitro. *Gene* **176:** 17–21.

Richter J.D. 1991. Translational control during early development. *Bioessays* **13:** 179–183.

Turner D.L. and Weintraub H. 1994. Expression of achaete-scute homolog 3 in *Xenopus* embryos converts ectodermal cells to a neural fate. *Genes Dev.* **8:** 1434–1447.

Cautions

GENERAL CAUTIONS

The following general cautions should always be observed.

- **The absence of a warning** does not necessarily mean that the material is safe, since information may not always be complete or available.

- **Proper disposal procedures** must be used for all chemical, biological, and radioactive waste.

- **Consult your local safety office** for specific guidelines on **appropriate gloves**.

- **Acids and bases** that are concentrated should be handled with great care. Wear goggles and appropriate gloves. A face shield should be worn when handling large quantities.

  Strong acids should not be mixed with organic solvents as they may react. Especially, sulfuric acid and nitric acid may react highly exothermic and cause fires and explosions.

  Strong bases should not be mixed with halogenated solvent as they may form reactive carbenes that can lead to explosions.

  For proper disposal of strong acids and bases, dilute them by placing the acid or base onto ice and neutralize them. Do not pour water into them. If the solution does not contain any other toxic compound, the salts can be flushed down the drain.

- **Never pipet solutions** using mouth suction. This method is not sterile and can be dangerous. Always use a pipette aid or bulb.

- **Halogenated and nonhalogenated** solvents should be kept separately (e.g., mixing chloroform and acetone can cause unexpected reactions in the presence of bases).

- **Laser radiation**, visible or invisible, can cause severe damage to the eyes and skin. Take proper precautions to prevent exposure to direct and reflected beams. Always follow manufacturers safety guidelines and consult your local safety office.

- **Photographic fixatives and developers** also contain chemicals that can be harmful. Handle them with care and follow manufacturer's directions

- **Power supplies and electrophoresis equipment** pose serious fire hazard and electrical shock hazards if not used properly.

- **The use of microwave ovens and autoclaves** in the lab requires certain precautions. Accidents have occurred involving their use (e.g., to melt agar or bactoagar stored in bottles or to sterilize). Often, the screw top is not completely removed and there is not enough space for the steam to vent. When the containers are removed from the microwave or autoclave, they can explode and cause severe injury. Always completely remove bottle caps before microwaving or autoclaving. An alternative method for routine agarose gels that do not require sterile agar is to weigh out the agar and place the solution in a flask.

HAZARDOUS MATERIALS

Acetic acid (concentrated) must be handled with great care. It is harmful by inhalation, ingestion, or skin absorption. Wear appropriate gloves and goggles and use in a chemical fume hood.

Acrylamide (unpolymerized) is a potent neurotoxin and is absorbed through the skin (the effects are cumulative). Avoid breathing the dust. Wear appropriate gloves and a face mask when weighing powdered acrylamide and methylene-bisacrylamide. Use in a chemical fume hood. Polyacrylamide is considered to be nontoxic, but it should be handled with care because it might contain small quantities of unpolymerized acrylamide.

Ammonium acetate may be harmful by inhalation, ingestion, or skin absorption. Wear appropriate gloves and safety glasses. Use in a chemical fume hood.

Ammonium persulfate is extremely destructive to tissue of the mucous membranes and upper respiratory tract, eyes, and skin. Inhalation may be fatal. Wear appropriate gloves, safety glasses, and protective clothing. Use only in a chemical fume hood. Wash thoroughly after handling.

Amyl alcohol is extremely flammable and may be harmful by inhalation, ingestion, or skin absorption. It may cause irritation to the skin, eyes, and respiratory tract and affects the central nervous system. Use only with adequate ventilation. Wear appropriate gloves and safety glasses. Keep away from heat, sparks, and open flame.

Animal treatment: Procedures for the humane treatment of animals must be observed at all times. Consult your local animal facility for guidelines.

BB/BA, see *also* **Benzyl alcohol; Benzyl benzoate.** BB/BA is a potent irritant. To avoid contamination, use it sparingly and clean up any spills promptly with wipes and 95% ethanol. It is useful to have an aspirator nearby to draw up any droplets of BB/BA that creep out of the wells.

BCIP, see **5-Bromo-4-chloro-3-indolyl-phosphate**

Benzyl alcohol is an irritant and may be harmful by inhalation, ingestion, or skin absorption. Wear appropriate gloves and safety glasses. Keep away from heat, sparks, and open flame.

Benzyl benzoate is an irritant and may be harmful by inhalation, ingestion, or skin absorption. Avoid contact with the eyes. Wear appropriate gloves and safety glasses.

Bisacrylamide is a potent neurotoxin and is absorbed through the skin (the effects are cumulative). Avoid breathing the dust. Wear appropriate gloves and a face mask when weighing powdered acrylamide and methylenebisacrylamide.

5-Bromo-4-chloro-3-indolyl-phosphate (BCIP) is hazardous. Handle with care.

Bromophenol blue may be harmful by inhalation, ingestion, or skin absorption. Wear appropriate gloves and safety glasses. Use in a chemical fume hood.

CaCl$_2$, see **Calcium chloride**

Calcium chloride (CaCl$_2$) is harmful by inhalation, ingestion, or skin absorption. Wear appropriate gloves and safety glasses. Use in a chemical fume hood.

Calcium nitrate (CaNO$_3$), is a strong oxidizer and it reacts violently upon contact with many organic substances. Handle with great care. It is harmful by inhalation, ingestion, or skin absorption. Wear appropriate gloves and safety glasses. Keep away from heat, sparks, and open flame.

CaNO$_3$, see **Calcium nitrate**

Chloroform is irritating to the skin, eyes, mucous membranes, and respiratory tract. It is a carcinogen and may damage the liver and kidneys. Wear appropriate gloves and safety glasses and always use in a chemical fume hood.

Cysteine is an irritant to the eyes, skin, and respiratory tract. It may be harmful by inhalation and skin absorption. Wear appropriate gloves and safety glasses. Do not breathe the dust.

DAB, see **3,3 -Diaminobenzidine tetrahydrochloride**

DEPC, see **Diethyl pyrocarbonate**

3,3 -Diaminobenzidine tetrahydrochloride (DAB) is a carcinogen. Handle with extreme care. Avoid breathing vapors. Wear appropriate gloves and safety glasses and use in a chemical fume hood.

Diethyl pyrocarbonate (DEPC) is a potent protein denaturant and is a suspected carcinogen. Aim bottle away from you when opening it; internal pressure can lead to splattering. Wear appropriate gloves, lab coat, and use in a chemical fume hood.

N,N-dimethylformamide (DMF) is irritating to the eyes, skin, and mucous membranes. It can exert its toxic effects through inhalation, ingestion, or skin absorption. Chronic inhalation can cause liver and kidney damage. Wear appropriate gloves and safety glasses. Use in a chemical fume hood.

Dimethyl sulfoxide (DMSO) is harmful by inhalation or skin absorption. Wear appropriate gloves and safety glasses. Use in a chemical fume hood. DMSO is also combustible. Store in a tightly closed container. Keep away from heat, sparks, and open flame.

Dithiothreitol (DTT) is a strong reducing agent that emits a foul odor. Wear lab coat and safety glasses and use in a chemical fume hood when working with the solid form or highly concentrated stocks.

DMF, see *N,N*-**dimethylformamide**

DMSO, see **Dimethyl sulfoxide**

DTT, see **Dithiothreitol**

Ethanol may be harmful by inhalation, ingestion, or skin absorption. Wear appropriate gloves and safety glasses.

Formaldehyde is highly toxic and volatile. It is also a carcinogen. It is readily absorbed through the skin and is irritating or destructive to the skin, eyes, mucous membranes, and upper respiratory tract. Avoid breathing the vapors. Wear appropriate gloves and safety glasses. Always use in a chemical fume hood. Keep away from heat, sparks, and open flame.

Formamide is teratogenic. The vapor is irritating to the eyes, skin, mucous membranes, and upper respiratory tract. It may be harmful by inhalation, ingestion, or skin absorption. Wear appropriate gloves and safety glasses. Always use in a chemical fume hood when working with concentrated solutions of formamide. Keep working solutions covered as much as possible.

Glutaraldehyde is toxic. It is readily absorbed through the skin and is irritating or destructive to the skin, eyes, mucous membranes, and upper respiratory tract. Wear appropriate gloves and safety glasses. Always use in a chemical fume hood.

Glycine may be harmful by inhalation, ingestion, or skin absorption. Wear gloves and safety glasses. Avoid breathing the dust.

HCl, see **Hydrochloric acid**

H₂O₂, see **Hydrogen peroxide**

Hydrochloric acid (HCl) is volatile and may be fatal if inhaled, ingested, or absorbed through the skin. It is extremely destructive to mucous membranes, upper respiratory tract, eyes, and skin. Wear appropriate gloves and safety glasses and use with great care in a chemical fume hood. Wear goggles when handling large quantities.

Hydrogen peroxide (H_2O_2) is corrosive, toxic, and extremely damaging to the skin. It is harmful by inhalation, ingestion, and skin absorption. Wear appropriate gloves and safety glasses and use only in a chemical fume hood.

Isopropanol is irritating and may be harmful by inhalation, ingestion, or skin absorption. Wear appropriate gloves and safety glasses. Do not breathe the vapor. Keep away from heat, sparks, and open flame.

KCl, see **Potassium chloride**

KOH, see **Potassium hydroxide**

Leupeptin (or its **hemisulfate**) may be harmful by inhalation, ingestion, or skin absorption. Wear appropriate gloves and safety glasses. Use in a chemical fume hood.

LiCl, see **Lithium chloride**

Lithium chloride (LiCl) is an irritant to the eyes, skin, mucous membranes, and upper respiratory tract. It may be harmful by inhalation, ingestion, or skin absorption. Wear appropriate gloves, safety goggles, and use in a chemical fume hood. Do not breathe the dust.

MAB (Maleic acid buffer) see **Maleic acid**

Magnesium chloride ($MgCl_2$) is harmful by inhalation, ingestion, or skin absorption. Wear appropriate gloves and safety glasses, and use in a chemical fume hood.

Magnesium sulfate ($MgSO_4$) may be harmful by inhalation, ingestion, or skin absorption. Wear appropriate gloves and safety glasses.

Maleic acid is toxic and harmful by inhalation, ingestion, or skin absorption. Reaction with water or moist air can release toxic, corrosive, or flammable gasses. Do not breathe the vapors or dust. Wear appropriate gloves and safety glasses.

MBS, see **$MgSO_4$; CaCl; KCl**

MEMFA (MOPS, EGTA, Magnesium sulfate, Formaldehyde) Buffer, see **MOPS; Magnesium sulfate; Formaldehyde**

Methanol is poisonous and can cause blindness. It is harmful by inhalation, ingestion, or skin absorption. Adequate ventilation is necessary to limit exposure to vapors. Avoid inhaling these vapors. Wear appropriate gloves and goggles. Use only in a chemical fume hood

Methylene chloride is toxic if inhaled, ingested, or absorbed through the skin. It is also an irritant and is suspected to be a carcinogen. Wear appropriate gloves and safety glasses and do not breathe the vapors. Use in a chemical fume hood.

MgCl$_2$, see **Magnesium chloride**

MgSO$_4$, see **Magnesium sulfate**

MMR, see **MgCl$_2$; CaCl$_2$; KCl**

3-(*N*-Morpholino)-propanesulfonic acid (MOPS) may be harmful by inhalation, ingestion, or skin absorption. It is irritating to mucous membranes and upper respiratory tract. Wear appropriate gloves and safety glasses and use in a chemical fume hood.

MOPS, see **3-(*N*-Morpholino)-propanesulfonic acid**

Na$_2$CO$_3$, see **Sodium carbonate**

Na$_2$HPO$_4$, see **Sodium hydrogen phosphate**

NaH$_2$PO$_4$, see **Sodium dihydrogen phosphate**

NaOH, see **Sodium hydroxide**

NBT, see **4-Nitro blue tetrazolium chloride**

4-Nitro blue tetrazolium chloride (NBT) is hazardous. Handle with care.

Paraformaldehyde is highly toxic. It is readily absorbed through the skin and is extremely destructive to the skin, eyes, mucous membranes, and upper respiratory tract. Avoid breathing the dust. Wear appropriate gloves and safety glasses, and use in a chemical fume hood. Paraformaldehyde is the undissolved form of formaldehyde.

Pepstatin A may be harmful by inhalation, ingestion, or skin absorption. Wear appropriate gloves and safety glasses. Use in a chemical fume hood.

Phenol is extremely toxic, highly corrosive, and can cause severe burns. Wear appropriate gloves, goggles, and protective clothing. Always use in a chemical fume hood. Rinse any areas of skin that come in contact with phenol with a large volume of water and wash with soap and water; do not use ethanol!

Phenyl-methyl-sulfonyl fluoride (PMSF) is a highly toxic cholinesterase inhibitor. It is extremely destructive to the mucous membranes of the respiratory tract, eyes, and skin. It may be fatal if inhaled, ingested, or absorbed through the skin. Wear appropriate gloves and safety glasses and always use in a chemical fume hood. In case of contact, immediately flush eyes or skin with copious amounts of water and discard contaminated clothing.

Picric acid powder is caustic and potentially explosive if it is dissolved and then allowed to dry out. Care must be taken to ensure that stored solutions do not dry out. Handle all concentrated acids with great care. It is also highly toxic and may be harmful by inhalation, ingestion, or skin absorption. Wear appropriate gloves and goggles.

PMSF, see **Phenyl-methyl-sulfonyl fluoride**

Polyvinyl alcohol may be harmful by inhalation, ingestion, or skin absorption. Wear appropriate gloves and safety glasses.

Polyvinylpyrrolidone may be harmful by inhalation, ingestion, or skin absorption. Wear appropriate gloves and safety glasses. Use in a chemical fume hood.

Potassium chloride (KCl) may be harmful by inhalation, ingestion, or skin absorption. Wear appropriate gloves and safety glasses.

Potassium hydroxide (KOH and KOH/methanol) can be highly toxic. Solutions are caustic and should be handled with great care. Wear appropriate gloves.

Radioactive substances: Wear appropriate gloves when handling. Consult the local safety office for further guidance in the appropriate use and disposal of radioactive materials. Always monitor thoroughly after using radioisotopes.

Retinoic acid poses a possible risk to the unborn child. It may be harmful by inhalation, ingestion, and skin absorption. Avoid prolonged or repeated exposure. Wear appropriate gloves and safety glasses. Do not breathe the dust.

SDS, see **Sodium dodecyl sulfate**

Sodium acetate, see **Acetic acid**

Sodium carbonate (Na_2CO_3) may be harmful by inhalation, ingestion, or skin absorption. Wear appropriate gloves and safety glasses. Use in a chemical fume hood.

Sodium dihydrogen phosphate (NaH$_2$PO$_4$) (sodium phosphate, monobasic) may be harmful by inhalation, ingestion, or skin absorption. Wear appropriate gloves and safety glasses. Use in a chemical fume hood.

Sodium dodecyl sulfate (SDS) is harmful if inhaled. Wear a face mask when weighing SDS.

Sodium hydrogen phosphate (Na$_2$HPO$_4$) (sodium phosphate, dibasic) may be harmful by inhalation, ingestion, or skin absorption. Wear appropriate gloves and safety glasses. Use in a chemical fume hood.

Sodium hydroxide (NaOH) and solutions containing NaOH are highly toxic and caustic and should be handled with great care. Wear appropriate gloves and a face mask. All other concentrated bases should be handled in a similar manner.

Triethylamine is highly toxic and flammable. It is extremely corrosive to the mucous membranes, upper respiratory tract, eyes, and skin. It may be harmful by inhalation, ingestion, or skin absorption. Wear appropriate gloves and safety glasses. Use in a chemical fume hood. Keep away from heat, sparks, and open flame.

UV light and/or **UV radiation** is dangerous and can damage the retina of the eyes. Never look at an unshielded UV light source with naked eyes. View only through a filter or safety glasses that absorb harmful wavelengths. UV radiation is also mutagenic and carcinogenic. To minimize exposure, make sure that the UV light source is adequately shielded. Wear protective appropriate gloves when holding materials under the UV light source.

Xylene must always be used in a chemical fume hood. It is flammable and may be narcotic at high concentrations. Keep away from heat, sparks, and open flame.

APPENDIX 5

Suppliers

With the exception of those suppliers listed in the text with their addresses, all suppliers mentioned in this manual can be found in the BioSupplyNet Source Book and on the Web site at:

htpp://www.biosupplynet.com

If a copy of BioSupplyNet Source Book was not included in this manual, a free copy can be ordered by using any of the following methods:

- Complete the Free Source Book Request Form found at the Web site at:

  http://www.biosupplynet.com

- E-mail a request to info@biosupplynet.com

- Fax a request to 516-349-5598.

Trademarks

The following trademarks and registered trademarks are accurate to the best of our knowledge at the time of printing. Please consult individual manufacturers and other resources for specific information.

Isopac	Sigma Chemical Co.
Netwell	Corning Costar Corp.
Adams Nutator	Becton, Dickinson & Co.
Benchkote	Whatman International Ltd.
Ektachrome	Eastman Kodak Co.
Ethicon	Johnson & Johnson
GENECLEAN	BIO 101, Inc.
Gyrotory	New Brunswick Scientific Co., Inc.
Instant Ocean	Aquarium Systems, Inc.
Kimwipes	Kimberly-Clark Corp.
Labquake	Barnstead Thermolyne Corp.
Maroxy	Mardel Laboratories, Inc.
Nasco-Guard	Nasco Industries, Inc.
Netwell	Corning Costar Corp.
Nitex	Tetko Inc.
Nonidet P-40	Shell International Petroleum Co. Ltd. UK
Parafilm	American National Can Co.
pBluescript	Stratagene
Permoplast	American Art Clay Co., Inc.
Permount	Fisher Scientific Co.
pGEM	Promega Corp.
Picospritzer	General Valve Corp.

Plexiglas	Rohm and Haas Co.
Pronase	Calbiochem-Novabiochem Corp.
QIAGEN	QIAGEN, Inc.
RNasin	Promega Corp.
SeaKem	FMC BioProducts
Sephadex	Pharmacia Biotech AB
Sigmacote	Sigma Chemical Co.
Sorvall	Kendro Laboratory Products
Sylgard	Dow Corning Corp.
Trout Chow	Ralston Purina Company
Tygon	Norton Company
VECTASTAIN	Vector Laboratories Inc.
Versilube	N.S.C.G., Inc.

Index

natural mating, 7, 100
FGF. *See* Fibroblast growth factor
Fibroblast growth factor (FGF), 43, 45, 48, 62
Fibroblast growth factor receptor, dominant-negative mutants, 50, 61–62
Film, 87
Fixation
  lineage labeling specimens, 165–166
  paraffin embedding of embryos, 277
  plastic embedding of embryos, 280
  whole-mount in situ hybridization, 259–260
Fluorescent dextran
  fixation of specimens, 165
  lineage tracing, 153, 157–159, 162
  preparation, 163
Forceps, 83–84, 105
Forebrain, 23, 30, 33, 116
Formaldehyde, 232, 238–239, 317
Formamide, 317
Fungal infection, 8, 10
Fusion protein. *See* Hormone-inducible fusion protein

Gain-of-function mutant, 39–40
GAL4, 60
β-Galactosidase
  fixation of specimens, 166
  histochemical staining, 167
  lineage tracing, 157–159, 161–162
  vector preparation, 164
Gall bladder, 35
Gap junction, tracer permeability, 157
Gastrulation, 16–17, 113
Gatenby and Cowdry adhesive, 276
Gene trap, 199
Genomic DNA isolation
  red blood cells
    lysis and extraction, 310–311
    nuclei isolation, 309–310
  single embryo extraction, 308–309
  solutions, 292

Gentamycin, 8, 284
GFP. *See* Green fluorescent protein
Gills, 23, 25
Glass needles, 86
Glass vials, immunohistochemical staining, 233, 241
Glossopharyngeal nerve, 31, 33
Glutaraldehyde, 165, 280, 317
Glycine, 317
Glycogen synthase kinase-3 (GSK-3), 64, 114
Goat serum, 290
Goosecoid, 41
Green fluorescent protein (GFP)
  fixation of specimens, 165
  lineage tracing, 157–159, 161–162
  transgenic frogs, 199
  vector preparation, 164
GSK-3. *See* Glycogen synthase kinase-3

Habenular commissure, 156
Hair loop, 84–85, 185
Hammerhead ribozyme, 69
hCG. *See* Human chorionic gonadotropin
Head mesoderm, 186
Heart, 20, 24, 35, 116, 155
Heparin, 309
Hindbrain, 23, 27, 30, 33, 116
Hindgut, 23
Histology
  JB-4 kit, 280
  paraffin embedding, 276–279
  plastic embedding, 280–281
  scoring for manipulated gene expression, 41–42
HNK-1, 33
Hoescht 33258, 145
Hoescht 33342, 214
Holtfreter's solution, 286
Hormone-inducible fusion protein
  ligand-binding domain insertion, 58
  protein selection, 56–57

www.ingramcontent.com/pod-product-compliance
Lightning Source LLC
Chambersburg PA
CBHW042313210326
41599CB00038B/7114